Emerging Advances in Petrophysics

Emerging Advances in Petrophysics

Porous Media Characterization and Modeling of Multiphase Flow

Special Issue Editors

Jianchao Cai
Shuyu Sun
Ali Habibi
Zhien Zhang

MDPI • Basel • Beijing • Wuhan • Barcelona • Belgrade

MDPI

Special Issue Editors
Jianchao Cai
University of Geosciences
China

Shuyu Sun
King Abdullah University of Science and Technology
Kingdom of Saudi Arabia

Ali Habibi
University of Alberta
Canada

Zhien Zhang
The Ohio State University
USA

Editorial Office
MDPI
St. Alban-Anlage 66 4052
Basel, Switzerland

This is a reprint of articles from the Special Issue published online in the open access journal *Energies* (ISSN 1996-1073) from 2018 to 2019 (available at: https://www.mdpi.com/journal/energies/special_issues/petrophysics)

For citation purposes, cite each article independently as indicated on the article page online and as indicated below:

LastName, A.A.; LastName, B.B.; LastName, C.C. Article Title. *Journal Name* **Year**, *Article Number*, Page Range.

ISBN 978-3-03897-794-0 (Pbk)
ISBN 978-3-03897-795-7 (PDF)

Contents

About the Special Issue Editors

Jianchao Cai received his B.Sc in Physics from Henan Normal University in 2005 and his MSc and Ph.D in Condensed Matter Physics from Huazhong University of Science and Technology in 2007 and 2010, respectively. He joined the Institute of Geophysics and Geomatics at the China University of Geosciences (Wuhan) in 2010. From July 2013 to July 2014, he acted as a Visiting Scholar at University of Tennessee-Knoxville, USA. He currently is Professor of Geological Resources and Geological Engineering (since 2015). Meanwhile, he is the founder and Editor-in-Chief of *Advances in Geo-Energy Research* and serves as Associate Editor for some international journals. Dr. Cai focuses on petrophysical characterization and micro-transport phenomena in porous media, and fractal theory as well as its application. He has published more than 100 peer-refereed journal articles and numerous book chapters.

Shuyu Sun is currently the Principal Investigator of the Computational Transport Phenomena Laboratory (CTPL) at King Abdullah University of Science and Technology (KAUST) and a Co-Director of the Center for Subsurface Imaging and Fluid Modeling consortium (CSIM) at KAUST. He is a founding faculty member jointly appointed by the program of Earth Sciences and Engineering (ErSE) and the program of Applied Mathematics and Computational Science (AMCS) at KAUST since 2009. He also holds a number of adjunct faculty positions across the world, including Adjunct Professorship in Xi'an Jiao Tong University, Adjunct Professorship in China University of Petroleum at Beijing, Adjunct Professorship in China University of Petroleum at Qingdao, and Adjunct Professorship in China University of Geosciences at Wuhan. Before joining KAUST, Dr. Sun served as an (tenure-tracked) Assistant Professor of Mathematical Sciences at Clemson University in the United States. Professor Sun's research includes the modeling and simulation of flow and reactive transport in porous media from molecular scales, pore scales to Darcy scales, as well as the numerical analysis of relevant algorithms. Professor Sun has published 300+ articles or book chapters, including 190+ refereed journal articles and numerous conference papers and technical reports.

Ali Habibi is a postdoctoral fellow in the Department of Civil and Environmental Engineering at the University of Alberta. His primary research interests include understanding rock-fluid interactions in tight rocks, evaluating the wetting affinity of tight rocks using DLVO theory, investigating the application of EOR techniques in tight oil formations, and investigating the application of nanotechnology in the formation damage reduction. Habibi has authored or coauthored more than 25 peer-reviewed and conference papers, and two book chapters. Habibi holds a BS degree from the University of Tehran in chemical engineering, and an MS degree from the University of Tehran and a PhD degree from the University of Alberta, both in petroleum engineering.

Zhien Zhang is currently a postdoctoral researcher in William G. Lowrie Department of Chemical and Biomolecular Engineering at Ohio State University. His research interests include advanced processes and materials (i.e., membranes) for CO2 capture, CCUS processes, gas separation, and gas hydrates. Dr Zhang has published 65+ journal papers and 10+ editorials in the high-impact journals, e.g., Renewable & Sustainable Energy Reviews. He has authored two "hot papers" (top 0.1%) and six "highly cited papers" (top 1%). He is an Editor or Guest Editor for several international journals, e.g., *Applied Energy, Fuel*, and *Journal of Natural Gas Science and Engineering*, and he is also a visiting Professor at University of Cincinnati.

energies

MDPI

Editorial

Emerging Advances in Petrophysics: Porous Media Characterization and Modeling of Multiphase Flow

Jianchao Cai [1,*], Shuyu Sun [2], Ali Habibi [3] and Zhien Zhang [4]

[1] Hubei Subsurface Multi-Scale Imaging Key Laboratory, Institute of Geophysics and Geomatics, China University of Geosciences, Wuhan 430074, China

[2] Computational Transport Phenomena Laboratory, Division of Physical Science and Engineering, King Abdullah University of Science and Technology, Thuwal 23955-6900, Saudi Arabia; shuyu.sun@kaust.edu.sa

[3] Department of Civil & Environmental Engineering, University of Alberta, Edmonton, AB T6G 2W2, Canada; ahabibi@ualberta.ca

[4] William G. Lowrie Department of Chemical and Biomolecular Engineering, The Ohio State University, Columbus, OH 43210, USA; zhienzhang@hotmail.com

* Correspondence: caijc@cug.edu.cn

Received: 28 December 2018; Accepted: 4 January 2019; Published: 17 January 2019

Abstract: With the ongoing exploration and development of oil and gas resources all around the world, applications of petrophysical methods in natural porous media have attracted great attention. This special issue collects a series of recent studies focused on the application of different petrophysical methods in reservoir characterization, especially for unconventional resources. Wide-ranging topics covered in the introduction include experimental studies, numerical modeling (fractal approach), and multiphase flow modeling/simulations.

Keywords: petrophysics; fractal porous media; unconventional reservoirs; multiphase flow

1. Introduction

A subsurface reservoir, as an important natural porous media, usually has a complex pore/fracture structure. Several key parameters such as pore-throat size distribution, pore connectivity, and micro-scale fractures strongly affect transport, distribution, and residual saturation of fluids in porous media.

Petrophysics, defined as the study of the physical and chemical properties of rock and its interactions with fluids, has many applications in different industries, especially in the oil and gas industries. Key parameters studied in petrophysics are lithology, porosity, water saturation, permeability, and density. Petrophysics is closely related to different areas such as petroleum engineering, geology, mineralogy, and exploration geophysics. Petrophysicists work closely with reservoir engineers and geoscientists to understand different phenomena in subsurface reservoirs, i.e. how pores are interconnected in a subsurface reservoir, and how this connectivity affects the migration and accumulation of hydrocarbon.

From a broader point of view, petrophysics, especially in terms of characterization of and multiphase flow in porous media, covers a wide range of research studies including hydrocarbon extraction, geosciences, environmental issues, hydrology, and biology. The relevant stakeholders in this issue are authorities and service companies working in the petroleum, subsurface water resources, air and water pollution, environmental, and bio-material industries. Implementing reliable methods for the characterization of multiphase flow in porous media is crucial in many fields, including the characterization of residual water or oil in hydrocarbon reservoirs and long-term storage of supercritical CO_2 in geological formations.

This collection associated with the special issue in *Energies* emphasizes the fundamental innovations and gathers together 15 recent papers on novel applications of petrophysics in unconventional reservoirs.

2. Overview of Work Presented in This Special Issue

The papers published in this special issue present new advancements in the characterization of porous media and the modeling of multiphase flow in porous media. These research studies are divided into four categories.

Studies in the first category focus on the experimental approaches for the characterization of porous media. Through selecting 12 rock samples from the Bakken formation, Liu et al. [1] conducted a set of experiments including X-ray diffraction analysis, total organic carbon analysis, vitrinite reflectance, and low-temperature nitrogen adsorption experiments to study pore structures and the main controlling factors in this formation. The Bakken formation has micro-, meso-, and macro-pores. Total organic carbon and maturity are the main parameters controlling the pore structure of the upper and lower Bakken formation. However, clays and quartz are the controlling parameters for the middle Bakken formation.

Using pressure-controlled mercury injection, casting sheet images and scanning electron microscopy analysis, Wang et al. [2] studied the pore structure of the mouth bar sand bodies in Guan195 area, China. Three types of pores exist in this area, including intergranular pores, dissolution pores, and micro fractures. Pore structure heterogeneity of single mouth bar sand bodies in the short-, middle- and long-terms base-level were respectively analyzed and compared.

By employing a computer-controlled creep setup, Zhou et al. [3] conducted a set of creep tests on salt rock under a constant uniaxial stress. They studied the acoustic emission space–time evolution and energy-releasing characteristics. A new creep-damage model was proposed based on a fractional derivative of combined acoustic emission statistical regularities. This could provide a precise description of full creep regions in salt rock. The acoustic emission data in the non-decay creep process of salt rock could be divided into three stages: the initial transient creep period, the steady-state period, and the tertiary stage.

Based on the functional relationships between porosity, true density, and bulk density, Zhang et al. [4] proposed a new experimental method for characterizing the porosity of loose media subjected to overburden pressure. This method was used to test the total porosity of loose coal particles including the influence of pressure and particle size. The total porosity and pressure obey an attenuated exponential function, while the total porosity and particle size obey a power function. The sensitivity of total porosity and particle size to pressure were also analyzed.

Studies in the second category focus on a fractal-based approach for the characterization of porous media. In addition to pores, fractures can be characterized by fractal geometry [5–8]. Gong et al. [9] utilized the fractal method to evaluate the behavior of fractures in tight conglomerate reservoirs. Three types of fractures were identified in these reservoirs: intra-gravel, gravel edge, and trans-gravel fractures. The fractal dimension of the fractures is in the range of 1.20–1.50. The fractal dimensions are exponentially correlated with the fracture areal density, porosity and permeability. The cumulative frequency distribution of both fracture apertures and areal densities obeys the power law distribution. The fracture parameters at different scales could be predicted by extrapolating their power law distributions. Jiang et al. [10] used mercury injection capillary pressure (MICP) data to characterize the heterogeneity of pore structures. The multifractal analysis based on the MICP data was conducted to investigate the heterogeneity of tight sandstone reservoirs. The relationships among physical properties, MICP data and multifractal parameters were analyzed in detail. Tao et al. [11] presented an efficient fractal-based approach to investigate the effects of initial void ratio on the soil–water characteristic curve (SWCC) in a deformable unsaturated soil. This approach included only two parameters: fractal dimension and air-entry value. The SWCCs are mainly controlled by the air-entry value, while the fractal dimension can be assumed constant.

Studies in the third category focus on the modeling of multiphase flow in porous media. To accurately find flow patterns in high-yield wells with different inclined angles, Qi et al. [12] conducted multiphase pipe flow tests in mid–high yield and highly deviated wells under different: (i) inclined angles; (ii) liquid flow rates; and (iii) gas flow rates. A pressure prediction model with new coefficients and higher accuracy for wellbores in selected blocks was developed. In addition, the effects of inclination, output, and gas flow rate on the flow pattern, liquid holdup, and friction during multiphase flow were comprehensively analyzed. Six types of classical flow regimes were verified with the developed model.

Microfractures (natural and induced) have a great significance on reservoir development. Yang et al. [13] utilized the Lattice Boltzmann method to calculate the equivalent permeability of artificially induced three-dimensional fractures. The fractal dimensions, geometrical parameters and porosity of induced fractures in Berea sandstone were calculated based on digital cores of fractures. The relations between permeability and fractal dimension, geometrical parameters and the porosity of factures were summarized and discussed.

To evaluate the producing degree of commingled production, Shen et al. [14] presented a one-dimensional linear flow model and a planar radial flow model for water-flooded multilayer offshore heavy oil reservoirs based on the Buckley–Leverett theory. A dynamic method was used to evaluate seepage resistance, sweep efficiency, and oil recovery factor. An analysis with field data was also presented.

Scientifically determining a heating scheme requires an understanding of the behavior of oil when the temperature declines during tanker transportation. Yu et al. [15] investigated the free liquid surface movement and the temperature drop characteristics of crude oil in cargo when the tanker was subjected to rotational motion. It was found that the oscillating motion significantly enhanced the temperature decline rate and it was positively related to the rotational angular velocity.

The behavior of flow resistance with velocity is still poorly defined for post-laminar flow through coarse granular media. Banerjee et al. [16] investigated the effect of flow resistance on independently varying media size and porosity subjected to parallel post-laminar flow through granular media. They then simulated the post-laminar flow conditions with the help of a computational fluid dynamic model. Their output advocated the importance and applicability of computational fluid dynamic modelling in better understanding post-laminar flow through granular media.

Zhang et al. [17] developed a generalized method to determine the diffusion coefficient for supercritical CO_2 diffusing into porous media saturated with oil under reservoir conditions. They established a mathematical model to describe the mass transfer process. The results showed that oil with lower viscosity and lighter oil components can enhance the mass transfer process.

Studies in the fourth category focus on marine gas hydrates. Molecular and isotopic analysis of marine gas hydrate samples and potential hazards are associated with their production and development. Ye et al. [18] analyzed the isotopic gas composition of 300 samples from China's first gas hydrate production test in the South China Sea. All gas samples were predominated by methane. However, no H_2S was detected. The methane in all samples was of microbial origin and derived from CO_2 reduction. Moreover, Wang et al. [19] analyzed the potential hazards associated with the production and development of marine gas hydrates. Marine geo-hazards, greenhouse gas emissions, marine ecological hazards, and marine engineering hazards are four reported hazard categories. They proposed the concept of life-cycle management for the prevention of hazards during the development and production from marine gas hydrates. This concept was divided into three steps including preparation, production control, and post-production protection. A production test in the Shenhu area of the South China Sea showed that marine gas hydrate exploration and development could be planned using the three-step methodology.

3. Conclusions

Many researchers around the world from different areas, ranging from natural sciences to engineering fields, have been working on: (i) the characterization of petrophysical properties for unconventional resources; and (ii) simulating multiphase flow in these resources. The aim of this special issue is to provide new ideas related to petrophysics, and then to advance this multidisciplinary effort. Clearly, the modeling of multiphase flow in porous media continues to be helpful for oil and gas reservoirs. Moreover, the molecular and isotopic analysis of gas hydrates in marine fields are beneficial for minimizing the hazardous and polluting gas emissions.

Author Contributions: The authors contributed equally to this work.

Funding: This research was funded by National Natural Science Foundation of China, Grant Numbers 41722403 and 41572116, and King Abdullah University of Science and Technology (KAUST), Grant Numbers BAS/1/1351-01, URF/1/2993-01, and REP/1/2879-01.

Acknowledgments: The guest editors would like to acknowledge MDPI for the invitation to act as the guest editor of this special issue of *"Energies"* with the kind cooperation and support of the editorial staff. The guest editors are also grateful to the authors for their inspiring contributions and the anonymous reviewers for their tremendous efforts. The first guest editor, Jianchao Cai, would like to thank the National Natural Science Foundation of China for supporting his series of studies on flow and transport properties in porous media. The second guest editor, Shuyu Sun, gratefully acknowledges that his research on porous media modeling has been supported by funding from King Abdullah University of Science and Technology (KAUST).

Conflicts of Interest: The authors declare no conflict of interest.

References

1. Liu, Y.; Shen, B.; Yang, Z.; Zhao, P. Pore structure characterization and the controlling factors of the bakken formation. *Energies* **2018**, *11*, 2879. [CrossRef]
2. Wang, X.; Hou, J.; Liu, Y.; Ji, L.; Sun, J.; Gong, X. Impacts of the base-level cycle on pore structure of mouth bar sand bars: A case study of the Paleogene Kongdian formation, Bohai bay basin, China. *Energies* **2018**, *11*, 2617. [CrossRef]
3. Zhou, H.; Liu, D.; Lei, G.; Xue, D.; Zhao, Y. The creep-damage model of salt rock based on fractional derivative. *Energies* **2018**, *11*, 2349. [CrossRef]
4. Zhang, C.; Zhang, N.; Pan, D.; Qian, D.; An, Y.; Yuan, Y.; Xiang, Z.; Wang, Y. Experimental study on sensitivity of porosity to pressure and particle size in loose coal media. *Energies* **2018**, *11*, 2274. [CrossRef]
5. Mandelbrot, B.B. *The Fractal Geometry of Nature*; W. H. Freeman: New York, NY, USA, 1982.
6. Bonnet, E.; Bour, O.; Odling, N.E.; Davy, P.; Main, I.; Cowie, P.; Berkowitz, B. Scaling of fracture systems in geological media. *Rev. Geophys.* **2001**, *39*, 347–383. [CrossRef]
7. Barton, C.C.; la Pointe, P. *Fractals in the Earth Sciences*; Plenum: New York, NY, USA, 1995.
8. Cai, J.; Wei, W.; Hu, X.; Liu, R.; Wang, J. Fractal characterization of dynamic fracture network extension in porous media. *Fractals* **2017**, *25*, 1750023. [CrossRef]
9. Gong, L.; Fu, X.; Gao, S.; Zhao, P.; Luo, Q.; Zeng, L.; Yue, W.; Zhang, B.; Liu, B. Characterization and prediction of complex natural fractures in the tight conglomerate reservoirs: A fractal method. *Energies* **2018**, *11*, 2311. [CrossRef]
10. Jiang, Z.; Mao, Z.; Shi, Y.; Wang, D. Multifractal characteristics and classification of tight sandstone reservoirs: A case study from the Triassic Yanchang formation, Ordos basin, China. *Energies* **2018**, *11*, 2242. [CrossRef]
11. Tao, G.; Chen, Y.; Kong, L.; Xiao, H.; Chen, Q.; Xia, Y. A simple fractal-based model for soil-water characteristic curves incorporating effects of initial void ratios. *Energies* **2018**, *11*, 1419. [CrossRef]
12. Qi, D.; Zou, H.; Ding, Y.; Luo, W.; Yang, J. Engineering simulation tests on multiphase flow in middle- and high-yield slanted well bores. *Energies* **2018**, *11*, 2591. [CrossRef]
13. Yang, Y.; Liu, Z.; Yao, J.; Zhang, L.; Ma, J.; Hejazi, S.; Luquot, L.; Ngarta, T. Flow simulation of artificially induced microfractures using digital rock and lattice Boltzmann methods. *Energies* **2018**, *11*, 2145. [CrossRef]
14. Shen, F.; Cheng, L.; Sun, Q.; Huang, S. Evaluation of the vertical producing degree of commingled production via waterflooding for multilayer offshore heavy oil reservoirs. *Energies* **2018**, *11*, 2428. [CrossRef]
15. Yu, G.; Yang, Q.; Dai, B.; Fu, Z.; Lin, D. Numerical study on the characteristic of temperature drop of crude oil in a model oil tanker subjected to oscillating motion. *Energies* **2018**, *11*, 1229. [CrossRef]
16. Banerjee, A.; Pasupuleti, S.; Singh, M.K.; Pradeep Kumar, G.N. An investigation of parallel post-laminar flow through coarse granular porous media with the Wilkins equation. *Energies* **2018**, *11*, 320. [CrossRef]
17. Zhang, C.; Qiao, C.; Li, S.; Li, Z. The effect of oil properties on the supercritical CO_2 diffusion coefficient under tight reservoir conditions. *Energies* **2018**, *11*, 1495. [CrossRef]

18. Ye, J.; Qin, X.; Qiu, H.; Xie, W.; Lu, H.; Lu, C.; Zhou, J.; Liu, J.; Yang, T.; Cao, J.; et al. Data report: Molecular and isotopic compositions of the extracted gas from China's first offshore natural gas hydrate production test in South China Sea. *Energies* **2018**, *11*, 2793. [CrossRef]

19. Wang, F.; Zhao, B.; Li, G. Prevention of potential hazards associated with marine gas hydrate exploitation: A review. *Energies* **2018**, *11*, 2384. [CrossRef]

energies

MDPI

Article

Pore Structure Characterization and the Controlling Factors of the Bakken Formation

Yuming Liu [1,*] , Bo Shen [2,*], Zhiqiang Yang [3] and Peiqiang Zhao [4]

[1] State Key Laboratory of Petroleum Resources and Prospecting, China University of Petroleum, Beijing 102249, China
[2] Geophysics and Oil Resource Institute, Yangtze University, Wuhan 430100, China
[3] Exploration & Development Research Institute of Liaohe Oilfield Company, Petrochina, Panjin 124000, China; yangzhiqiang2@petrochina.com.cn
[4] Institute of Geophysics and Geomatics, China University of Geosciences, Wuhan 430074, China; zhaopq@cug.edu.cn
* Correspondence: liuym@cup.edu.cn (Y.L.); boshen151@gmail.com (B.S.);
   Tel.: +86-10-8973-9071 (Y.L.); +86-27-6911-1036 (B.S.)

Received: 20 September 2018; Accepted: 23 October 2018; Published: 24 October 2018

Abstract: The Bakken Formation is a typical tight oil reservoir and oil production formation in the world. Pore structure is one of the key factors that determine the accumulation and production of the hydrocarbon. In order to study the pore structures and main controlling factors of the Bakken Formation, 12 samples were selected from the Bakken Formation and conducted on a set of experiments including X-ray diffraction mineral analysis (XRD), total organic carbon (TOC), vitrinite reflectance (R_o), and low-temperature nitrogen adsorption experiments. Results showed that the average TOC and R_o of Upper and Lower Bakken shale is 10.72 wt% and 0.86%, respectively. The Bakken Formation develops micropores, mesopores, and macropores. However, the Upper and Lower Bakken shale are dominated by micropores, while the Middle Bakken tight reservoir is dominated by mesopores. The total pore volume and specific surface area of the Middle Bakken are significantly higher than those of the Upper and Lower Bakken, indicating that Middle Bakken is more conducive to the storage of oil and gas. Through analysis, the main controlling factors for the pore structure of the Upper and Lower Bakken shale are TOC and maturity, while those for Middle Bakken are clay and quartz contents.

Keywords: Bakken Formation; pore structure; controlling factors; low-temperature nitrogen adsorption

1. Introduction

The unconventional reservoirs such as shale gas and tight oil have been paid more attention since the production of conventional oil and gas decreased [1–3]. It is difficult to obtain economic oil flows without horizontal well drilling and fracturing technology, as these unconventional reservoirs are impermeable or extremely low permeable [4,5]. In unconventional reservoirs, the development may be more important than the exploration. However, the exploration of unconventional resources cannot be neglected. A series of geochemical and petrophysical properties are required to characterize and evaluate [5,6]. Among these parameters, pore structures are important factors affecting the fluid transport both in the hydrocarbon accumulation and production in porous media and are also key parameters for reservoir grading and productivity evaluation [7–10]. Therefore, it is necessary to study the pore structure characteristics such as pore size distribution and pore types, and the controlling factors of the unconventional reservoirs.

According to pore sizes, the pores can be classified into three types: Micropore (<2 nm), mesopore (2–50 nm), and macropore (>50 nm) [11]. Compared with the conventional reservoirs, the unconventional shale oil and gas reservoirs always develop more micropores and mesopores, and show more complex pore systems with strong heterogeneity [4]. Thus, it is necessary to utilize proper methods to characterize the pore structure. A variety of advanced experimental methods can be used to analyze the pore structure of porous media. The pore types and sizes can be qualitatively observed by CT scan [12], scanning electron microscope (SEM) [13], field emission scanning electron microscope (FESEM) [14], and transmission electron microscope (TEM) [15]. Meanwhile, quantitative characterization methods include high pressure mercury injection (PMI) [16], constant rate mercury injection (RMI) [17], nuclear magnetic resonance (NMR) [18,19], low-temperature N_2 and CO_2 adsorption [20,21], small-angle and ultra-small-angle neutron scattering (SANS and USANS) [22], etc. Each method has its own advantages and disadvantages. For example, it is difficult for qualitative methods providing pore size distributions. NMR can capture almost of all sizes of pores but it requires other quantitative methods (PMI) to scale [18]. Low-temperature N_2 adsorption (LTNA) is suitable for capturing the pores with a size of approximately 1 to 150 nm. Hence, it is suitable for obtaining nano-scale pore structure parameters. In recent years, a growing number of studies have applied low-pressure N_2 adsorption to explore the pore structure characteristics of unconventional shale and tight sandstone, such as pore volume, specific area, and pore size distributions [23,24]. Kuila and Prasad [25] investigated the specific surface area and pore-size distributions in shales with a nitrogen gas-adsorption technique. Wang et al. [26] studied the pore structure of shale gas of Longmaxi Formation, Sichuan Basin, China using LTNA data. Su et al. [27] combined the LTNA and low-temperature CO_2 adsorption method to characterize the pore structure of shale oil reservoirs in the Zhanhua Sag, Bohai Bay Basin, China.

The Bakken formation is a typical tight oil reservoir which underlies parts of Northern USA and Southern Canada. This formation is divided into three members: The Upper Bakken, the Middle Bakken, and the Lower Bakken. The previous geological studies include petroleum source rocks [28], systemic petroleum geology [29], lithofacies and paleoenvironments [30], diagenesis and fracture [31], and petrophysics properties [32,33]. Studies on the pore structure of the Bakken Formation have been reported. Liu and Ostadhassan [34] characterized the microstructures of Upper/Lower Bakken shales with the aid of SEM images and observed that micropores developed extensively in those shale samples. Li et al. [35] measured the permeability of Middle Bakken samples including Kinkenberg permeability and PMI-based permeability. Kinkenberg permeability is commonly higher than PMI-based permeability, indicating smaller pores cannot be detected by the PMI method. Saidian and Prasad [25] reported the pore size distribution of Middle Bakken and Three Forks formations provided by PMI and LTNA methods. Pore size distributions are affected by clay content. The greater the clay content, the higher the amplitude of the pore size distribution. Liu et al. [36] investigated the fractal and multifractal characteristics of pore-throats of Upper/Lower Bakken shale using the PMI method. Liu et al. [1] reported the pore structure and fractal characteristics of the Bakken formation in North Dakota, USA. It was observed that the pore structure of the Middle Bakken and the Upper/Lower Bakken are significantly different. However, the differences in the pore structure of the three Bakken members and their main controlling factors are not thoroughly investigated.

In this paper, 12 samples were derived from the Bakken Formation for experimental tests. The X-ray diffraction (XRD), total organic carbon (TOC) analysis, vitrinite reflectance (R_o) and low-temperature N_2 adsorption experiments were conducted on these samples. Pore structure parameters, such as pore morphology, specific surface area, and pore size distribution, were then analyzed based on low-pressure N_2 adsorption curves. The controlling factors of pore structures were determined by analyzing the correlations of the pore structures with the mineral composition, TOC, and thermal maturity.

2. Samples and Methods

2.1. Geological Background and Samples

The Williston Basin is located in the Northern United States and in Southern Canada with an area of $40,000 \times 10^4$ m^2, see Figure 1 [37]. This basin occupies portions of North Dakota, South Dakota, Montana USA, and Alberta, Saskatchewan, and Manitoba Canada [38,39]. It was deposited on the Superior and Wyoming Craton, and Trans-Hudson orogenic belt from the Cambrian to Carboniferous (mainly Mississippian) system. Subsidence and basin filling were most intense during the Ordovician, Silurian, and Devonian Periods, when thick accumulations of limestone and dolomite, with lesser thicknesses of sandstones, siltstones, shales, and evaporites were laid down [31]. The Bakken Formation overlies the Upper Devonian Three Forks Formation and underlies the Lower Mississippian Lodgepole Formation. However, the Bakken Petroleum System (BPS) included the Bakken, lower Lodgepole and upper Three Forks formations [35].

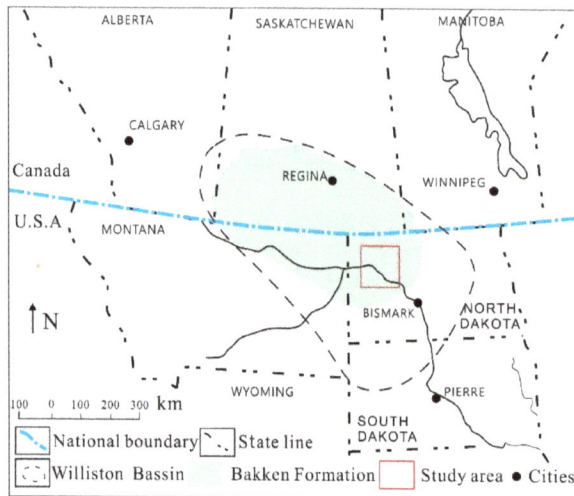

Figure 1. Location of the Williston Basin and study area.

As mentioned in the introduction to this paper, the Bakken Formation in the Williston Basin is divided into three different members: The Upper, Middle, and Lower Bakken, see Figure 2 [1]. The lithology of Upper and Lower Bakken members mainly consists of dark-gray to brownish-black to black, fissile, slightly calcareous, organic-rich shale, which was deposited in an offshore marine environment during periods of sea-level rise [31]. They serve as the main source rocks and seal rocks and have a thickness of about 8 m and 13 m, respectively. The total organic carbon (TOC) content of the Upper and Lower Bakken is between 12% and 36% with an average of 11.33% [29]. The Middle Bakken member consists of the products of the continental shelf and the shallow foreshore environment. The lithology of the middle member is highly variable and consists of a light-gray to medium-dark-gray, interbedded sequence of siltstones and sandstones with lesser amounts of shale, dolostones, and limestones rich in silt, sand, and oolites [31,39]. The oil and gas of the Bakken Formation are mainly produced from the middle member with a thickness of about 15 m. It is easy to identify the three members using well logs. For the black shale, the gamma ray (GR) log curve shows a large value, generally greater than 200 API, while the resistivity logs are generally higher than 100 Ω·m. For the Middle Bakken, the GR log values are very small and resistivity log readings are lower than those in Upper and Lower Bakken shale.

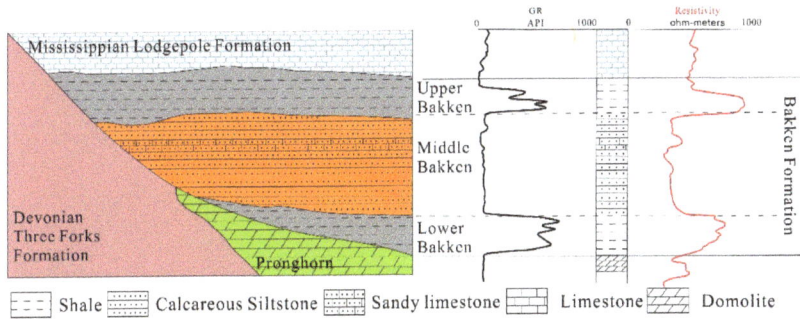

Figure 2. Basin schematic diagram and the typical log curves (modified after Liu et al. [1]).

The study area is located in North Dakota, near the center of the Williston Basin. In this part of North Dakota, the Bakken Formation reaches its maximum thickness of approximately 46 m, which is conducive to our research. In this study, we respectively chose four samples from the Upper, Middle, and Lower Bakken members (a total of 12 samples) for pore structure and main controlling factors. A series of experiments including: (1) Low-pressure N_2 adsorption; (2) X-ray diffraction (XRD); (3) total organic carbon (TOC) analysis; and (4) vitrinite reflectance (R_o) were conducted.

2.2. Experimental Methods

We used the Dmax-2500 X-ray diffraction analyzer for the X-ray diffraction (XRD) experiments following the Chinese national standard SY/T5163-2010 (SY/T5163-2010). The Cornerstone™ carbon-sulfur analyzer was used for the TOC analysis and the MPV-SP microphotometer was used for the vitrinite reflectance (R_o) measurement under the condition of 22 °C and 35% humidity.

We used the JWBK-200C surface area analyzer for the low-pressure N_2 adsorption experiments following the Chinese national standard GB/T21652-2008. Prior to the adsorption measurement, approximately 3 g of 40 to 80 mesh samples were first dried under vacuum for 12 h at high temperature (110 °C) to remove bound water and residual volatile compounds. The nitrogen adsorption was performed with a surface area and pore structure analyzer at 77 K. The adsorbed volume was measured at different relative equilibrium adsorption pressure (P_0/P) which ranges from 0.01 to 0.99, where P is the gas vapor pressure in the system and P_0 is the saturation pressure of gas. Brunauer-Emmett-Teller (BET) method [40] by multipoint calculation was used to calculate the surface area (SA). For the pore volume determination, it was calculated as the total volume of nitrogen adsorbed at the relative pressure of 0.99. The pore size distribution (PSD) was determined based on Barrett-Joyner-Halenda (BJH) model [41].

These experiments were performed in the Wuxi Department of Petroleum Geology, Research Institute of Petroleum Exploration and Development, SINOPEC.

3. Results

3.1. Mineralogical Compositions and Geochemical Characteristics

Table 1 shows the mineral composition and geochemical characteristics of the studied samples. The minerals in the samples of the Upper and Lower Bakken are dominated by clay minerals and quartz, followed by feldspar and pyrite. A small amount of dolomite and calcite is contained. The clay content ranges from 24.07% to 50.59%, with an average value of 42.23%, and the quartz content varies between 38.04% and 63.59%, with an average of 45.28%. The mineral composition of the Middle Bakken samples is also dominated by clay minerals and quartz. However, the difference between Middle Bakken and Upper/Lower Bakken is that the dolomite and calcite contents of Middle Bakken are higher, and pyrite and feldspar content of Middle Bakken are relatively smaller. The clay content is in the range of 13.27% to 36.18% with an average of 25.76%, which is less than that of the Upper/Lower Bakken samples. The Upper and Lower Bakken are abundant in organic matter, and the TOC content is between 6.58% and 15.86 wt%, with a mean value of 10.72 wt%. According to previous studies, the kerogen type of this area is mainly type II [39,42]. The R_o ranges between 0.62% and 1.11%, with an average of 0.86%, indicating the Bakken shale belongs to the low mature and mature stages in thermal maturity.

Table 1. The mineral composition and geochemical characteristics of the Bakken samples.

Sample	Members	Ro (%)	TOC (wt%)	Mineral Composition (wt%)						
				Clay	Quartz	Pyrite	Potassium Feldspar	Albite	Dolomite	Calcite
#1	Upper Bakken	0.94	13.57	44.87	39.57	4.50	6.11	3.35	1.60	-
#2	Upper Bakken	0.62	7.41	42.29	45.08	4.33	5.39	1.57	1.34	-
#3	Upper Bakken	0.88	9.14	36.24	48.90	4.32	6.65	3.34	0.23	0.32
#4	Upper Bakken	0.93	13.33	24.07	63.59	5.73	3.58	2.20	0.68	0.15
#5	Middle Bakken	-	-	27.45	31.43	3.61	7.57	5.89	15.56	8.49
#6	Middle Bakken	-	-	36.18	34.49	5.77	4.25	4.25	5.24	9.82
#7	Middle Bakken	-	-	26.16	43.28	2.65	1.31	2.61	16.44	7.55
#8	Middle Bakken	-	-	13.27	33.14	3.31	2.66	19.54	20.72	7.36
#9	Lower Bakken	0.82	6.58	49.23	38.36	5.06	2.05	3.85	1.12	0.33
#10	Lower Bakken	0.85	11.34	53.92	38.04	2.29	2.16	2.63	0.31	0.65
#11	Lower Bakken	1.11	15.86	36.63	50.27	4.58	3.20	4.05	0.74	0.45
#12	Lower Bakken	0.73	8.59	50.59	38.44	4.57	4.60	1.65	0.15	-

3.2. Pore Structure Characteristics

The adsorption/desorption models of studied samples can be divided into two stages, see Figure 3. The first stage is the single-multiple layer adsorption stage. In this stage, the adsorption curve coincides with the desorption curve, and the adsorption quantity increases gradually with the increase of relative pressure, which suggests the presence of micropores (<2 nm) in the samples. The second stage is the capillary condensation stage, i.e., separation of adsorption and desorption curves, namely, the hysteresis loop exists. The initial point of the hysteresis loop indicates that the minimum capillary begins to show capillary condensation, and the end point of the hysteresis loop suggests that the maximum capillary is filled with nitrogen. Additionally, the adsorption/desorption curves of the samples have no obvious "platform segment" at high relative pressure, indicating that the Bakken samples contain the macropores (>50 nm) beyond the measurement range of low-pressure N_2 adsorption [43].

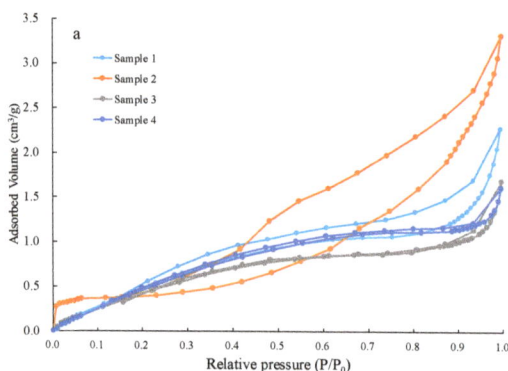

Figure 3. *Cont.*

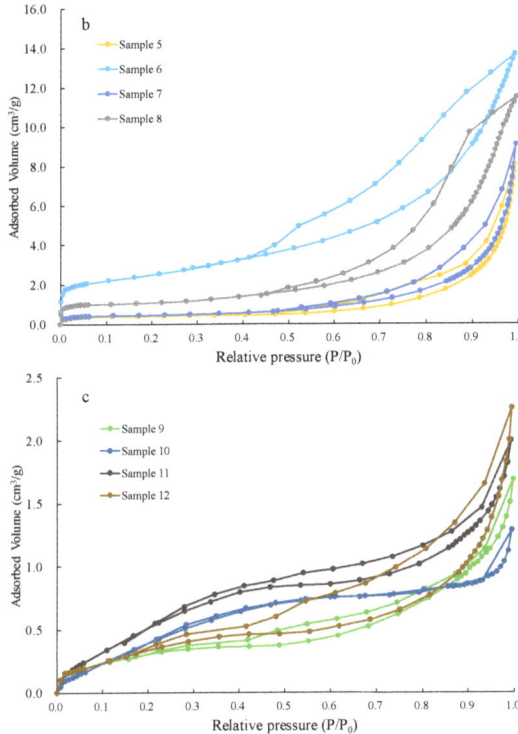

Figure 3. The nitrogen adsorption-desorption curves of the Bakken samples. (**a**) Upper Bakken; (**b**) Middle Bakken; (**c**) Lower Bakken.

The adsorption/desorption isotherms of porous media often separate under certain conditions, producing the hysteresis loop. The morphology of the hysteresis loop is closely related to the pore structure of the samples [44]. The adsorption/desorption curves and pore types of Bakken samples can be divided into three models, according to whether the hysteresis loop exists or not and the morphology of the hysteresis loop, see Figure 4. The first type has no hysteresis loop. This kind of adsorption/desorption curves results because the relative pressure of the same pore is identical during capillary condensation and evaporation and the adsorption curve coincides with the desorption curve. It generally corresponds to cylindrical pores, parallel plate-like pores, or wedge pores. The second type has a hysteresis loop without an inflection point, which belongs to types H1 or H3 based on the classification of IUPAC (International Union of Pure and Applied Chemistry) [11]. The types H1 or H3 adsorption/desorption isotherms correspond to cylindrical pores with openings at both ends or plate-shaped pores with openings on all sides, namely the open, permeable pores [45]. The hysteresis loop of the third type has an inflection point and belongs to type H2 based on the classification of IUPAC [11]. There is a sharp drop of the inflection point in the desorption curve in type H2, corresponding to ink bottle-like pores.

The adsorption/desorption curves of sample 3 and sample 10 belong to the first type, indicating the development of impermeable pores dominated in the two samples. The adsorption/desorption curves of sample 1, 4, 5, 6, 7, 8, 9, 11, and 12 belong to the second type, developing the open, permeable pores that are conducive to the flow of oil and gas, see Table 2. The curve of sample 2 belongs to the third type. It indicates that the sample contains more ink bottle-like pores but cannot deny the existence of the impermeable pores closed at one end and the open, permeable pores. That may be because the effect of these two kinds of pores on the adsorption/desorption curve is obscured by the ink bottle-like pores. The existence of ink bottle-like pores is beneficial to the adsorption of shale oil and gas. However, the closeness of the pores is not conducive to the desorption and diffusion of shale oil and gas. The condensed liquid in the bottle would evaporate and flow out quickly as the relative pressure drops to a certain extent. Therefore, it is necessary to prevent shale gas from bursting out when the relative pressure drops to a critical pressure.

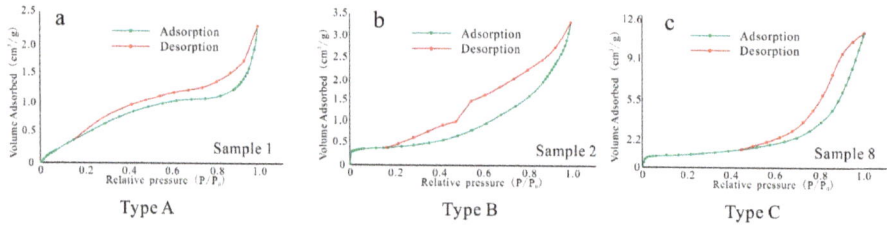

Figure 4. Three types of adsorption-desorption curves of Bakken samples. (**a**) Type A; (**b**) Type B; (**c**) Type C.

Table 2. Ideal pore model division of Bakken formation samples.

Type	Samples	Pore Morphology	Hysteresis Loop	Pore Model
I	3, 10	Pores closed at one end	No	Cylindrical pores closed at one end, wedge-shaped pores
II	1, 4, 5, 6, 7, 8, 9, 11, 12	Open pores	Yes	Cylindrical pores, parallel-plate pores
III	2	Ink bottle-like pores	Yes	Ink bottle-like pores

Table 3 depicts the pore structure parameters including pore volume, specific surface area, and mean pore diameter obtained from N_2 adsorption isotherms. We computed the total pore volume based on the Barrett–Joyner–Halenda (BJH) model [41]. The total pore volume of the Upper/Lower Bakken shale samples is relatively small with an average of 3.92 cm³/kg, while that of the Middle Bakken ranges from 13.52 cm³/kg to 24.95 cm³/kg, with an average of 18.52 cm³/kg. The average pore diameter of the Upper/Lower Bakken shale samples varies between 3.5 and 9.84 nm, with an average of 6.03 nm. However, the average pore diameter of the Middle Bakken sample is between 9.05 and 25.72 nm, with an average of 18.04 nm, which is significantly greater than that of the Upper and Lower Bakken samples. The specific surface area of the Middle Bakken samples varies between 1.28 and 8.67 m²/g, with an average of 3.73 m²/g, higher than that of the Upper and Lower Bakken samples, which is between 1.156 and 3.107 m²/g with a mean value of 2.12 m²/g. All these features indicate the samples taken from the Upper and Lower Bakken formations have greater micropores percentage contents than the Middle Bakken samples.

Table 3. Pore structure parameters of the Bakken samples.

Sample	Micropore Volume (cm³/kg)	Mesopore Volume (cm³/kg)	Macropore Volume (cm³/kg)	Total Pore Volume (cm³/kg)	BET Specific Surface Area (m²/g)	Average Pore Diameter (nm)
#1	0.48	3.06	1.07	4.61	2.886	4.67
#2	0.57	4.52	1.06	6.15	1.328	9.842
#3	0.42	2.17	0.84	3.43	2.435	4.221
#4	0.43	2.53	0.61	3.57	3.107	3.598
#5	0.56	6.38	6.58	13.52	1.276	25.716
#6	3.4	18.34	3.21	24.95	8.67	9.053
#7	0.66	7.6	7.11	15.37	1.449	23.166
#8	1.57	15.14	3.53	20.24	3.55	14.256
#9	0.37	1.79	0.83	2.99	1.156	8.804
#10	0.38	1.76	0.55	2.69	2.159	3.765
#11	0.52	2.67	0.75	3.94	2.522	4.549
#12	0.38	2.24	1.33	3.95	1.406	8.864

The pore size distributions are obtained from the N_2 adsorption isotherms using the BJH model. Figure 5 shows the pore size distribution of sample 1 and sample 8, representing the characteristics of the Upper/Lower Bakken and Middle Bakken, respectively. For the sample 1 of the Upper/Lower Bakken shale, the pore size distribution is characterized by double peaks, consisting of a (left) half peak and a (right) whole peak. The left peak occurs between 2 and 3 nm and the right wave crest is smaller, located between 20 and 60 nm, which suggests the pores of shale samples are dominated by micropores and mesopores, and with only a small amount of macropores. Nevertheless, the pore size distribution of the Middle Bakken samples is featured by a single peak distributed between 5 nm and 10 nm, indicating that the pores of tight sandstone samples mainly develop as mesopores.

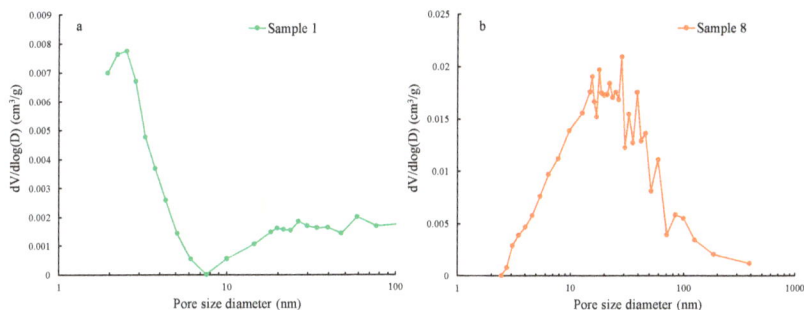

Figure 5. The pore size distributions of the samples 1 and 8. (**a**) Sample 1 of the Upper/Lower Bakken; (**b**) Sample 8 of the Middle Bakken.

4. Discussion

4.1. Correlation of the Pore Structure Parameters

We analyze the correlations of the average pore diameter, the specific surface area, and the total pore volume. In order to describe the correlation, the degree of correlation defined in this study is depicted in Table 4. Figure 6a shows the relationship between the average pore diameter and the total pore volume. The average pore diameter of the Middle Bakken sample is strongly correlated with the total pore volume with the coefficient of determination (R^2) of 0.99. The total pore volume decreases as the average pore diameter increases. In contrast, the average pore diameter of the Upper and Lower Bakken samples was positively correlated with the total pore volume, but the correlation was low, and the coefficient of determination was only 0.22. Figure 6b shows that the average pore diameter has a high and negative correlation with the specific surface area. The correlation coefficient between the two parameters for Upper and Lower Bakken samples is 0.81, and for the Middle Bakken sample is 0.86. As the average pore diameter increases, the specific surface area of the sample gradually becomes smaller, which indicates that the greater the small pore content in the Bakken sample is, the larger the specific surface area. From Figure 6c, a strong positive correlation between the total pore volume and the specific surface area in the Middle Bakken samples, with a correlation coefficient of 0.92 can also be observed. However, no correlation between the two parameters of the Upper and Lower Bakken samples was found. These observations suggest that there is a significant difference in the pore structure between the Upper/Lower Bakken and the Middle Bakken, and the control factors are also obviously different.

Table 4. The defined degree of correlation.

Ranges for the Coefficient of Determination (R²)	Degree of Correlation
0.9–1	Strong
0.7–0.9	High
0.4–0.7	Medium
0.2–0.4	Low
0.0–0.2	Not exist

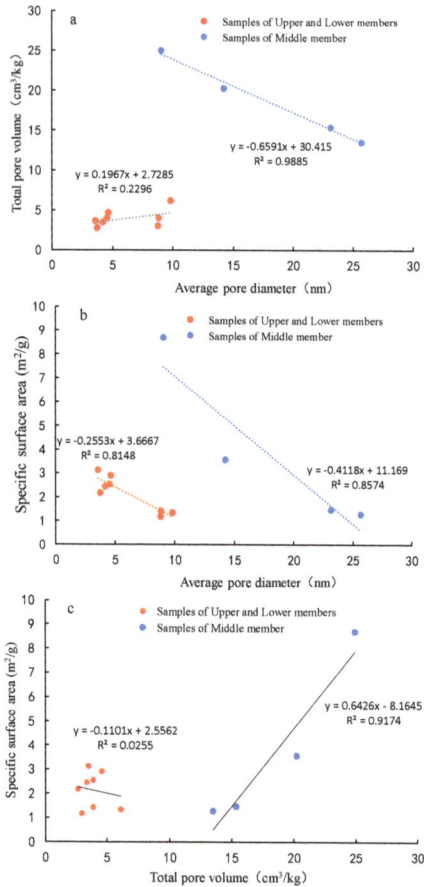

Figure 6. The interrelation of pore structure parameters of the Bakken samples. (**a**) Total pore volume vs. average pore diameter; (**b**) Specific surface area vs. average pore diameter; (**c**) Specific surface area vs. total pore volume.

4.2. Control of Mineral Composition on Pore Structure

Figure 7 shows the relationships between the quartz content and the pore structure of samples of the Bakken formation. It can be seen that the specific surface area is moderately correlated with the quartz content. For the samples of the Upper and Lower Bakken, the specific surface area decreases with the increase of quartz content, see Figure 7a. However, the specific surface area increases as the quartz content increases for the Middle Bakken samples, as shown in Figure 7d. The quartz content has a weak control over the total pore volume in the Upper and Lower Bakken samples but has an obvious control over that of the Middle Bakken samples, see Figure 7b,e. In the Upper and Lower Bakken samples, the greater the quartz content, the larger the average pore diameter, while there is an opposite trend in the Middle Bakken samples, see Figure 7c,f. These results demonstrate that the quartz has an obvious control over the pore structure of the Middle Bakken tight rock, but its effect on the Upper and Lower Bakken shale is different.

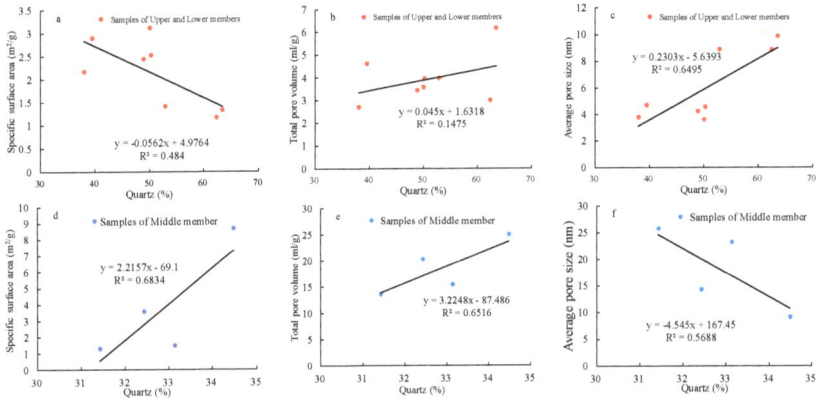

Figure 7. The relationship between pore structure parameters and quartz content. (**a**) Specific surface area vs. quartz content for Upper/Lower Bakken samples; (**b**) Total pore volume vs. quartz content for Upper/Lower Bakken samples; (**c**) Average pore size vs. quartz content for Upper/Lower Bakken samples; (**d**) Specific surface area vs. quartz content for Middle Bakken samples; (**e**) Total pore volume vs. quartz content for Middle Bakken samples; (**f**) Average pore size vs. quartz content for Middle Bakken samples.

Figure 8 displays the relationships between clay content and pore structures of the Bakken samples. Upon increasing the clay content, the average pore diameter reduces while the specific surface area increases. The tiny intergranular pores between clay minerals increase as the clay content becomes high, which leads to the reduction of the average pore diameter. According to the above discussion, the smaller pores have a larger specific surface area. Thus, more intergranular pores result in a larger total specific surface area. In the Upper and Lower Bakken samples, the lack of any correlation between the clay content and the total pore volume is due to a coefficient of determination of 0.15. However, a positive correlation exists between the clay content and total pore volume in the Middle Bakken samples. That is to say, the total pore volume raises with the increasing of the clay content, as shown in Figure 8e.

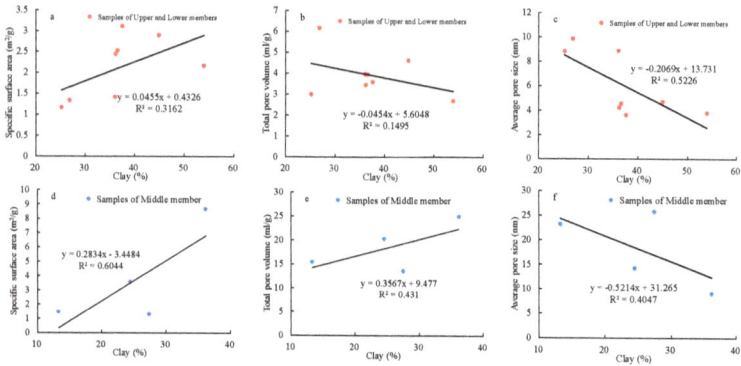

Figure 8. The relationship between pore structure parameters and clay content. (**a**) Specific surface area vs. clay content for Upper/Lower Bakken samples; (**b**) Total pore volume vs. clay content for Upper/Lower Bakken samples; (**c**) Average pore size vs. clay content for Upper/Lower Bakken samples; (**d**) Specific surface area vs. clay content for Middle Bakken samples; (**e**) Total pore volume vs. clay content for Middle Bakken samples; (**f**) Average pore size vs. clay content for Middle Bakken samples.

4.3. Relationship between TOC Content and Pore Structure

The TOC content of shale can not only demonstrate the hydrocarbon potential of source rock but is also an important parameter to control the shale pore structure. Generally, the average pore diameter of the organic pore is far less than that of the inorganic pore [46]. Figure 9 displays the relationships between the TOC and the pore structure parameters in the Upper and Lower Bakken shale samples. It can be seen that the TOC is linearly correlated with the specific surface area and the average pore diameter. The coefficients of determination are 0.7 and 0.58, respectively. The specific surface area gradually increases and the average pore diameter gradually reduces as the TOC content increases. This indicates that the TOC is one of the main factors controlling the average pore diameter and the specific surface area of shale.

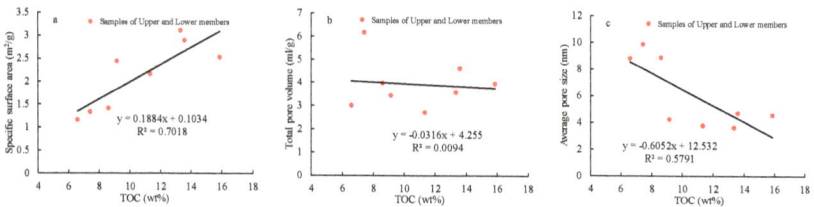

Figure 9. The relationship between total organic carbon content and pore structure parameters in the Upper/Lower Bakken samples. (**a**) Specific surface area vs. TOC; (**b**) Total pore volume vs. TOC; (**c**) Average pore size vs. TOC.

4.4. Relationship between Thermal Maturity and Pore Structure

The thermal maturity has a complicated effect on the shale pore structure. The degree of thermal evolution not only controls the change of the pore structure but also controls the transformation between the clay minerals, causing the pore structure changes of the clay pores. For example, the montmorillonite can transform into illite with maturity. Thus, the pore structure changes as different clay minerals have different pore shapes [47,48]. The thermal evolution of organic matter in the Upper and Lower Bakken shale is relatively low and belongs to early catagenesis, according to the vitrinite reflectance R_o ranges in Table 1. The relationships between the R_o and the pore structure in the Upper and Lower Bakken shale are shown in Figure 10. It can be seen that the R_o has a poor correlation with the total pore volume of shale. In addition, the specific surface area gradually increases and the average pore diameter gradually decreases with the increase of R_o. It suggests that a large number of micropores are developed in organic matter during the thermal evolution. Micropores usually have a larger specific surface area compared with macropores. Consequently, the degree of thermal evolution is one of the main factors controlling the pore structure of the Upper and Lower Bakken shale.

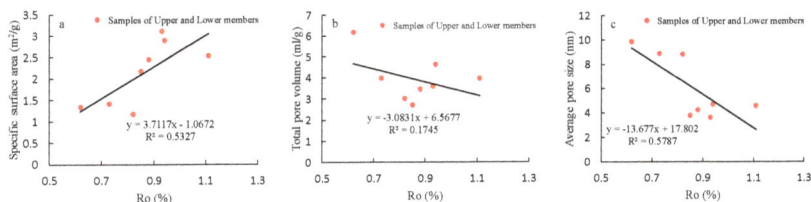

Figure 10. The relationship between pore structure parameters and vitrinite reflectance (R_o) in the Upper/Lower Bakken samples. (**a**) Specific surface area vs. R_o; (**b**) Total pore volume vs. R_o; (**c**) Average pore size vs. R_o.

As for the effect of depositional environment on the pore structure, Zhai et al. [49] reported that in the reduced marine sedimentary environment, larger pores and micro-fracture can be observed in shale. Source rocks in a salinized strong reduction terrestrial sedimentary environment developed laminated layers, while in weak reduction terrestrial sedimentary environment, the rock structure is dominated by blocks and the amount of micro-fracture is less. Bakken Formation was deposited in the reduced marine sedimentary environment. According to Table 3, large pores of Bakken Formation account for an average of 25.7%, which is conducive to oil and gas storage. In different sedimentary environments, the lithology and organic matter contents are different, which results in differences in the pore structure. However, the sedimentary environments vary slightly in a specific study area; hence, the effect of depositional environment on the pore structure should be weak.

5. Conclusions

(1) The Upper and Lower Bakken shale are characterized by rich organic matter and low maturity. TOC content ranges from 6.58% to 15.86%, with an average value of 10.72%. The R_o varies between 0.62% and 1.11%, with an average value of 0.86%.

(2) The pores of the Upper and Lower Bakken shale are dominated by micropores. The specific surface area is distributed from 1.156 to 3.107 m^2/g, with an average of 2.12 m^2/g. The total pore volume is distributed from 2.69 to 6.15 cm^3/kg, with a mean of 3.92 cm^3/kg. The average pore diameter is distributed from 3.76 to 9.84 nm, with a mean of 6.03 nm.

(3) The pores of the Middle Bakken formation are dominated by mesopores. The specific surface area ranges from 1.276 to 8.67 m^2/g, and the average value is 3.73 m^2/g, which is significantly higher than that of the Upper and Lower Bakken shale. The total pore volume is distributed from 13.52 to 24.95 cm^3/kg, with an average value of 18.52 cm^3/kg, which is also higher than that of the Upper and Lower Bakken shale. The average pore diameter distribution varies between 9.05 nm and 25.72 nm with a mean of 18.04 nm.

(4) The main controlling factors of the pore structure of the Upper and Lower Bakken shale are total organic carbon content and thermal maturity. With the increase of total organic carbon content, the specific surface area of Bakken shale gradually increases, and the average pore radius gradually decreases. With the increase of thermal maturity, the specific surface area of Bakken shale gradually increases, and the average pore radius gradually decreases.

(5) The main controlling factor of the pore structure of the Middle Bakken sample is the content of clay and quartz. With the increase of clay and quartz content, the specific surface area in the Middle Bakken gradually increases, and the average pore radius gradually decreases.

Author Contributions: Conceptualization, Y.L. and B.S.; Methodology, Y.L. and B.S.; Investigation, Z.Y. and P.Z.; Writing-Original Draft Preparation, Y.L.; Writing-Review & Editing, Z.Y. and P.Z.

Funding: This research is was funded by China National Science and Technology Major Project (Grant No. 2016ZX05010-001), the National Basic Research Program of China (Grant No. 2015CB250901) and the National Natural Science Foundation of China (Grant No. 41574121).

Acknowledgments: The authors wish to thank North Dakota Geological Survey, Core Library, for giving us access to the Bakken core samples. The authors thank the Chunxiao Li at the University of North Dakota for helping drill the samples.

Conflicts of Interest: The authors declare no conflict of interest.

References

1. Liu, K.; Ostadhassan, M.; Zhou, J.; Gentzis, T.; Rezaee, R. Nanoscale pore structure characterization of the Bakken shale in the USA. *Fuel* **2017**, *209*, 567–578. [CrossRef]
2. Wood, D.A.; Hazra, B. Characterization of organic-rich shales for petroleum exploration & exploitation: A review—Part 1: Bulk properties, multi-scale geometry and gas adsorption. *J. Earth Sci.* **2017**, *28*, 739–757. [CrossRef]
3. Mendhe, V.A.; Mishra, S.; Khangar, R.G.; Kamble, A.D.; Kumar, D.; Varma, A.K.; Singh, H.; Kumar, S.; Bannerjee, M. Organo-petrographic and pore facets of Permian shale beds of Jharia Basin with implications to shale gas reservoir. *J. Earth Sci.* **2017**, *28*, 897–916. [CrossRef]
4. Jia, C.Z.; Zou, C.N.; Li, J.Z.; Li, D.H.; Zheng, M. Assessment criteria, main types, basic features and resource prospects of the tight oil in China. *Acta Pet. Sin.* **2012**, *33*, 343–350.
5. Zhao, P.; Ma, H.; Rasouli, V.; Liu, W.; Cai, J.; Huang, Z. An improved model for estimating the TOC in shale formations. *Mar. Pet. Geol.* **2017**, *83*, 174–183. [CrossRef]
6. Dullien, F.A. *Porous Media: Fluid Transport and Pore Structure*; Academic Press: New York, NY, USA, 2012.
7. Cai, J.; Wei, W.; Hu, X.; Liu, R.; Wang, J. Fractal characterization of dynamic fracture network extension in porous media. *Fractals* **2017**, *25*, 1750023. [CrossRef]
8. Verweij, J.M. *Hydrocarbon Migration Systems Analysis*; Elsevier: Amsterdam, The Netherlands, 1993.
9. Kong, L.; Ostadhassan, M.; Li, C.; Tamimi, N. Pore characterization of 3D-printed gypsum rocks: A comprehensive approach. *J. Mater. Sci.* **2018**, *53*, 5063–5078. [CrossRef]
10. Cai, J.; Yu, B.; Zou, M.; Mei, M. Fractal analysis of invasion depth of extraneous fluids in porous media. *Chem. Eng. Sci.* **2010**, *65*, 5178–5186. [CrossRef]
11. International Union of Pure and Applied Chemistry. Physical chemistry division commission on colloid and surface chemistry, subcommittee on characterization of porous solids: recommendations for the characterization of porous solids (technical report). *Pure Appl. Chem.* **1994**, *66*, 1739–1758. [CrossRef]
12. Bai, B.; Zhu, R.; Wu, S.; Yang, W.; Gelb, J.; Gu, A.; Zhang, X.; Su, L. Multi-scale method of Nano (Micro)-CT study on microscopic pore structure of tight sandstone of Yanchang Formation, Ordos Basin. *Pet. Explor. Dev.* **2013**, *40*, 354–358. [CrossRef]
13. Tüysüz, H.; Lehmann, C.W.; Bongard, H.; Tesche, B.; Schmidt, R.; Schüth, F. Direct imaging of surface topology and pore system of ordered mesoporous silica (MCM-41, SBA-15, and KIT-6) and nanocast metal oxides by high resolution scanning electron microscopy. *J. Am. Chem. Soc.* **2008**, *130*, 11510–11517. [CrossRef] [PubMed]
14. Chalmers, G.R.; Bustin, R.M.; Power, I.M. Characterization of gas shale pore systems by porosimetry, pycnometry, surface area, and field emission scanning electron microscopy/transmission electron microscopy image analyses: Examples from the Barnett, Woodford, Haynesville, Marcellus, and Doig units. *AAPG Bull.* **2012**, *96*, 1099–1119. [CrossRef]
15. Wang, Z.L. Transmission electron microscopy of shape-controlled nanocrystals and their assemblies. *J. Phys. Chem. B* **2000**, *104*, 1153–1175. [CrossRef]

16. Wang, X.; Hou, J.; Liu, Y.; Zhao, P.; Ma, K.; Wang, D.; Ren, X.; Yan, L. Overall PSD and fractal characteristics of tight oil reservoirs: A case study of Lucaogou Formation in Junggar Basin, China. *Fractals* **2019**, *27*, 1940005. [CrossRef]

17. Wang, X.; Hou, J.; Song, S.; Wang, D.; Gong, L.; Ma, K.; Liu, Y.; Li, Y.; Yan, L. Combining pressure-controlled porosimetry and rate-controlled porosimetry to investigate the fractal characteristics of full-range pores in tight oil reservoirs. *J. Pet. Sci. Eng.* **2018**, *171*, 353–361. [CrossRef]

18. Wang, L.; Zhao, N.; Sima, L.; Meng, F.; Guo, Y. Pore structure characterization of the tight reservoir: systematic integration of mercury injection and nuclear magnetic resonance. *Energy Fuels* **2018**, *32*, 7471–7484. [CrossRef]

19. Zhao, P.; Sun, Z.; Luo, X.; Wang, Z.; Mao, Z.; Wu, Y.; Xia, P. Study on the response mechanisms of nuclear magnetic resonance (NMR) log in tight oil reservoirs. *Chin. J. Geophys.* **2016**, *29*, 1927–1937.

20. Kruk, M.; Jaroniec, M.; Sayari, A. Application of large pore MCM-41 molecular sieves to improve pore size analysis using nitrogen adsorption measurements. *Langmuir* **1997**, *13*, 6267–6273. [CrossRef]

21. Yang, F.; Ning, Z.; Liu, H. Fractal characteristics of shales from a shale gas reservoir in the Sichuan Basin, China. *Fuel* **2014**, *115*, 378–384. [CrossRef]

22. Yang, R.; He, S.; Hu, Q.; Sun, M.; Hu, D.; Yi, J. Applying SANS technique to characterize nano-scale pore structure of Longmaxi shale, Sichuan Basin (China). *Fuel* **2017**, *197*, 91–99. [CrossRef]

23. Ferrero, G.A.; Preuss, K.; Fuertes, A.B.; Sevilla, M.; Titirici, M.M. The influence of pore size distribution on the oxygen reduction reaction performance in nitrogen doped carbon microspheres. *J. Mater. Chem. A* **2016**, *4*, 2581–2589. [CrossRef]

24. Sandoval-Díaz, L.E.; Aragon-Quiroz, J.A.; Ruíz-Cardona, Y.S.; Domínguez-Monterroza, A.R.; Trujillo, C.A. Fractal analysis at mesopore scale of modified USY zeolites by nitrogen adsorption: A classical thermodynamic approach. *Microporous Mesoporous Mater.* **2017**, *237*, 260–267. [CrossRef]

25. Saidian, M.; Prasad, M. Effect of mineralogy on nuclear magnetic resonance surface relaxivity: A case study of Middle Bakken and Three Forks formations. *Fuel* **2015**, *161*, 197–206. [CrossRef]

26. Wang, L.; Fu, Y.; Li, J.; Sima, L.; Wu, Q.; Jin, W.; Wang, T. Mineral and pore structure characteristics of gas shale in Longmaxi formation: A case study of Jiaoshiba gas field in the southern Sichuan Basin, China. *Arab. J. Geosci.* **2016**, *9*, 733. [CrossRef]

27. Su, S.; Jiang, Z.; Shan, X.; Zhang, C.; Zou, Q.; Li, Z.; Zhu, R. The effects of shale pore structure and mineral components on shale oil accumulation in the Zhanhua Sag, Jiyang Depression, Bohai Bay Basin, China. *J. Pet. Sci. Eng.* **2018**, *165*, 365–374. [CrossRef]

28. Webster, R.L. Petroleum source rocks and stratigraphy of the Bakken Formation in North Dakota. *AAPG Bull.* **1984**, *68*, 953.

29. Meissner, F.F. Petroleum geology of the Bakken Formation Williston Basin, North Dakota and Montana. In *1991 Guidebook to Geology and Horizontal Drilling of the Bakken Formation: Billings*; Hanson, W.B., Ed.; Montana Geological Society: Billings, MT, USA, 1991.

30. Smith, M.G.; Bustin, R.M. Lithofacies and paleoenvironments of the Upper Devonian and Lower Mississippian Bakken Formation, Williston Basin. *Bull. Can. Pet. Geol.* **1996**, *44*, 495–507.

31. Pitman, J.K.; Price, L.C.; LeFever, J.A. *Diagenesis and Fracture Development in the Bakken Formation, Williston Basin: Implications for Reservoir Quality in the Middle Member (No. 1653)*; U.S. Department of the Interior, U.S Geological Survey: Denver, CO, USA, 2001.

32. Havens, J.B.; Batzle, M.L. Minimum horizontal stress in the Bakken Formation. In Proceedings of the 45th U.S. Rock Mechanics/Geomechanics Symposium, San Francisco, CA, USA, 26–29 June 2011.

33. Sayers, C.M.; Dasgupta, S. Elastic anisotropy of the Middle Bakken Formation. *Geophysics* **2014**, *80*, D23–D29. [CrossRef]

34. Liu, K.; Ostadhassan, M. Quantification of the microstructures of Bakken shale reservoirs using multi-fractal and lacunarity analysis. *J. Nat. Gas Sci. Eng.* **2017**, *39*, 62–71. [CrossRef]

35. Li, H.; Hart, B.; Dawson, M.; Radjef, E. Characterizing the Middle Bakken: Laboratory measurement and rock typing of the Middle Bakken Formation. In Proceedings of the Unconventional Resources Technology Conference, San Antonio, TX, USA, 20–22 July 2015.

36. Liu, K.; Ostadhassan, M.; Kong, L. Fractal and multifractal characteristics of pore throats in the Bakken Shale. *Transp. Porous Med.* **2018**, 1–20. [CrossRef]

37. Sonnonberg, S.A.; Jin, H.; Sarg, J.F. Bakken mudrocks of the Williston Basin, world class source rocks. In Proceedings of the AAPG Annual Convention and Exhibition, Houston, TX, USA, 11–13 April 2011.

38. Zhao, P.; Ostadhassan, M.; Shen, B.; Liu, W.; Abarghani, A.; Liu, K.; Luo, M.; Cai, J. Estimating thermal maturity of organic-rich shale from well logs: Case studies of two shale plays. *Fuel* **2019**, *235*, 1195–1206. [CrossRef]

39. Li, C.; Ostadhassan, M.; Gentzis, T.; Kong, L.; Carvajal-Ortiz, H.; Bubach, B. Nanomechanical characterization of organic matter in the Bakken formation by microscopy-based method. *Mar. Pet. Geol.* **2018**, *96*, 128–138. [CrossRef]

40. Brunauer, S.; Emmett, P.H.; Teller, E.J. Adsorption of gases in multimolecular layers. *J. Am. Chem. Soc.* **1938**, *60*, 309–319. [CrossRef]

41. Barrett, E.P.; Joyner, L.G.; Halenda, P.P. The determination of pore volume and area distributions in porous substances. I. Computations from nitrogen isotherms. *J. Am. Chem. Soc.* **1951**, *73*, 373–380. [CrossRef]

42. Abarghani, A.; Ostadhassan, M.; Gentzis, T.; Carvajal-Ortiz, H.; Bubach, B. Organofacies study of the Bakken source rock in North Dakota, USA, based on organic petrology and geochemistry. *Int. J. Coal Geol.* **2018**, *188*, 79–93. [CrossRef]

43. Groen, J.C.; Peffer, L.A.; Pérez-Ramírez, J. Pore size determination in modified micro-and mesoporous materials. Pitfalls and limitations in gas adsorption data analysis. *Microporous Mesoporous Mater.* **2003**, *60*, 1–17. [CrossRef]

44. Sing, K.S. Reporting physisorption data for gas/solid systems with special reference to the determination of surface area and porosity (Recommendations 1984). *Pure Appl. Chem.* **1985**, *57*, 603–619. [CrossRef]

45. Jiang, F.; Chen, D.; Wang, Z.; Xu, Z.; Chen, J.; Liu, L.; Huyan, Y.; Liu, Y. Pore characteristic analysis of a lacustrine shale: A case study in the Ordos Basin, NW China. *Mar. Pet. Geol.* **2016**, *73*, 554–571. [CrossRef]

46. Guo, X.; Li, Y.; Liu, R.; Wang, Q. Characteristics and controlling factors of micropore structures of the Longmaxi Shale in the Jiaoshiba area, Sichuan Basin. *Nat. Gas Ind. B* **2014**, *1*, 165–171. [CrossRef]

47. Wei, X.; Liu, R.; Zhang, T.; Liang, X. Micro-pores structure characteristics and development control factors of shale gas reservoir: A case of Longmaxi formation in XX area of southern Sichuan and northern Guizhou. *Nat. Gas Geosci.* **2013**, *24*, 1048–1059.

48. Liu, Y.; Zhang, B.; Dong, Y.; Qu, Z.; Hou, J. The determination of variogram in the presence of horizontal wells—An application to a conglomerate reservoir modeling, East China. *J. Pet. Sci. Eng.* **2018**. [CrossRef]

49. Zhai, Z.; Wang, X.; Li, Z.; Li, J.; Liu, Q. The influence of shale sedimentary environments on oil and gas potential: Examples from China and North America. *Geol. J. China Univ.* **2016**, *22*, 690–697.

energies

MDPI

Article

Impacts of the Base-Level Cycle on Pore Structure of Mouth Bar Sand Bars: A Case Study of the Paleogene Kongdian Formation, Bohai Bay Basin, China

Xixin Wang [1,2], Jiagen Hou [1,2,*], Yuming Liu [1,2], Ling Ji [3], Jian Sun [3] and Xun Gong [3]

[1]   State Key Laboratory of Petroleum Resource and Prospecting-Beijing, Beijing 102249, China;
     xixin.wang@und.edu (X.W.); liuym@cup.edu.cn (Y.L.)
[2]   College of Geosciences, China University of Petroleum-Beijing, Beijing 102249, China
[3]   CNPC Da Gang Oilfield, Tianjin 061100, China; dg_jiling@petrochina.com.cn (L.J.);
     dg_sunjian@petrochina.com.cn (J.S.); c3_gongxun@petrochina.com.cn (X.G.)
*   Correspondence: houjg63@cup.edu.cn

Received: 23 August 2018; Accepted: 29 September 2018; Published: 1 October 2018

Abstract: The pore structure of rocks can affect fluid migration and the remaining hydrocarbon distribution. To understand the impacts of the base-level cycle on the pore structure of mouth bar sand bodies in a continental rift lacustrine basin, the pore structure of the mouth bar sand bodies in the ZVC (ZV4 + ZV5) of the Guan195 area was studied using pressure-controlled mercury injection (PMI), casting sheet image and scanning electron microscopy (SEM). The results show that three types of pores exist in ZVC, including intergranular pores, dissolution pores, and micro fractures. The porosity is generally between 1.57% and 44.6%, with a mean value of 19.05%. The permeability is between 0.06 μm^2 and 3611 μm^2, with a mean value of 137.56 μm^2. The pore structure heterogeneity of a single mouth bar sand body in the early stage of the falling period of short-term base-level is stronger than that in the late stage. During the falling process of the middle-term base level, the pore structure heterogeneity of a late single mouth bar sand body is weaker than that of an early single mouth bar sand body. In the long-term base-level cycle, the pore structure heterogeneity of mouth bar sand bodies becomes weaker with the falling of the base-level.

Keywords: base-level cycle; pore structure; mouth bar sand body; Huanghua Depression

1. Introduction

Pore structure refers to the shape, size, distribution and the connective relationship of pores and pore throats [1,2]. Pore structure is an important factor controlling the porosity and permeability of rock, and hence the potential of remaining oil and gas recovery [3–5]. Investigations of the factors controlling pore structure are critical to the exploration and development of an oil field [6–10]. Many scholars have conducted extensive research work from different perspectives, using various methods such as rate-controlled mercury injection testing [11–13], pressure-controlled mercury injection (PMI) testing [14–16], scanning electron microscopy (SEM) [17–21] and multiscale computed tomography (CT) imaging technology [22–24].

The pore structure of the reservoir is not only controlled by tectonic movement and diagenesis but also by the influence of sedimentary factors (e.g., Ju et al. [25], Tang et al. [26], Wang et al. [27], Wang et al. [28]). The controlling effects of sedimentation on reservoir characteristics are mainly reflected by the changes of the base level. The influence of base level cycle on the reservoir characteristics can be divided into autocycles and allocycles [29]. Autocycles refer to the sedimentary cycles formed by the change of the internal factors of the basin under the condition of a relatively stable sedimentary background. Allocycles refer to the sedimentary cycles formed by the change of the source supply conditions, the tectonic movement of the ocean floor or eustatic sea level change. Short-term base-level cycles (SSC) is the minimal genetic stratigraphic unit and is also the basis for high-resolution sequence stratigraphy, which can be identified in outcrop, core and well logging curve data [30,31]. Middle-term base-level cycles (MSC) correspond to multiple mouth bar sand bodies in this study. Such factors as stratum erosion surface and regional deposition scour surface are taken as the basis upon which to divide the long-term base-level cycle (LSC), which is equivalent to three order sequence stratigraphic units [32]. Miall commented that the base level cycle process determines the rock types, lithological characteristics, geometric shape and spatial continuity of reservoir rock, and controls pore structure, size, arrangement, and sorting of detritus [33].

The pore structure of rock can be obtained from mercury injection testing, which is one of the most common methods [12,34–37]. Toledo et al. [38] investigated pore-space statistics based on capillary pressure curves. Zhao et al. [39] explored the pore structure of tight oil reservoirs using pressure-controlled porosimetry. Xi et al. [12] studied the influence of pore size distribution on the oiliness and the porosity and permeability of tight sandstones based on PMI. Wang et al. [5] calculated the fractal dimension of the pore structure using the combination of PMI and rate-controlled mercury injection.

Many studies were devoted to the controlling effect of base level cycles on macroscopic characteristics of reservoirs [40,41]. However, very few studies have focused on the effect of base level cycles on the pore structure of a reservoir [42]. This study took mouth bar sand bodies of Guan195 area as an example, investigated the effects of different levels of base level cycles on the pore structure.

2. Geological Setting

The Bohai Bay Basin, located in eastern China, has an area of approximately 1.7×10^4 km^2 (Figure 1a), in which there are nine first class tectonic zones (Figure 1a). The Wang guantun oilfield, located in the Huanghua depression, is one of the most important oil-gas production areas. More than 1700 wells have been drilled in Wang guantun oilfield. Wang guantun oilfield is divided into two main tectonic zones by Kongdong fault, and some secondary faults make the structure of the oilfield more complicated (Figure 1b). Sediments comprise the Paleogene Kongdian (Ek), Shahejie (Es), Dongying (Ed) formations, the Neogene Guantao (Ng), Minghuazhen (Nm) Formations, the Quaternary Pingyuan (Q) formations (Figure 2). Oil shale of Ek2 (the second member of the Kongdian Formation) is the main source rock and oil-gas has migrated into suitable trap locations of Ek1 along connecting faults [43,44].

The studied interval is the six segments (ZV) in Ek1 (the first member of the Kongdian Formation); the ZV can be divided into seven layers (from ZV1 to ZV7). This research is focused on ZV4 and ZV5, called the ZVC (Figure 2). The ZVC is 40–80 m thick and mainly includes light gray fine sandstone, silty sandstone and mudstone (Figure 2). The reservoir sand body is mainly quartz sandstone, and the cementation type is mainly cement and calcareous cementation. The reservoir is characterized by high component maturity and relatively high textural maturity, and the particles are mainly characterized by point contact and point line contact; the sediment was mainly in the early diagenetic stage. The buried depth of the ZVC in this area and the spans of the buried depth are small (1820 m–1900 m), so diagenesis had a limited effect on reservoir characteristics. Therefore, the macroscopic and microscopic characteristics of the reservoir were mainly controlled by sedimentation [45]. Sediments including delta plain subfacies, delta front subfacies and front delta subfacies are deposited in the ZV in the study area with stronger water energy [46], where many mouth bar sand bodies are deposited in the horizontal and vertical, and mouth bar sand bodies are of cardinal significance for reservoir rock.

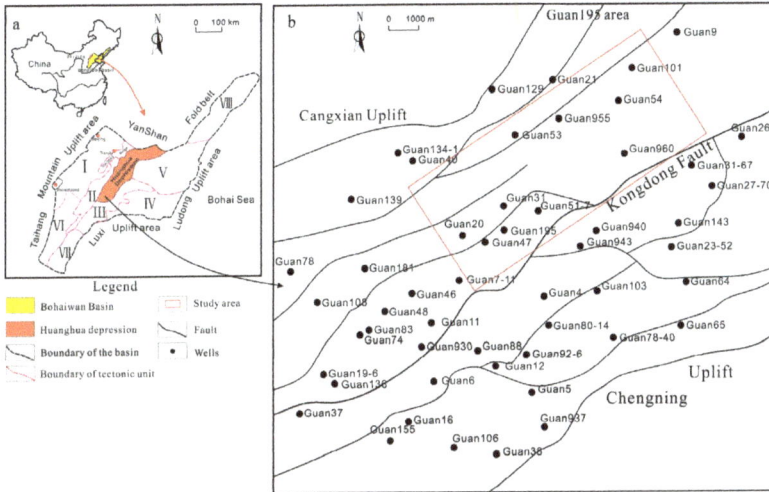

Figure 1. (**a**) Location map of the study area and sub-tectonic units of the Bohaiwan Basin. (I) JiZhong depression zone, (II) CangXian uplift zone, (III) ChengNing uplift zone, (IV) JiYang depression zone, (V) BoZhong depression zone, (VI) XingHeng uplift zone, (VII) LinQing depression zone, (VIII) LiaoHe uplift zone; (**b**) The sub-tectonic units of the study area and well locations.

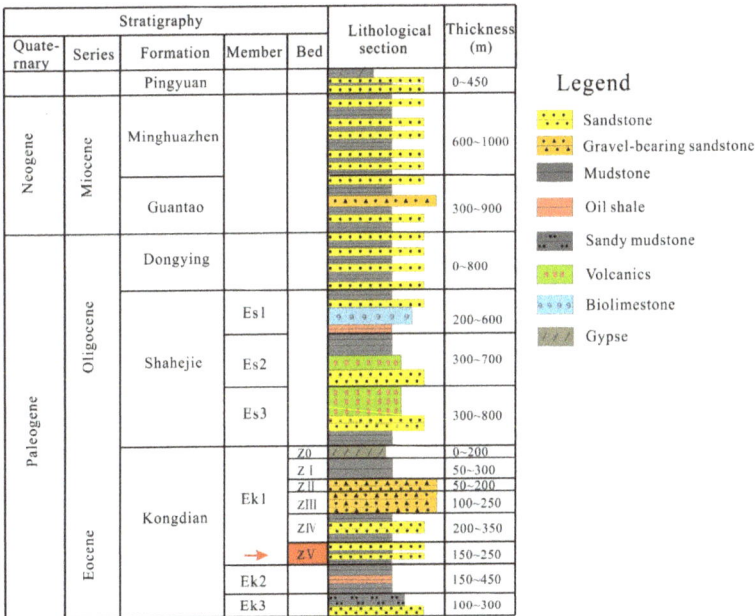

Figure 2. Stratigraphy, lithology, and thickness of Kongdian Formation in the study area (modified from Neng et al. [44]).

3. Data and Experimental Method

The EK1 member of the Wang guantun oilfield is one of the most oil-gas production members; there are more than 1700 wells, including vertical wells and inclined wells, have been drilled through the member. Among the wells, there are about 220 core wells, and consequently, the data necessary for this research are available, such as core, wireline-log, mud logging, oil test, and mercury injection data. Reservoir porosity and permeability of 4547 data points were collected from Dagang Oilfield, China National Petroleum Corporation (CNPC) (Beijing, China). 744 mercury injection testing data were tested in Research Institute of Petroleum Exploration & Development, Dagang Oilfield, CNPC.

One of the most commonly used methods to characterize pore structure characteristics is PMI testing. To obtain multifarious microscopic parameters, mercury, a non-wetting phase fluid, is forced into all the effective space and mercury intake volume is measured by a mercury porosimeter [47]. The pressure gradually increased to the highest value and slowly decreased thereafter, and the mercury extruded from the samples. Thus, test curves and different characterization parameters, including maximum pore-throat radius (R_{max}), average pore-throat radius (R_{ave}), residual injection saturation (S_r), maximum mercury injection saturation (S_{max}), sorting coefficient of pore throat, coefficient of variation of pore throat, homogeneity coefficient, characteristic structure coefficient, lithology coefficient, etc. were obtained according to the amount of mercury and the corresponding pressure.

4. Geological and Laboratory Modelling

4.1. Stratigraphic Correlation and Cycle Identification

According to the formation characteristics of ZVC bed of EK1 (the first member of the Kongdian Formation), eight single layers were divided within the ZVC bed. The thickness of a single sand body ranges from 2 to 8 m and most of them are between 3.5 m and 5 m. The thickness difference of a given sand body in the horizontal is mainly controlled by the depositional environment and the location of the sequence, while the thickness difference in the vertical is mainly controlled by such factors as the expansion and contraction of the lake, provenance supply, climate, and tectonic movement. A total of one LSC, three MSC and eight SSC were further divided within the ZVC according to the characteristics of core, lithology and well logging curve (Figure 3). Based on those data, stratigraphic correlation and cycle identification were carried out on the profile (Figure 4).

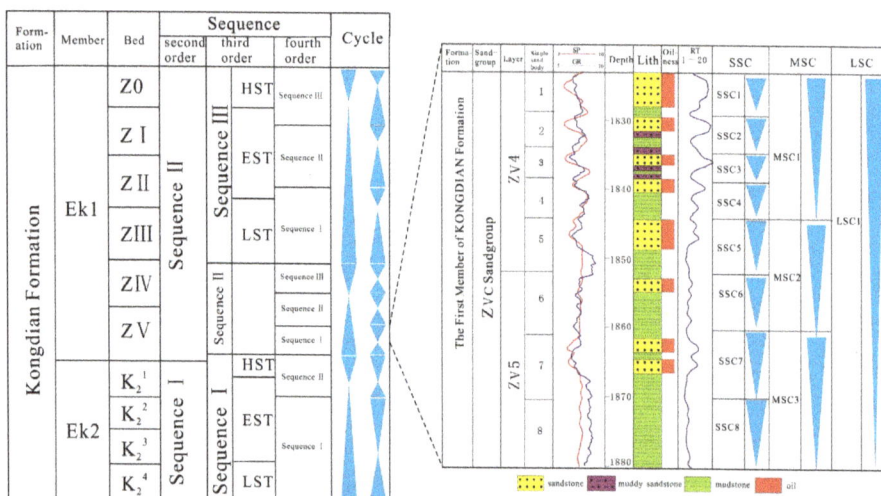

Figure 3. The synthetic pattern of the base-level cycle of the study area. SP, spontaneous potential curve; GR, gamma-ray curve; RT, resistivity curve.

4.2. Sedimentary Microfacies Characteristics

The source of Wang guantun area comes mainly from two directions, the Cangxian uplift zone and Xuhei uplift zone [48]. Sediments from the north formed fan delta deposits in the Guan195 area, and sediments from the southeast formed an alluvial fan deposit in the ZV bed [46]. Based on a high resolution sequence stratigraphic framework, mouth bar, underwater distributary channel, sheet sand, and distributary bay micro-facies were distinguished in the delta front subfacies in Guan195 area according to the lithological characteristics, sedimentary structure and logging curve characteristics (Table 1). Braided rever, channel bar, interchannel and alluvial plain micro-facies were distinguished in the alluvial fan subfacies [49]. The sedimentary facies map in ZV of the Wang guantun area is shown in Figure 5.

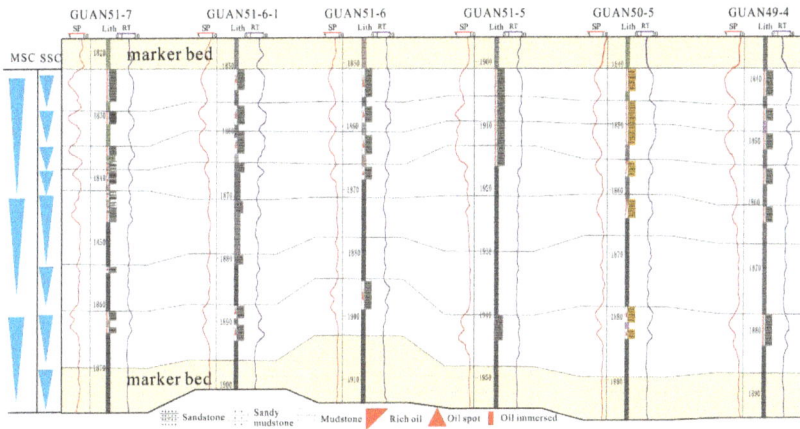

Figure 4. Cycle identification and stratigraphic correlation. SP, spontaneous potential curve; RT, resistivity curve.

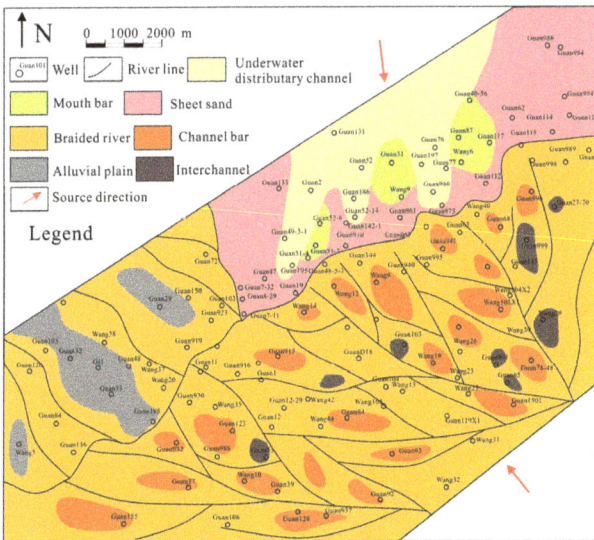

Figure 5. Sedimentary facies map in ZV of the Wang guantun area.

25

Table 1. Lithological and logging curve characteristics of different micro-facies.

Micro-Facies Type	Curve Model	Lithological Characteristics	Logging Curve Characteristics	Sedimentary Structure
Mouth bar micro-facies		Medium sand; Fine sand	Funnel shaped; Box shaped	Cross bedding Wavy bedding Parallel bedding Lenticular bedding
Underwater distributary channel micro-facies		Gravel bearing sandstone; Medium sand; Fine sand	Bell shaped; Box shaped	Graded bedding Wavy bedding Parallel bedding Cross-bedding
Sheet sand		Medium sand; Fine sand	Finger shaped	Parallel bedding Cross-bedding
Distributary bay micro-facies		Mainly mudstone	Low amplitude tooth curve or straight section	Horizontal bedding Wormhole

4.3. Petrophysical Properties and Pore Types

Under laboratory pressure conditions, the porosity of core samples is generally between 1.57% and 44.6% with an average of 19.05%, 58.6% of samples are between 15% and 25% (Figure 6a). The permeability is between 0.06 μm^2 and 3611 μm^2, 62.2% of samples are between 0.06 μm^2 and 50 μm^2, with an average of 137.56 μm^2 (Figure 6b).

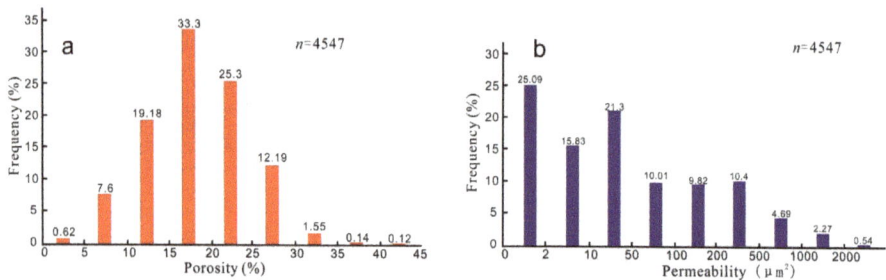

Figure 6. Reservoir properties of sandstones: (**a**) porosity distribution; (**b**) permeability distribution. *n* is the sample quantity.

The micrographs of the casting sheet image and SEM show that the main pore types in the ZVC sandstone reservoirs are intergranular pores, dissolution pores and micro fractures (Figure 7). Intergranular pores are relatively large in size and mostly irregular-polygonal in shape. The size of intergranular pores generally ranges from about 30 μm to about 60 μm, which is mainly controlled by the compaction effect. Dissolution pores are mainly from feldspar and matrix dissolution, with sizes mainly between about 5 μm and about 40 μm. The size of micro fractures is approximately 5 μm in width. Different pore types have different pore structure characteristics (Figure 7).

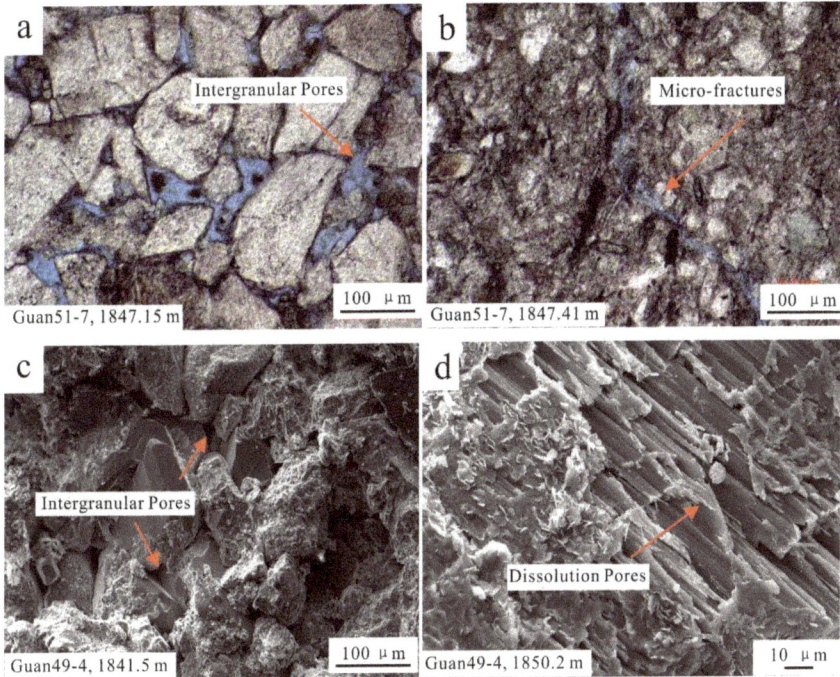

Figure 7. Typical pore types: (**a**) micrograph of casting sheet image showing intergranular pores; (**b**) micrograph of casting sheet image showing micro fractures; (**c**) micrograph of SEM showing intergranular pores; (**d**) micrograph of SEM showing dissolution pores.

4.4. The Results Obtained from Pressure-Controlled Mercury Injection Testing

The mercury injection and extrusion curves of the samples change at the threshold pressure, which is the pressure at which a wetting phase will begin to be displaced by a non-wetting phase from a porous medium [50]. The lower the threshold pressure (Pt), the larger the pore throat size. The intrusion and extrusion curves of typical samples can be classified into three patterns according to curve features (Figure 8). Pattern A includes samples with minimal Pt, the widest horizontal stage, and maximal intake of mercury, which represents a larger pore throat radius, better sorting, and less pore throat heterogeneity (Figure 8). The mercury intrusion curves of pattern B also have a horizontal stage, while their slopes are larger. The average Pl of pattern B is higher than that of pattern A (Figure 8). There is no horizontal stage in the samples of pattern C, in the process of mercury intrusion, mercury injection saturation increases with the increase of injection pressure. Although S_r is small, the extrusion efficiency is poor.

R_{ave} obtained from mercury intrusion saturation are mainly distributed between 0.0387 μm and 14.75 μm and the most common ones are from 9 μm to 12 μm (Figure 9a). R_{max} according to mercury intrusion saturation is generally between 0.0981 μm to 27.84 μm (mainly 0.1 μm −10 μm) (Figure 9b). S_{max} is mainly between 60% and 90% (Figure 9c), while S_r mainly ranges from 10% to 40% (Figure 9d). This indicates that much of the mercury was constrained by the pore throat, and the larger the discrepancy between pores and pore throats, the more residual mercury.

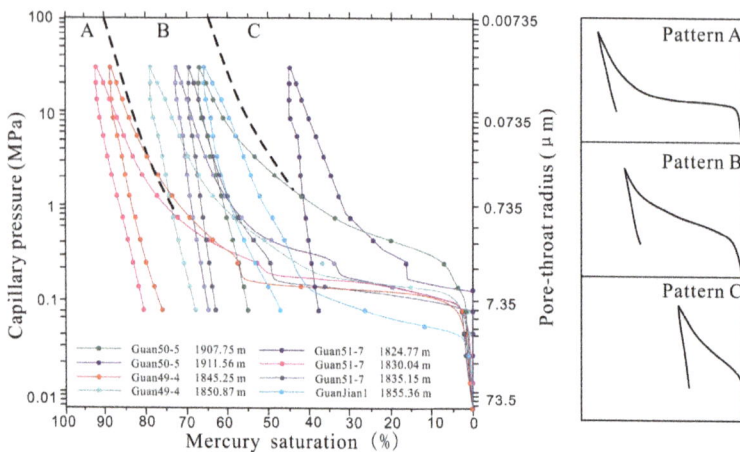

Figure 8. Intrusion and extrusion curves of the Pressure mercury method and the three patterns of intrusion and extrusion curves.

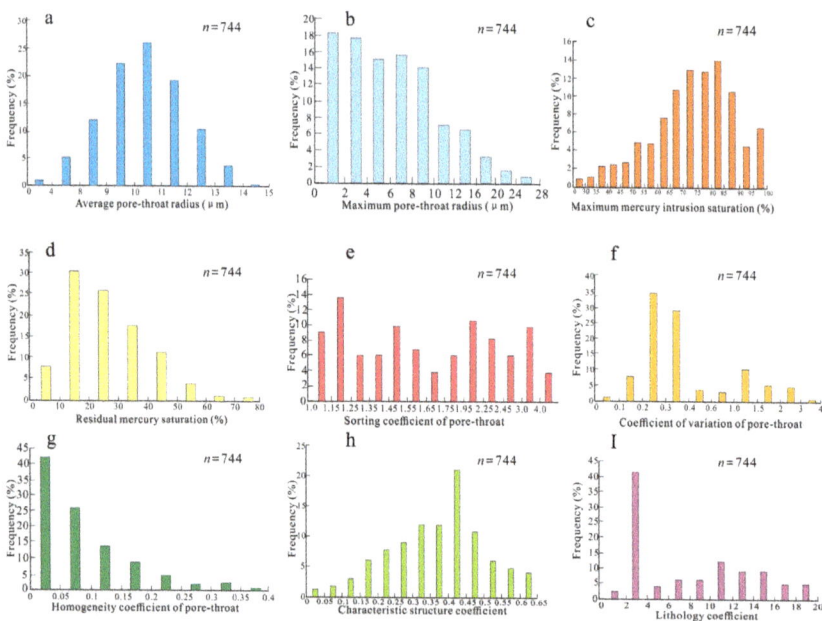

Figure 9. Illustrates the parameters derived from PMI experiments which are characterizing pore throat distribution. The "frequency" on the y-axis represents the sample ratio of each value intervals, n is the sample quantity. (**a**) Distribution of average pore-throat radius; (**b**) distribution of maximum pore-throat radius; (**c**) distribution of maximum mercury intrusion saturation; (**d**) distribution of residual mercury saturation; (**e**) distribution of sorting coefficient of pore-throat; (**f**) distribution of coefficient of variation of pore-throat; (**g**) distribution of homogeneity coefficient of pore-throat; (**h**) distribution of characteristic structure coefficient; (**i**) distribution of lithology coefficient.

Sorting coefficient refers to the uniformity of the pore throat radius distribution, which is calculated by Equation (1) [51]:

$$sc = \sqrt{\Delta S_i \sum_{i=1}^{n} (r_i - r_m)}$$ (1)

where sc is the sorting coefficient, r_i is the pore throat radius of an interval in the pore throat radius distribution function, nm, r_m is the average pore throat radius, nm, and ΔS_i is the saturation corresponding to r_i, %.

The more uniform the pore throat size, the better sorting, and the smaller sc. The coefficient of variation is the ratio of sorting coefficient and the average pore throat radius, which is calculated by Equation (2) [52]:

$$cv = \frac{\sqrt{\Delta S_i \sum_{i=1}^{n} (r_i - r_m)}}{r_m}$$ (2)

where cv is the coefficient of variation.

The better the sorting, the smaller the cv. The uniformity coefficient is used to quantify the uniformity of the pore throat radius distribution, which is calculated by Equation (3) [53], whose value ranges from 0 to 1, and the bigger the value, the more uniform pore-throat distribution. The characteristic structure coefficient reflects the connectivity of throats, and the bigger the value, the better the connectivity. The lithology coefficient is the ratio of the measured permeability and the calculated permeability, which reflects the degree of pore throat curvature:

$$ue = \frac{\sum_{i=1}^{n} \frac{r_i \Delta S_i}{r_d}}{\sum_{i=1}^{n} \Delta S_i}$$ (3)

where ue is the uniformity coefficient, r_d is the maximum pore throat radius, nm.

Sorting coefficients of pore throats range from 1.08 to 4.95 and display no obvious concentration distribution (Figure 9e). Coefficients of variation of pore throats range from 0.0361 to 3.27 and the most common ones are between 0.2 and 0.4 (Figure 9f). Uniformity coefficients of pore throats range from 0.0067 to 0.622 (mainly 0.2–0.4) (Figure 9g). This shows that the pore-throat distribution of the different samples is very different. Characteristic structure coefficients are mainly distributed between 0.3 and 0.5 (Figure 9h). Lithology coefficients are generally between 0.07 to 0.92 μm (mainly 0.1–0.2) (Figure 9i). This finding may be a manifestation of a complex tortuous shape in the pore throats. Overall, these relatively dispersed frequency distributions corroborate the complexity and heterogeneity of pore throat sizes.

With lower base-levels, the above parameters exhibit different behaviors. We intend to study and summarize these differences in order to further study the controlling effect of the base-level on the pore structure characteristics of sandstone.

5. Discussion

5.1. Relationship between Short-Term Base-Level Cycle and Pore Structure

The change of short-term base-level controls sedimentary sequence, stratigraphic structure, and superimposed pattern, and correspondingly controls the longitudinal variation of the physical properties and heterogeneity of the sand body [32,33]. An analysis of the relationship between short-term base-level cycles and pore structure of mouth bar sand bodies in the study area has shown that the pore structure of a mouth bar sand body is changed regularly with the falling of short-term base level.

The three samples in Figure 10 are from the SSC3 of well Guan51-7, and they belong to the same mouth bar sand body. The mercury intrusion curves of the three samples all display pattern B, and they all have a wider horizontal stage because of their shallower buried depth, though the horizontal stage of sample C is narrowest (Figure 10). The samples from late in the falling period of the SSC also have the lower Pt and larger S_{max}. This is an indication that the pore-throat distribution is better, and the effective porosity of the samples is larger in the late stage of the falling period of short-term base-level.

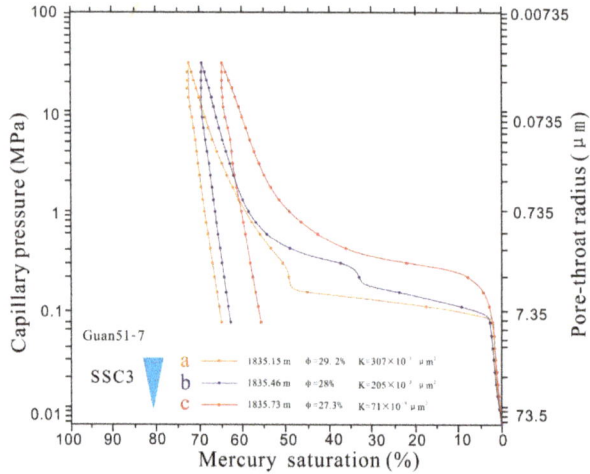

Figure 10. Intrusion and extrusion curves of PMI test of the samples in SSC3 (Well Guan51-7).

We can obtain the pore throat size distribution of samples from the test results of PMI [54]. The pore throat size distribution curves of Samples a, b and c are mainly distributed within 1.6–6.3 μm, 2.5–10 μm and 4–16 μm, respectively (Figure 11), which is gradually increasing with the falling of the short-term base level. This also shows that the pore throat size of the samples is larger in the late stage of the falling period of short-term base-level.

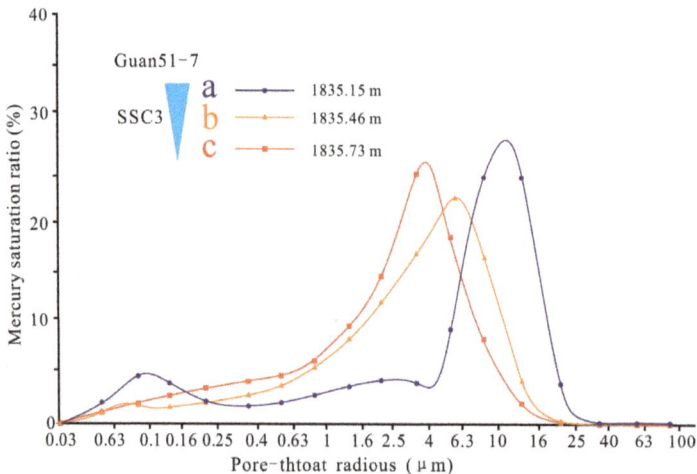

Figure 11. Pore throat size distributions by PMI of the samples in SSC3 (Well Guan51-7).

Because the developmental period of SSC was shorter, the allocyclic factors, such as geological structure and the change of lake level, were in a relatively static state, and the reduction rate of the accommodation space was approximately equivalent to the rate of sediment supply. As the short-term base-level fell, water energy was gradually strengthened, sediments could be fully washed; consequently, the porosity and permeability of the mouth bar sand body gradually increased upward, as did the radius of the pore throats, pore throat sorting, and the pore throat connectivity (Figure 12). In conclusion, the pore structure of single mouth bar sand bodies was mainly influenced by the autocycles and pore structure of single mouth bar sand bodies gradually become better with the falling of short-term base-level.

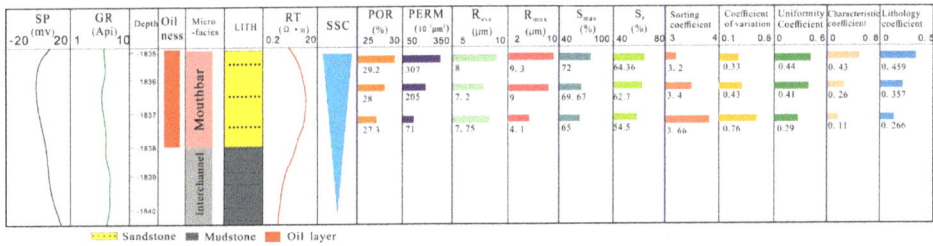

Figure 12. Pore structure of the samples in SSC3 (Well Guan51-7).

5.2. Relationship between Middle-Term Base-Level Cycle and the Pore Structure

The pore structure of the mouth bar of different SSCs in the same MSC has obvious differences. The five samples of the Guan 51-7 well were from 2 different short-term base-level cycles (SSC2, SSC3) but were all from the same middle-term base-level cycle (MSC1). It can be seen clearly in Figure 13 that the average width of the horizontal stage of samples is wider in SSC2 than in SSC3 and the samples in SSC2 also have the lower Pt and larger S_{max}. This indicates that the sorting of samples in SSC2 is better than in SSC3 and the interconnected pore space of the samples in SSC2 is larger under a given injection pressure. The two mercury curves of SSC2 belong to pattern A while the three mercury curves of SSC3 belong to pattern B, this may be an indication that the pattern of mercury curves changes from pattern B to pattern A with the falling of middle-term base-level (Figure 13).

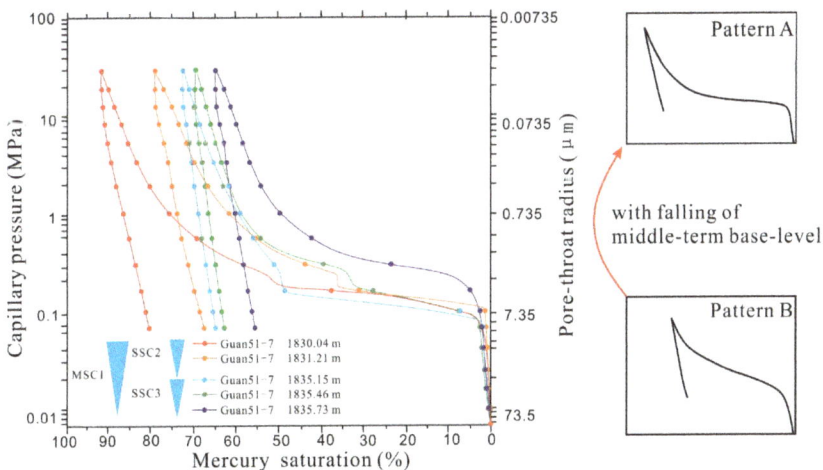

Figure 13. Intrusion and extrusion curves of PMI of the samples in MSC1.

The pore size distribution is shown in Figure 14; the pore throat radius of samples in SSC2 is mainly distributed between 6.3 μm and 30 μm; the samples in the SSC2 range from 1.6 μm to 10 μm, smaller than the samples in SSC2. This is an indication that the pore-throat radius of the single mouth bar sand body is larger in the late stage of the falling period of middle-term base-level.

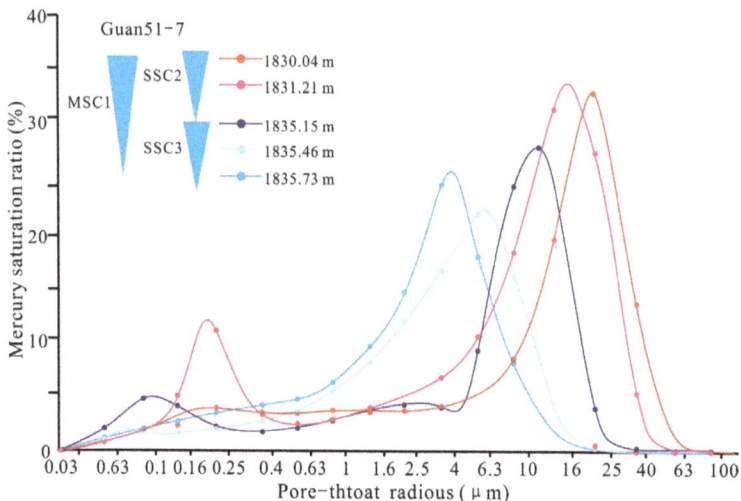

Figure 14. Pore throat size distributions by PMI of the samples in MSC1.

The pore structure of multiple mouth bar sand bodies in MSC is primarily controlled by the allocyclic factors, such as tectonic movement and lake level change. The ratio of the reduction rate of accommodation space and the supply rate of sediment (A/S) decreased, and the water energy increased continuously. As a result, the microscopic characteristics of mouth bar sand bodies present regular changes (Figure 15). With the falling of middle-term base-level, the porosity and permeability of the later single mouth bar sand body were larger, pore throat radii increased upward, pore throat sorting became better upward, and the connectivity of the pore throat increased, under the common influence of allocycles and autocycles. In brief, the pore throat characteristics of the later mouth bar sand bodies were better than those of the early mouth bar sand bodies.

LSC	MSC	SSC	POR (%) 10–30	PERM $(10^{-3}\ \mu m^2)$ 0–350	R_{ave} (μm) 6–10	R_{max} (μm) 0–10	S_{max} (%) 40–100	S_r (%) 40–80	Sorting coefficient 2–5	Coefficient of variation 0.1–0.8	Uniformity Coefficient 0–0.6	Characteristic coefficient 0–0.8	Lithology coefficient 0–0.5
		SSC1	28.39	322.5	7.89	6.78	82.95	73.66	2.585	0.2	0.45	0.69	0.445
	MSC1	SSC2	28.11	200	8.75	7.43	68.99	59.37	3.01	0.39	0.40	0.28	0.41
LSC1		SSC3	27.5	190.5	7.53	5.5	60.15	40.21	3.42	0.33	0.39	0.35	0.40
		SSC4	19.33	194.26	5.37	2.86	59.45	51.36	3.56	0.48	0.16	0.02	0.28
	MSC2	SSC5	22.5	84.5	7.5	7.515	79.7	68.95	3.48	0.35	0.31	0.15	0.28
		SSC6	17.1	63.48	6.7	4.2	57.11	50.32	3.53	0.52	0.15	0.0185	0.22

Figure 15. Relationship diagram between the pore structure of mouth-bar sand bodies and MSC.

5.3. Relationship between Long-Term Base-Level Cycle and the Pore Structure

The ZVC in the study area corresponds to the falling semi-cycle of the long-term base-level cycle (LSC1), including three middle-term base-level cycles (MSC1, MSC2, MSC3). Through comparative analyses, it was observed that although pore structure of single-period multi-mouth bar sand bodies tends to be of poor quality at the bottom and good quality at the top, the regularity of pore structure in multi-period multi-mouth bar sand bodies is not obvious. In other words, because the developmental period of LSC is fairly long. During this period, lake level and basin basement experience a series of rises and drops, the supply rate of sediment and the change rate of accommodation space increase and decrease repeatedly, the transport distance of sediment experience a number of discontinuity changes. In addition, the energy of water changes repeatedly under the influence of the allocycles, which also leads to the poor regularity of pore throat heterogeneity. The long-term base-level is in a downward trend, and the pore structure of multi-period multi-mouth bar sand bodies was gradually better overall, but the regularity is not obvious (Figure 14).

5.4. Summary

According to the above analysis, the main factors controlling the different levels' base-level cycles are different, which leads to different degrees of control of different levels base-level cycles on pore structure heterogeneity (Table 2). The pore structure of the sand bodies in the short-term base-level cycles and middle-term base-level cycles exhibit obvious changes. Therefore, the research on the pore structure of the sand bodies in short-term base-level cycles and middle-term base-level cycles can achieve better results. In summary, in the production process of oilfields, especially in the delta depositional environment of a continental fault depression basin, the pore structure of the sand bodies in the late stage of the falling period of base-level cycles are better, and those sand bodies are also the more favorable targets for production.

Table 2. Impacts of the base-level cycle on pore structure of mouth bar.

Cycles	With the Falling of Base-Level			
	Pore Size	Pore-Throat Distribution	Porosity	Permeability
SSC	Gradually increasing	Getting more uniform	Getting larger	Getting larger
MSC	Pore size of the sand body in the late period is larger	Getting more uniform	The porosity of the sand body in the late period is larger	The permeability of the sand body in the late period is larger

6. Conclusions

In this study, to investigate the effect of the base-level cycle on mouth bar pore structure, casting sheet image, SEM and PMI analyses were performed on samples from the Kongdian Formation. Several conclusions can be drawn as follows:

- Three types of pores (intergranular pores, dissolution pores, and micro fractures) exist in the Kongdian Formation of the Wang guantun oilfield. The pore structures of the different types of pores are different. Intrusion and extrusion curves of PMI were divided into three patterns: the curves of pattern A have minimum *Pt*, widest horizontal stages, and maximum intake of mercury, which altogether represent larger pore throat radius, better sorting, and weaker heterogeneity. The intrusion curves of most samples in mouth bar sand bodies belong to pattern A or pattern B.
- The pore structure heterogeneity in the early stage of the falling period of short-term base-level was stronger than that in the late stage, and pore size gradually became larger with the falling of short-term base-level; with the falling of middle-term base level, the pore size of late single mouth bar sand bodies were larger than those of early single mouth bar sand bodies, and the intrusion curves of samples tended to change from pattern B to pattern A; the pore size, porosity, and permeability in mouth bar sand bodies generally became larger with the falling of long-term base-level.

- In the oilfield production process, especially in the delta depositional environment of a continental fault depression basin, the pore structure of the sand bodies in the late stage of the falling period of base-level cycles are better, and those sand bodies are also the more favorable targets for production.

Author Contributions: Conceptualization, J.H. and Y.L.; Data curation, L.J.; Formal analysis, J.S.; Resources, X.G.; Writing—Original Draft Preparation, X.W. Writing—Review & Editing, J.H. and X.W.

Funding: This research was funded by [National Science and Technology Major Project] grant number [2016ZX05010-001] and [2016ZX05014002-008], and [National Basic Research Program of China] grant number [2015CB250901].

Conflicts of Interest: The authors declare no conflict of interest.

References

1. Durner, W. Hydraulic conductivity estimation for soils with heterogeneous pore structure. *Water. Resour. Res.* **1994**, *30*, 211–223. [CrossRef]
2. Nelson, P.H. Pore throat sizes in sandstones, tight sandstones, and shales. *AAPG Bull.* **2009**, *93*, 329–340. [CrossRef]
3. Ross, D.J.; Bustin, R.M. The importance of shale composition and pore structure upon gas storage potential of shale gas reservoirs. *Mar. Petrol. Geol.* **2009**, *26*, 916–927. [CrossRef]
4. Ge, X.; Fan, Y.; Li, J.; Zahid, M.A. Pore structure characterization and classification using multifractal theory—An application in Santanghu basin of western China. *J. Petrol. Sci. Eng.* **2015**, *127*, 297–304. [CrossRef]
5. Wang, X.; Hou, J.; Song, S.; Wang, D.M.; Gong, L.; Ma, K.; Liu, Y.M.; Li, Y.Q.; Lin, Y. Combining pressure-controlled porosimetry and rate-controlled porosimetry to investigate the fractal characteristics of full-range pores in tight oil reservoirs. *J. Petrol. Sci. Eng.* **2018**, *171*, 353–361. [CrossRef]
6. Wang, R. Characteristics of micro-pore throat in ultra-low permeability sandstone reservoir. *Acta Petrol. Sin.* **2009**, *30*, 560–563. [CrossRef]
7. Cai, J.C.; Wei, W.; Hu, X.Y.; Liu, R.C.; Wang, J.J. Fractal characterization of dynamic fracture network extension in porous media. *Fractals* **2017**, *25*, 1750023. [CrossRef]
8. Li, C.; Ostadhassan, M.; Guo, S.; Gentzis, T.; Kong, L. Application of PeakForce tapping mode of atomic force microscope to characterize nanomechanical properties of organic matter of the Bakken Shale. *Fuel* **2018**, *233*, 894–910. [CrossRef]
9. Li, C.; Ostadhassan, M.; Gentzis, T.; Kong, L.; Carvajal-Ortiz, H.; Bubach, B. Nanomechanical characterization of organic matter in the Bakken formation by microscopy-based method. *Mar. Petrol. Geol.* **2018**, *96*, 128–138. [CrossRef]
10. Gong, L.; Fu, X.F.; Gao, S.; Zhao, P.Q.; Luo, Q.Y.; Zeng, L.B.; Yue, W.T.; Zhang, B.J.; Liu, B. Characterization and Prediction of Complex Natural Fractures in the Tight Conglomerate Reservoirs: A Fractal Method. *Energies* **2018**, *11*, 2311. [CrossRef]
11. Gu, M.W.; Lu, S.F.; Xiao, D.H.; Guo, S.; Zhang, L. Present Situation of Core Mercury Injection Technology of Unconventional Oil and Gas Reservoir. *Acta Geol. Sin.-Engl.* **2015**, *89* (Suppl. S1), 388–389. [CrossRef]
12. Xi, K.L.; Cao, Y.C.; Beyene, G.H.; Zhu, R.K.; Jens, J.; Knut, B.; Zhang, X.X.; Helge, H. How does the pore throat size control the reservoir quality and oiliness of tight sandstones? The case of the Lower Cretaceous Quantou Formation in the southern Songliao Basin, China. *Mar. Petrol. Geol.* **2016**, *76*, 1–15. [CrossRef]
13. Wang, X.; Hou, J.; Liu, Y.; Ji, L.; Sun, J. Studying reservoir heterogeneity by Analytic Hierarchy Process and Fuzzy Logic, case study of Es1x formation of the Wang guan tun oilfield, China. *J. Petrol. Sci. Eng.* **2017**, *156*, 858–867. [CrossRef]
14. Cerepi, A.; Humbert, L.; Burlot, R. Pore-scale complexity of a calcareous material by time-controlled mercury porosimetry. *Stud. Surf. Sci. Catal.* **2000**, *128*, 449–458. [CrossRef]
15. Zhao, J.J.; Liu, Z.W.; Xie, Q.C.; Zhou, J.P. Micro Pore Throat Structural Classification of Chang 7 Tight Oil Reservoir of Jiyuan Oilfield in Ordos Basin. China. *Petrol. Exploit.* **2014**, *19*, 73–79. [CrossRef]

16. Zhu, H.Y.; An, L.Z.; Jiao, C.Y. The difference between constant-rate mercury injection and constant-pressure mercury injection and the application in reservoir assessment. *Nat. Gas Geosci.* **2015**, *26*, 1316–1322. [CrossRef]

17. Curtis, M.E.; Ambrose, R.J.; Sondergeld, C.H. Transmission and Scanning Electron Microscopy Investigation of Pore Connectivity of Gas Shales on the Nanoscale. In Proceedings of the North American Unconventional Gas Conference and Exhibition, The Woodlands, TX, USA, 14–16 June 2011.

18. Mendhe, V.A.; Mishra, S.; Khangar, R.G.; Kamble, A.D.; Kumar, D.; Varma, A.K.; Bannerjee, M. Organo-petrographic and pore facets of Permian shale beds of Jharia Basin with implications to shale gas reservoir. *J. Earth Sci.* **2017**, *28*, 897–916. [CrossRef]

19. Singh, H. Representative Elementary Volume (REV) in spatio-temporal domain: A method to find REV for dynamic pores. *J. Earth Sci.* **2017**, *28*, 391–403. [CrossRef]

20. Zhao, P.; Cai, J.; Huang, Z.; Ostadhassan, M.; Ran, F. Estimating permeability of shale gas reservoirs from porosity and rock compositions. *Geophysics* **2018**, *83*, MR283–MR294. [CrossRef]

21. Kong, L.; Ostadhassan, M.; Li, C.; Tamimi, N. Pore characterization of 3D-printed gypsum rocks: A comprehensive approach. *J. Mater. Sci.* **2018**, *53*, 5063–5078. [CrossRef]

22. Bloomfield, J.; Gooddy, D.; Bright, M.; Williams, P. Pore throat size distributions in permo-triassic sandstones from the United Kingdom and some implications for contaminant hydrogeology. *Hydrogeol. J.* **2001**, *9*, 219–230. [CrossRef]

23. Bai, B.; Zhu, R.K.; Wu, S.T.; Yang, W.J.; Jeff, G. Multi- scale of Nano (Micro)-CT study on microscopic pore structure of tight sandstone of Yanchang Formation, Ordos Basin. *Petrol. Explor. Dev.* **2013**, *40*, 329–333. [CrossRef]

24. Kong, L.; Ostadhassan, M.; Li, C.; Tamimi, N. Can 3-D Printed Gypsum Samples Replicate Natural Rocks? An Experimental Study. *Rock Mech. Rock Eng.* **2018**, 1–14. [CrossRef]

25. Ju, Y.; Sun, Y.; Tan, J.; Bu, H.; Han, K.; Li, X.; Fang, L. The composition, pore structure characterization and deformation mechanism of coal-bearing shales from tectonically altered coalfields in eastern China. *Fuel* **2018**, *234*, 626–642. [CrossRef]

26. Tang, S.W.; Wang, L.; Cai, R.J. The evaluation of electrical impedance of three-dimensional fractal networks embedded in a cube. *Fractals* **2017**, *25*, 1740005. [CrossRef]

27. Wang, L.; Zhao, N.; Sima, L.Q.; Meng, F.; Guo, Y.H. Pore Structure Characterization of the Tight Reservoir: Systematic Integration of Mercury Injection and Nuclear Magnetic Resonance. *Energy Fuel* **2018**, *32*, 7471–7484. [CrossRef]

28. Wang, X.X.; Hou, J.G.; Liu, Y.M.; Zhao, P.Q.; Ma, K.; Wang, D.M.; Ren, X.X.; Yan, L. Overall PSD and fractal characteristics of tight oil reservoirs: A case study of lucaogou formation in junggar basin, China. *Fractals* **2019**, *1*, 1940005. [CrossRef]

29. Read, W.A.; Forsyth, I.H. Allocycles and Autocycles in the Upper Part of the Limestone Coal Group (Pendleian E1) in the Glasgow–Stirling Region of the Midland Valley of Scotland. *Geol. J.* **1989**, *24*, 121–137. [CrossRef]

30. Deng, H.W.; Wang, H.L.; Li, X.Z. Identification, comparison technique and application of sequence stratigraphic datum. *Petrol. Nat. Gas Geol.* **1996**, *17*, 177–184. [CrossRef]

31. Zheng, R.C.; Yin, S.M.; Peng, J. Sedimentary dynamics analysis of base level cycle structure and superimposed style. *J. Sediment.* **2000**, *18*, 369–375. [CrossRef]

32. Deng, H.W.; Wang, H.L.; Ning, N. Principle of sediment volume distribution: Basic theory of high resolution stratigraphy. *Front. Earth Sci.* **2000**, *7*, 305–313. [CrossRef]

33. Miall, A.D. Reconstructing the architecture and sequence stratigraphy of the preserved fluvial record as a tool for reservoir development: A reality check. *AAPG Bull.* **2006**, *90*, 989–1002. [CrossRef]

34. Pittman, E.D. Relationship of porosity and permeability to various parameters derived from mercury injection-capillary pressure curves for sandstone (1). *AAPG Bull.* **1992**, *76*, 191–198. [CrossRef]

35. Yao, Y.; Liu, D. Comparison of low-field NMR and mercury intrusion porosimetry in characterizing pore size distributions of coals. *Fuel* **2012**, *95*, 152–158. [CrossRef]

36. Clarkson, C.R.; Solano, N.; Bustin, R.M.; Bustin, A.; Chalmers, G.; He, L. Pore structure characterization of North American shale gas reservoirs using USANS/SANS, gas adsorption, and mercury intrusion. *Fuel* **2013**, *103*, 606–616. [CrossRef]

37. Wang, S.; Wu, T.; Cao, X.; Zheng, Q.; Ai, M. A Fractal Model for Gas Apparent Permeability in Microfractures of Tight/shale Reservoirs. *Fractals* **2017**, *25*, 1750036. [CrossRef]

38. Toledo, P.G.; Scriven, L.E.; Davis, H.T. Pore-space statistics and capillary pressure curves from volume-controlled porosimetry. *SPE Form. Eval.* **1994**, *9*, 46–54. [CrossRef]

39. Zhao, H.; Ning, Z.; Wang, Q.; Zhang, R.; Zhao, T.; Niu, T.; Zeng, Y. Petrophysical characterization of tight oil reservoirs using pressure-controlled porosimetry combined with rate-controlled porosimetry. *Fuel* **2015**, *154*, 233–242. [CrossRef]

40. Wang, X.; Hou, J.G.; Liu, Y.M.; Dou, L.X.; Sun, J.; Gong, X. Quantitative characterization of vertical heterogeneity of mouth bar based on the analitic hierarchy process and the fuzzy mathematics: A case study of Guan195 fault block of Wangguantun Oilfield. *Nat. Gas Geosci.* **2017**, 1914–1923. [CrossRef]

41. Liang, H.W.; Wu, S.H.; Yue, D.L. Base-level cyclic controls on the reservoir macro-heterogeneity: A case study of the mouth bar in 2block of the Shengtuo oilfield. *J. Chin. Univ. Min. Technol.* **2013**, *42*, 413–420. [CrossRef]

42. Tang, M.A.; Sun, B.L. Relationship between base-level cycles and hiberarchy of reservoir flow units of Lower Shihezi Formation in Daniudi Gas field. *Acta Sedimentol. Sin.* **2007**, *1*, 39–47. [CrossRef]

43. Xu, E.A.; Li, X.L.; Wang, Q.K.; Xia, G.C.; Dong, L.Y.; Li, Z.W. Main controlling factor and model of Kongdian Formation hydrocarbon accumulation in Kongnan Area of Huanghua Depression. *Fault Oil Gas Field* **2014**, *21*, 278–281. [CrossRef]

44. Neng, Y.; Qi, J.F.; LI, T.H.; Zhang, G.; Li, M.G.; Shi, K.T. Characteristics of Cenozoic Fault System and Its Significance in Petroleum Geology in Kongnan Area, Huanghua Depression. *Geoscience* **2009**, *23*, 1077–1084. [CrossRef]

45. Sung, K.; Jin, Z.K.; Wang, X.W.; Zhang, A.H.; Duan, Z.Q.; Wang, C.S.; Zhang, G.H. Control of sedimentary facies on reservoir quality—An example from Zao II and III oil sets in Wangguantun Oilfield, Huanghua Depression. *Petrol. Explor. Dev.* **2006**, *33*, 335–339. [CrossRef]

46. Shi, Z.Z.; Ji, Y.L. Fan-delta Sedimentation Formed under the Environment of Lake Level Frequently Varying—Taking the First Member of Kongdian Formation, Huanghua Depression as an Example. *J. Xi'an Petrol. Inst.* **2002**, *17*, 24–28. [CrossRef]

47. Ziarani, A.S.; Aguilera, R. Pore throat radius and tortuosity estimation from formation resistivity data for tight-gas sandstone reservoirs. *J. Appl. Geophys.* **2012**, *83*, 65–73. [CrossRef]

48. Yang, Q.; Qi, J.F.; Chang, D.S.; Li, M.G. Tectonpalaeogeography during the deposition of Paleogene Kongdian Formation in Kongnan area, Huanghua Depression of Bohai Bay Bain. *J. Palaeogeogr.* **2009**, *3*, 306–313. [CrossRef]

49. Xue, L.F.; Sun, J.; Chen, C.W.; Xu, X.K.; Meng, Q.L.; Yan, Q.H. Tectonic sedimentary evolution of the first and second members of the Kongdian Formation in southern Kongdian, Huanghua depression. *Sediment. Geol. Tethyan Geol.* **2006**, *28*, 62–68. [CrossRef]

50. Thomas, L.K.; Katz, D.L.; Tek, M.R. Threshold pressure phenomena in porous media. *Soc. Petrol. Eng. J.* **1968**, *8*, 174–184. [CrossRef]

51. Rogers, J.J.; Head, W.B. Relationships between porosity, median size, and sorting coefficients of synthetic sands. *J. Sediment. Res.* **1961**, *31*, 467–470. [CrossRef]

52. Brown, C.E. Coefficient of variation. In *Applied Multivariate Statistics in Geohydrology and Related Sciences*; Springer Science & Business Media: Berlin, Germany, 1998; pp. 155–157.

53. Sepaskhah, A.R.; Ghahraman, B. The effects of irrigation efficiency and uniformity coefficient on relative yield and profit for deficit irrigation. *Biosyst. Eng.* **2004**, *87*, 495–507. [CrossRef]

54. Washburn, E.W. The dynamics of capillary flow. *Phys. Rev.* **1921**, *17*, 273–283. [CrossRef]

energies

MDPI

Article

The Creep-Damage Model of Salt Rock Based on Fractional Derivative

Hongwei Zhou [1,*], Di Liu [1], Gang Lei [2], Dongjie Xue [1] and Yang Zhao [1]

[1] State Key Laboratory of Coal Resources and Safe Mining, China University of Mining and Technology, Beijing 100083, China; liudi@student.cumtb.edu.cn (D.L.); xuedongjie@163.com (D.X.); zhaoyang@student.cumtb.edu.cn (Y.Z.)
[2] College of Engineering, Peking University, Beijing 100871, China; lg1987cup@126.com
* Correspondence: zhw@cumtb.edu.cn; Tel.: +86-139-1058-5080

Received: 30 July 2018; Accepted: 29 August 2018; Published: 6 September 2018

Abstract: The use of salt rock for underground radioactive waste disposal facilities requires a comprehensive analysis of the creep-damage process in salt rock. A computer-controlled creep setup was employed to carry out a creep test of salt rock that lasted as long as 359 days under a constant uniaxial stress. The acoustic emission (AE) space-time evolution and energy-releasing characteristics during the creep test were studied in the meantime. A new creep-damage model is proposed on the basis of a fractional derivative by combining the AE statistical regularity. It indicates that the AE data in the non-decay creep process of salt rock can be divided into three stages. Furthermore, the authors propose a new creep-damage model of salt rock based on a fractional derivative. The parameters in the model were determined by the Quasi-Newton method. The fitting analysis suggests that the new creep-damage model provides a precise description of full creep regions in salt rock.

Keywords: salt rock; creep; damage; fractional derivative; acoustic emission

---

1. Introduction

Salt rock is widely used in energy storage and radioactive waste disposal in underground engineering facilities. It is hard to predict their mechanical behavior during a long design life [1–5]. Thus, research on the creep-damage of salt rock is significant to avoid the loss of effective storage space in underground cavities.

The full creep regions of salt rock can be divided into three stages: The transient creep region (the primary region), the steady-state creep region (the secondary region), and the accelerated creep region (the tertiary region) [6]. Many efforts have been expanded on analyzing creep-damage features through mathematical modeling. Passaris has studied the creep of salt rock by using a three element model. The results indicated that the three-element model provided a precise description of creep deformation [7]. Ghavidel performed axial creep experiments of salt rock under different temperatures. The theoretical analysis revealed that the Burgers model is accurate in describing the characteristics of salt rock in different temperatures [8]. Hou and Lux conducted a creep experiment of salt rock and proposed the Hou-Lux creep constitutive model under the theory of strain hardening and recovery, damage and damage healing [9]. The Hou-Lux model was applied to predict the time-dependent deformation of salt rock in the Excavation Disturbed Zone (EDZ) of a 37-year-old underground cavity [10].

In addition, fractional calculus is helpful to propose the creep constitutive models as it has advantages in explicating the accumulation process of internal stress, reducing the parameters in a constitutive model and representing the nonlinear characteristics. Zhou has proposed a creep constitutive model of salt rock based on a fractional derivative by replacing a Newtonian dashpot in the classical Nishihara model with the fractional derivative Abel dashpot and found that the predicted results were consistent with the experimental data [11]. Wu improved the Maxwell creep model and established a constitutive model of salt rock based on variable-order fractional derivatives [12]. By combining ultrasonic testing (UT), Zhou introduced a variable-viscosity Abel dashpot in a new creep constitutive model [13]. However, the UT could only reflect two-dimensional (2D) damage information inside of the salt rock.

The acoustic emission (AE) test provides a more precise three-dimensional (3D) description of damage during the creep process as it records information at any given point inside the salt rock. It is widely used for rock damage testing both in the laboratory and situ. Lavrov performed AE experiments in clay to study the damage evolution of a boom clay specimen during uniaxial compression. These experiments indicated that, as is the case with salt rock, clay also shows the Kaiser effect in AE with Felicity ratios around unity [14]. Hardy collected the AE signals during a uniaxial compression experiment. This research suggested that the number of accumulated AE signals follows a linear relationship with the axial strain [15]. Kong used AE location technology to study the multifractal characteristics of coal. These studies revealed the process of crack evolution of coal and the generation mechanism of AEs [16]. Zhang performed AE experiments with granite samples and studied the relationship between the AE temporal-spatial distribution and the fractal dimension. It suggested that the characteristics of the AE temporal-spatial distribution can be quantitatively described by the fractal dimension [17].

In this paper, the authors completed uniaxial compression experiments of salt rock, which lasted for 359 days, and analyzed the AE features. These experiments indicated that the AE data in the unsteady creep process of salt rock can be divided into three stages. Furthermore, the parameters of a new creep-damage model were determined by the Quasi-Newton method. The fitting analysis suggested that the creep-damage model, based on a fractional derivative in this paper, provided a precise description of the full creep regions in salt rock.

2. Methods

The salt rock specimen used in the uniaxial compression experiments was taken from a salt mine in Pingdingshan City, Henan Province, Central China. It was drilled from the PT Well No.1 at a depth of 1719 m below the ground's surface. The main surrounding rock mass had a salt content up to 98.14–98.71%. The cylindrical salt rock specimen (No. 7-27-17) used in the uniaxial experiments was processed on a dry lathe and prepared with the required dimensions of 80 mm in diameter and 160 mm in length. According to other experiments using salt samples drilled at the same place, the long-term strength is about 12 to 14 MPa. Then, the axial load was set to be a constant value of 17 MPa so that the accelerated creep curve can be obtained.

The uniaxial compression tests were carried out at Sichuan University using a computer-controlled creep setup, with a uniaxial load in the range of 0–600 kN [11]. A three-dimensional real-time monitor and display system (model: PCI-2), manufactured by the American Physical Acoustics Corporation, was used to monitor the AE signals (Figure 1). The preamplifier gain was 40 dB, and the threshold value was set at 35 dB in order to eliminate background noise [18]. Eight AE sensors were installed symmetrically in the radial direction along the cylinder's surface and the distance from the sensor to the nearest end surface was about 1 cm (Figure 1). The system was able to capture and display the acoustic emissions during the whole rock damage and failure process. To get a better effect in receiving AE signals, the AE sensors were isolated from the specimen by thin Vaseline plates. The experiment lasted from 10 May 2013 to 4 May 2014. During the whole failure process of salt rock, the temperature was kept at 22 °C.

Figure 1. Experimental setup of the creep test of salt rock.

3. Experiment Results and Discussion

The salt rock exhibited expansion failure during the creep experiment without a confining pressure (Figure 2). The fractures in the salt rock specimen were mainly concentrated in the middle part of the specimen. A mass of penetrative cranny and discrete salt gains was clearly observed after failure.

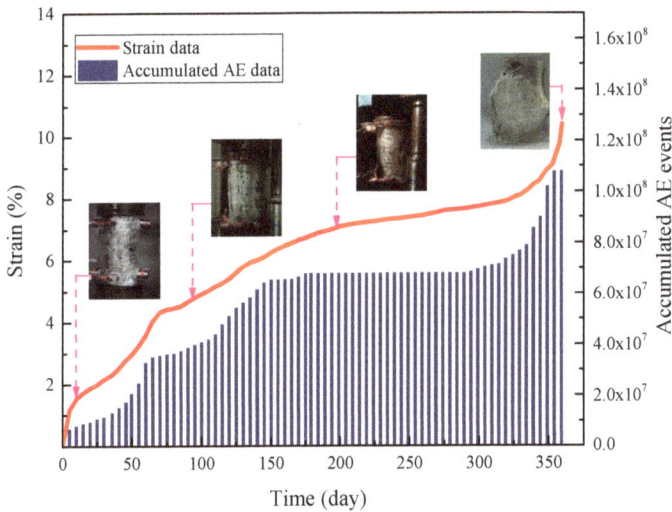

Figure 2. The deformation characteristics of salt rock during the creep process.

The authors analyzed the strain rate to further investigate the relationship between deformation and the AE spatial distribution during the creep process. The analysis indicated that the creep process can be divided into three stages (Figure 3). During Stage I, the strain hardening effect was stronger than the strain recovery effect at room temperature, so the strain rate curve showed a downtrend. This stage ends after about 179 days. Then, the strain rate remained steady at a relatively low level of 10^{-9}/s until 319 days. This stage was considered to be the steady stage of creep. After 319 days, the axial strain rate sharply increased, which means that the creep process entered the accelerated stage (Stage III). The salt rock quickly deformed until failure.

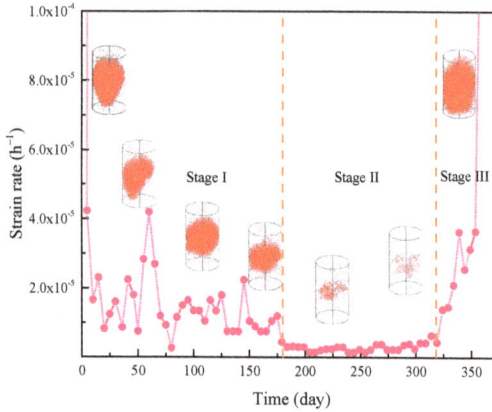

Figure 3. The strain rate and spatial distribution of AE events. The inset images show the spatial distribution characteristics of AE events in the different creep stages.

The inset images in Figure 3 also illustrate the spatial distribution of AE events during the full creep process. The red points represent the locations of acoustic emission events and each point corresponds to a fracture surface or volume in physical space. It revealed that the AE events were mainly focused on the middle part of the specimen during the loading process. The AE spatial distribution corresponded to the damaged parts of the salt rock, shown in Figure 2. During Stage I, AE events were relatively extensive in the middle portion and covered a comparatively large volume of the specimen. Then, the number of AE events markedly decreased and remained stable at a low level for about 90 days. Finally, the AE events rapidly increased and spread through the whole specimen.

The creep curves of full regions and their corresponding AE events can be observed in Figure 4. During Stage I, the AE events number and the released energy from the salt rock sample was at a relatively high level and fluctuated greatly during the primary region (Figure 4a,b). At 64 days, the AE rate reached a peak of 8.34×10^4/h. Meanwhile, the released energy also reached a peak of 3.16×10^{-6} J. The irregular fluctuation of the AE rate in this stage was mainly caused by the inhomogeneous distribution of different components in natural salt rock.

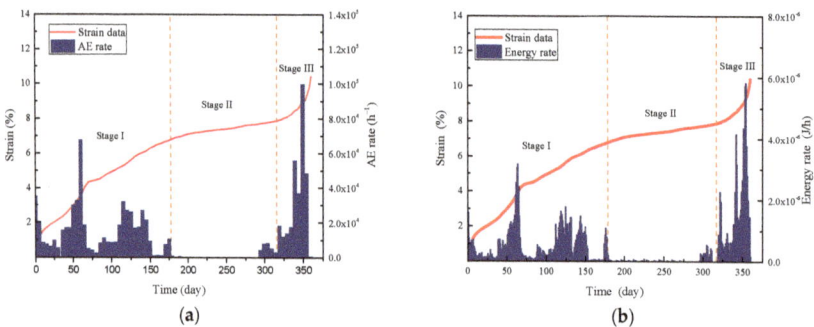

Figure 4. The curves for the AE count rate and released energy: (**a**) AE count rate vs. strain-time curve; (**b**) Released energy rate vs. strain-time curve.

The AE numbers and released energy of salt rock in the steady stage (Stage II) was rather small in contrast to the initial stage. It appeared that the fracture hardly grew so that only a few AE signals could be generated and received. During Stage III, the number of AE events and released energy rapidly increased to the maximum value. The sample continued to be damaged as the strain energy was released and the strain rate accelerated until failure. The ultimate deformation value of salt rock was 10.396 mm. Further, it was be found that the number of AE events and released energy of salt rock almost had the same trend (Figure 4).

4. Creep-Damage Model of Salt Rock Based on a Fractional Derivative

The creep behavior of salt rock can be divided into the steady creep process and the unsteady creep process. The former usually happens when an external load is under the long-time strength of the salt rock. The creep rate gradually decreased to a small constant close to zero under ideal conditions. The unsteady creep process happens when an external load is greater than or equal to the long-time strength of the salt rock. A typical unsteady creep process is usually divided into three stages, which are named initial transient creep period, the steady-state period, and the tertiary stage (an unsteady state period). The constitutive model in this paper aimed to study the unsteady creep process of salt rock.

4.1. The Maxwell Model

The Maxwell model is the simplest viscoelasticity model and the Maxwell model is set up by putting a spring and a dashpot together in a series. The Maxwell constitutive model is given as

$$\varepsilon(t) = \frac{\sigma_0}{E} + \frac{\sigma_0}{\eta}t \tag{1}$$

where σ_0 = const, E is the elasticity modulus, η is the viscosity coefficient, the initial condition is $\varepsilon_0 = \sigma_0/E$ when $t = 0$, ε_0 represents the instantaneous strain of salt rock. It can only describe the ideal fluid features of salt rock.

However, the creep behavior of salt rock is the interactions among elastic, viscoelastic and viscoplastic behavior. Thus, the Maxwell model should be improved in order to get a better explanation on creep behavior of natural salt rocks.

4.2. The Abel Dashpot: A Fractional Derivative Element

A typical application of fractional calculus is the Abel dashpot, which is a fractional derivative description of the Newtonian dashpot. The constitutive relation of the Abel dashpot is given by [11,13]

$$\sigma = \eta^\beta D^\beta[\varepsilon(t)] \quad (0 \le \beta \le 1) \tag{2}$$

where η^β is the viscosity coefficient and D^β indicates the fractional derivative. The Abel dashpot in Equation (2) can be used to describe both the Newtonian dashpot in the special case of $\beta = 1$, representing an ideal fluid and a spring in the special case of $\beta = 0$, representing an ideal solid. The Abel dashpot exhibits characteristics of both a spring and the Newtonian dashpot and eliminates the limitation of an element being solely either a spring or the Newtonian dashpot.

Considering $\sigma(t) = \sigma = $ const in Equation (3), taking the fractional integral calculation of Equation (2) on the basis of the Riemann-Liouville operator, we obtain

$$\varepsilon(t) = \frac{\sigma}{\eta^\beta} \frac{t^\beta}{\Gamma(1 + \beta)} \quad (0 \le \beta \le 1) \tag{3}$$

where Equation (3) denotes the creep strain characterized by the Abel dashpot. More details on the Abel dashpot can be found in Zhou [11,13].

4.3. Maxwell Model Based on a Fractional Derivative

The creep-damage behavior of salt rock has been studied through indoor creep experiments, which lasted for 359 days. By replacing the Newtonian dashpot in the Maxwell model with the Abel dashpot, the new constitutive relation of the Maxwell creep model (Figure 5), based on a fractional derivative, is given by

$$\varepsilon(t) = \frac{\sigma}{E} + \frac{\sigma}{\eta^\beta} \frac{t^\beta}{\Gamma(1+\beta)}. \tag{4}$$

Equation (4) denotes the creep strain characterized by the Abel dashpot. According to other experiments using salt samples drilled at the same place, the long-term strength is about 12 to 14 MPa. To make sure that the accelerated creep finally occurred during the experiment, the axial stress was greater than the long-term strength. So, according to accumulated experiences, substituting $\sigma = 20$ MPa and $\eta^\beta = 8$ GPa · h into Equation (4), one finds a series of creep curves under different derivative orders, β (Figure 5). It can be easily observed that the Maxwell creep model, based on a fractional derivative, has advantages in describing the characteristics of the decay creep process.

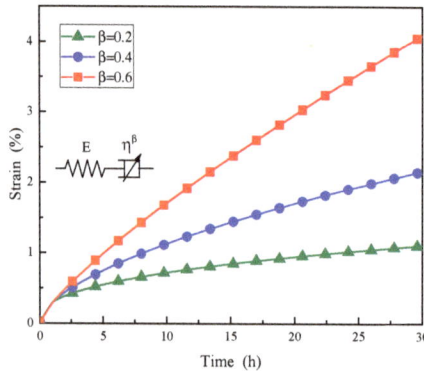

Figure 5. Creep strain of the Maxwell model based on a fractional derivative given by: σ = 20 MPa and $\eta^\beta = 8$ GPa · h.

4.4. The Creep-Damage Model Based on a Fractional Derivative

Many researchers have considered the creep of salt rock as a decay creep process, when the accelerating creep stage is ignored, which was caused by damage accumulation. Thus, they usually ignore the damage in the first two creep stages when establishing the constitutive model. But according to the discontinuity in natural salt rock, the damage actually existed since the beginning of the experiment and directly affected the whole creep process. As such, when considering damage during the full creep region, the total creep deformation can be concluded as the combination of two parts: primary and steady creep strains ε_{t+ss} (the decay creep process without damage) and the strain induced by the damage ε_d. Thus, the total creep deformation is given as

$$\varepsilon_c = \varepsilon_{t+ss} + \varepsilon_d \tag{5}$$

where ε_c is the total strain of creep.

By using the Maxwell creep model, based on a fractional derivative, to define the primary and steady creep strain, ε_{t+ss}, the constitutive relation of ε_{t+ss} and ε_d is given as [9]

$$\varepsilon_{t+ss} = \frac{\sigma}{E} + \frac{\sigma}{\eta^\beta} \frac{t^\beta}{\Gamma(1+\beta)} \tag{6}$$

$$\varepsilon_d = A\left(\frac{\sigma}{1-D}\right)^n t \tag{7}$$

where E is the elasticity modulus, η^β is the viscosity coefficient, A, n are material coefficients, and D is the damage factor, $0 \le D \le 1$.

The accumulated energy of AE events was used to evaluate the damage. Considering the data from the 5th sensor, the damage variable, d, can be defined as

$$d = \int_0^{\varepsilon_t} p_{i\varepsilon} d\varepsilon \Big/ \int_0^{\varepsilon_c} p_{i\varepsilon} d\varepsilon \qquad (8)$$

where ε_t is the strain of the specimen at time, t, ε_c is the total strain, and $p_{i\varepsilon}$ is the energy density of the 5th sensor at the loading level of ε.

By using acoustic emission data from the uniaxial compression test, the energy density $p_{i\varepsilon}$, relative to the strain curve of the specimen, is plotted in Figure 6. The damage variable, d, at any creep strain can be calculated by integration methods. According to the least square method, the relation between the damage variable, d, and strain at time, t, can be defined as

$$d = a \cdot [\varepsilon(t)]^b. \qquad (9)$$

All units are calculated using the International System of Units. So, the values of a and b from the fitted curve of damage variables are 0.0161 and 1.7572, respectively. It can be seen from the global evolution of damage that the damage increases slowly and steadily at the beginning of the loading procedure and then increases rapidly towards the end.

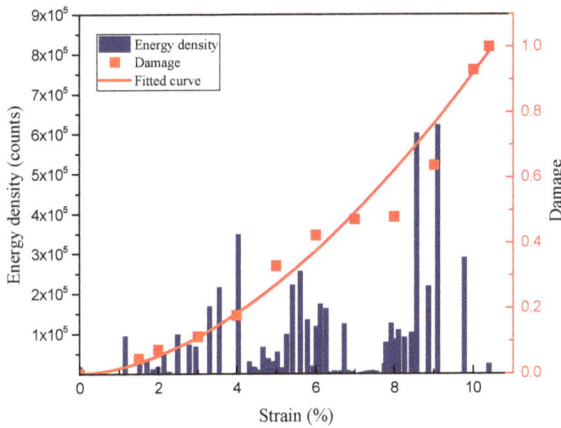

Figure 6. Curves of energy density and damage relative to the strain of the salt rock sample.

Synthetically, the creep-damage model, based on a fractional derivative, is given by

$$\varepsilon(t) = \frac{\sigma}{E} + \frac{\sigma}{\eta^\beta} \frac{t^\beta}{\Gamma(1+\beta)} + A \left(\frac{\sigma}{1 - a \times [\varepsilon(t)]^b} \right)^n t. \qquad (10)$$

4.5. Parameter Determination by Fitting Analysis

The efficacy of the creep-damage model, based on a fractional derivative, is dependent on its ability to adequately fit experimental data. Using the experimental data of salt rock creep under uniaxial compression, the parameters of the creep-damage model in Equation (10) were determined by the Quasi-Newton method (Figure 7, Table 1). It is indicated that the creep-damage model, based on a fractional derivative, proposed in this paper, can adequately represent the creep deformation of salt rock and accord with the experimental data better than the results estimated by the Maxwell model.

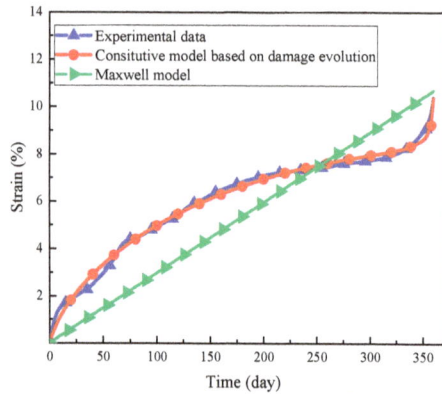

Figure 7. Experimental data and fitting curves.

Table 1. Parameters determined by a fitting analysis based on creep tests of salt rock.

Model	E (GPa)	η^β (GPa)	β	A	n	Correlation Coefficient
Creep-damage model	1.102	0.178	0.486	2.263×10^{-13}	0.998	0.989
Maxwell model	4.533	13.702	/	/	/	0.627

5. Conclusions

By replacing the Newtonian dashpot in the classical Maxwell model with the variable-viscosity Abel dashpot, a new creep-damage constitutive model was proposed with AE statistical regularity. The parameters of the new creep-damage model were determined by the Quasi-Newton method.

The results show that the creep deformation reflected the three stages of characteristics of a non-decay creep process. Meanwhile, the AE signals could also be divided into three obvious different stages. The distribution of the acoustic emission signal agreed well with the location where the deformation or growing fractures appeared in the salt rock. Furthermore, the cumulated AE events and released energy showed the same trend.

By considering the damage factor during the whole process of creep, a new mathematical model, which is named the creep-damage model based on a fractional derivative was proposed. In addition, the fitting analysis suggested that the creep-damage model based on a fractional derivative, in this paper, provided a precise description of the full creep regions in salt rock.

Author Contributions: H.Z. and D.L. contributed to the concept, results explanation and writing of the paper. D.L. and Y.Z. performed the experiments and analyzed the data. G.L. and D.X. contributed to the formal analysis. Supervision, Project Administration, and Funding Acquisition, H.Z.

Funding: This work was funded by the National Natural Science Foundation of China (51674266), and the State Key Research Development Program of China (2016YFC0600704) and the Yueqi Outstanding Scholar Program of CUMTB (2017A03). The financial supports are gratefully acknowledged.

Conflicts of Interest: The authors declare no competing interests.

References

1. Li, Y.P.; Yang, C.H.; Daemen, J.J.K.; Yin, X.Y. A new Cosserat-like constitutive model for bedded salt rocks. *Int. J. Numer. Anal. Meth. Geomech.* **2009**, *33*, 1691–1720. [CrossRef]
2. Adolfsson, K.; Enelund, M.; Olsson, P. On the fractional order model of viscoelasticity. *Mech. Time-Depend. Mater.* **2005**, *9*, 15–34. [CrossRef]
3. Bagley, R.L.; Torvik, P.J. Fractional calculus in the transient analysis of viscoelastically damped structures. *AIAA J.* **1985**, *23*, 918–925. [CrossRef]

4. Jin, J.; Cristescu, N.D. An elastic/viscoplastic model for transient creep of rock salt. *Int. J. Plast.* **1988**, *14*, 85–107. [CrossRef]

5. Li, Y.P.; Liu, W.; Yang, C.H.; Daemen, J.J.K. Experimental investigation of mechanical behavior of bedded rock salt containing inclined interlayer. *Int. J. Rock Mech. Min. Sci.* **2014**, *69*, 39–49. [CrossRef]

6. Cai, M.F. *Rock Mechanics and Engineering*; Bejing Science Press: Beijing, China, 2006.

7. Passaris, E.K.S. The rheological behavior of rock salt as determined in an in situ pressurized test cavity. In Proceedings of the 4th Conference on International Society for Rock Mechanics, Montreux, Switzerland, 2–8 September 1979; pp. 257–264.

8. Ghavidel, N.A.; Nazem, A.; Heidarizadeh, M.; Moosavi, M.; Memarian, H. Identification of rheological behavior of salt rock at elevated temperature, case study: Gachsaran evaporative formation. In Proceedings of the SRM European Regional Symposium EUROCK 2014, Vigo, Spain, 27–29 May 2014.

9. Hou, Z.M.; Lux, K.H. A constitutive model for rock salt including structural damages as well as practice-oriented applications. In Proceedings of the 5th Conference on Mechanical Behaviour of Salt, Bucharest, Romania, 9–11 August 1999; pp. 151–169.

10. Hou, Z.M. Mechanical and hydraulic behaviour of salt in the excavation disturbed zone around underground facilities. *Int. J. Rock Mech. Min. Sci.* **2003**, *40*, 725–738. [CrossRef]

11. Zhou, H.W.; Wang, C.P.; Han, B.B.; Duan, Z.Q. A creep constitutive model for salt rock based on fractional derivatives. *Int. J. Rock Mech. Min. Sci.* **2011**, *48*, 116–121. [CrossRef]

12. Wu, F.; Liu, J.F.; Wang, J. An improved maxwell creep model for rock based on variable-order fractional derivatives. *Environ. Earth Sci.* **2015**, *73*, 6965–6971. [CrossRef]

13. Zhou, H.W.; Wang, C.P.; Mishnaevsky, L., Jr.; Duan, Z.Q.; Ding, J.Y. A fractional derivative approach to full creep regions in salt rock. *Mech. Time-Depend. Mater.* **2013**, *17*, 413–425. [CrossRef]

14. Lavrov, A.; Vervoort, A.; Filimonov, Y.; Wevers, M.; Mertens, J. Acoustic emission in host-rock material for radioactive waste disposal: Comparison between clay and rock salt. *Bull Eng. Geol. Environ.* **2002**, *61*, 379–387. [CrossRef]

15. Hardy, H.R. Acoustic emission in salt during incremental creep tests. In Proceedings of the 5th Conference on Mechanical Behaviour of Salt, Bucharest, Romania, 9–11 August 1999.

16. Kong, X.G.; Wang, E.Y.; He, X.Q.; Li, Z.; Li, D.; Liu, Q. Multifractal characteristics and acoustic emission of coal with joints under uniaxial loading. *Fractals* **2017**, *25*, 1750045. [CrossRef]

17. Zhang, R.; Dai, F.; Gao, M.Z.; Xu, N.W. Fractal analysis of acoustic emission during uniaxial and triaxial loading of rock. *Int. J. Rock Mech. Min. Sci.* **2015**, *79*, 241–249. [CrossRef]

18. Xie, H.P.; Liu, J.F.; Ju, Y.; Li, J.; Xie, L.Z. Fractal property of spatial Distribution of acoustic emissions during the failure process of bedded rock salt. *Int. J. Rock Mech. Min. Sci.* **2011**, *48*, 1344–1351. [CrossRef]

energies

MDPI

Article

Experimental Study on Sensitivity of Porosity to Pressure and Particle Size in Loose Coal Media

Chenghao Zhang [1], Nong Zhang [1,*], Dongjiang Pan [2], Deyu Qian [1,*], Yanpei An [1], Yuxin Yuan [1], Zhe Xiang [1] and Yang Wang [1,3]

[1] Key Laboratory of Deep Coal Resource Mining, Ministry of Education of China, School of Mines, China University of Mining and Technology, Xuzhou 221116, China; zhangchenghao421@126.com (C.Z.); anyanpei@cumt.edu.cn (Y.A.); yuanyuxin@cumt.edu.cn (Y.Y.); xiangzhe@cumt.edu.cn (Z.X.); TB15020024B0@cumt.edu.cn (Y.W.)

[2] State Key Laboratory of Shield Machine and Boring Technology, China Railway Tunnel Group Co., Ltd., Zhengzhou 450001, China; cumtpdj@163.com

[3] School of Public Policy and Urban Affairs, College of Social Sciences and Humanities, Northeastern University, Boston, MA 02115, USA

* Correspondence: zhangnong@126.com (N.Z.); qian@cumt.edu.cn (D.Q.); Tel.: +86-136-0521-0567 (N.Z.); Tel.: +86-152-5201-6098 (D.Q.)

Received: 8 June 2018; Accepted: 27 August 2018; Published: 29 August 2018

Abstract: A new experimental method for characterizing the porosity of loose media subjected to overburden pressure is proposed based on the functional relationships between porosity, true density, and bulk density. This method is used to test the total porosity of loose coal particles from the Guobei coal mine in Huaibei mining area, China, in terms of the influence of pressure and particle size on total porosity. The results indicate that the total porosity of loose coal under 20 MPa in situ stress is about 10.22%. The total porosity and pressure obey an attenuated exponential function, while the total porosity and particle size obey a power function. The total porosity of the loose coal is greatly reduced and the sensitivity is high with increased pressure when stress levels are low (shallow burial conditions). However, total porosity is less sensitive to pressure at higher stress when burial conditions are deep. The effect of particle size on the total porosity reduction rate in loose coal is not significant, regardless of low- or high-pressure conditions; i.e., the sensitivity is low. The total porosity remains virtually unchanged as particle size changes when pressure exceeds 20 MPa. Overall, the sensitivity of total porosity to pressure is found to be significantly higher than sensitivity to particle size.

Keywords: loose media; coal; porosity; true density; bulk density; overburden pressure; particle size

1. Introduction

Porosity is one of the most important physical parameters used to characterize geological materials, including loose media which consists of a large number of particles basically belonging to the same order of magnitude, with voids between the particles and low mechanical strength, such as broken coal which has been subjected to tectonic stress failure, soil, and sand [1]. It is widely used in the field of mining and geotechnical engineering, as porosity is a primary control on material strength and behavior. The porosity of loose coal can be used to evaluate, among other things, the properties of coal reservoirs, the selection of grouting materials, the design of grouting parameters and the evaluation of grouting effects.

Mercury intrusion porosimetry (MIP), low-temperature nitrogen adsorption, and scanning electron microscopy (SEM) are the conventional methods of porosity testing [2–4]. Recently, some advanced techniques—such as nuclear magnetic resonance (NMR), computed tomography (CT), and

3D or 4D X-ray microscopes (XRM)—have also been applied to porosity testing [5–8]. However, these methods are mainly carried out at room temperature and ambient pressure, neglecting the key factor of overburden pressure for in situ geological materials. Although XRM is possible to measure pore structure parameters of subsurface rock systems at reservoir temperatures and pressures, it can only image mm-scale samples and it is ideal for resolving µm-scale processes [9,10]. To study the influence of pressure on porosity of cm-scale samples, some scholars have used automated porosity and permeability instruments or similar gas expansion principles to hold drilled core samples, and then apply different confining pressures and temperatures to study porosity variation [11,12], especially under different confining pressures. These core samples have typically consisted of low-permeability rocks and lump coal [13–22].

The above methods use drilled cores to test porosity of coal and other rock samples, but this is not possible for loose media that is pulverized and broken, in which drilling cores and press forming is not practicable. As such, it is difficult to study the influence of pressure and other factors on the porosity of loose media using the above-mentioned automated instruments, and some scholars tried to compress briquette and then testing porosity with porosity and permeability instruments [11,23–25]. There have also been very few experimental studies on the effects of particle size on the porosity of loose media [26–29] based on fractal theory or seepage experimental instruments.

In this paper, the research object of loose coal which is formed by tectonic stress failure, also called tectonically deformed coal [30], consists of a large number of broken particles, a type of loose media. Results are presented from the use of a true density tester and the use of a UTM5504 microcomputer-controlled electronic universal testing machine (Shenzhen Suns Technology Co., Ltd., Shenzhen, China) to accurately measure bulk density under overburden pressure. From this work, a suitable porosity testing method for loose media subjected to overburden pressure is proposed. It is based on the functional relationship between porosity, true density, and bulk density. This work also reveals the influence laws of different pressures and particle sizes on porosity.

2. Materials and Methods

2.1. Materials

Loose coal samples were collected from the haulage gateway of working face 8105-1 in coal seam 8_1 of the Guobei Coal Mine in the Huaibei mining area. The altitude of working face 8105-1 is between -683.6 m and -890.0 m, and the corresponding ground surface altitude varies from 29.6 m to 31.5 m. The structure of coal seam 8_1 is complex, as shown in Figure 1.

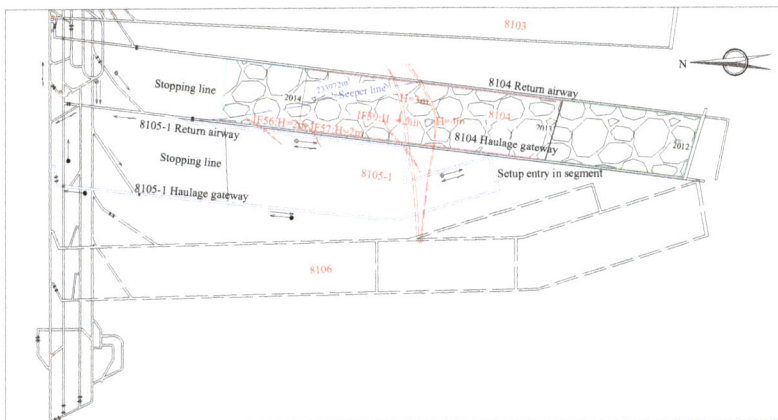

Figure 1. Position, production system, and geological structure of 8105-1 working face.

The coal body is grayish black to black, powdery to lumpy, partially scaly, with black streaks and a dull luster. It is a semi-bright coal. Detailed description and rock strata information were shown in Figure 2.

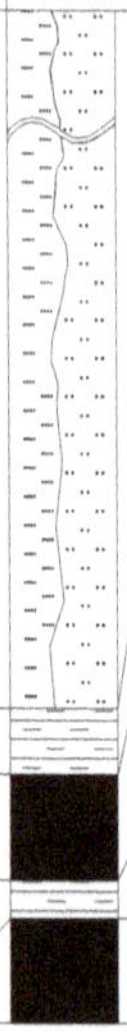

Rock strata	Borehole column	Thickness(m)	Lithological description
siltstone~ medium sandstone		15.59~29.26 22.02	Inner section: gray to light gray sandstone, continuous horizontal bedding. Outer section: grayish white, fine grained structure, local grain size is mainly composed of coarse siliceous, a small amount of dark minerals, parallel bedding, occasionally veined bedding, longitudinal oblique fracture.
		0.92~3.04 2.08	Gray to dark gray, flat fracture, muddy structure, partially silty, containing more plant fossil fragments.
mudstone		1.53~4.37 3.35	Black. Streak brown-brown black-black, powdery structure, glass luster-asphalt luster,endogenous fissures at the top and bottom are slightly developed, semi-bright coal dominated, occasionally bright coal, staggered-stepped fracture.
8₁coal		0.83~1.50 1.19	Light gray to grayish black, blocky structure. Partially containing siltstone, containing more plant fossil fragments.
mudstone			
8₂coal		1.87~4.75 3.29	Black,powder to broken structure, streaks brown to black, asphalt glass luster. Semi-dark to semi-light coal.

Figure 2. Borehole columnar section of coal seam.

To visually examine the pore spaces of the loose coal samples, the microstructure of the collected coal particles was observed using a SIGMA 300 field emission scanning electron microscope (Carl Zeiss, Oberkochen, Germany). The results were shown in Figure 3.

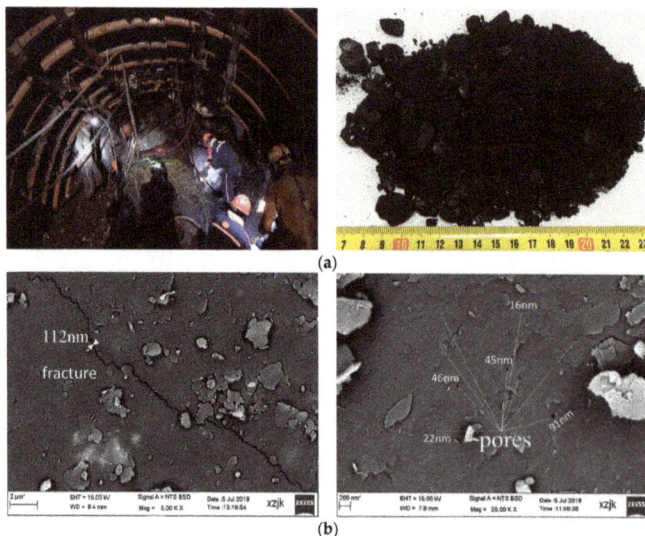

(a)

(b)

Figure 3. *Cont.*

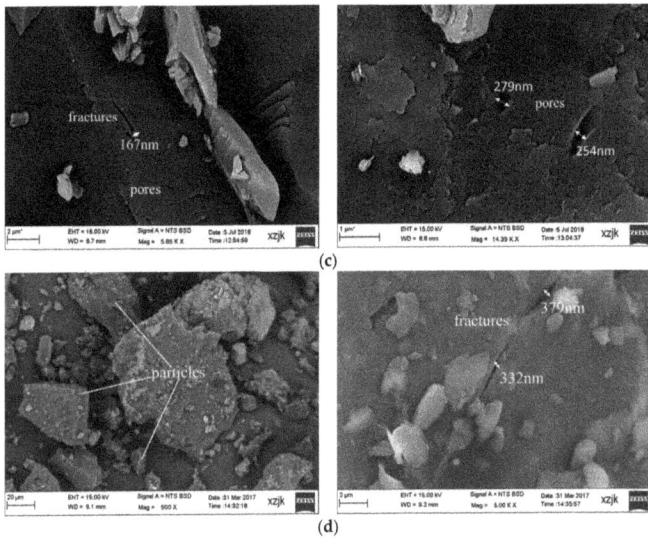

Figure 3. (a) On-site sampling photographs and (**b**–**d**) surface micromorphology of 5–15 mm, 2–5 mm, 0–2 mm coal particles from the 8_1 coal seam.

The sampling photographs and SEM images show that the loose coal body is composed of particles of different sizes, and the surfaces of the particles are further adhered to by tiny coal pieces. There are voids between the particles and when the coal particles are magnified, the microfractures and pores in the matrix are seen. The average labeled fracture width is about 247.5 nm, and the pore diameters focus on 16–45 nm.

The raw coal collected on-site from the mine is in the form of pulverized and fragmented coal. To understand the particle size distribution (PSD) of the raw coal and select the particle size gradient to study, initial sample screening (the mass of studied specimen was 2000 g) was performed, as shown in Table 1. Mesh diameter, grader retained percentage, and accumulated retained percentage in the table were used to plot the screening curve, as shown in Figure 4.

Table 1. Initial sample screening experiment data.

Mesh Diameter (mm)	Grader Retained Mass m_i (g)	Grader Retained Percentage (m_i/2000) (%)	Accumulated Retained Percentage (%)
31.5	0	0	0
16	124.77	6.2385	6.2385
9.5	131.25	6.5625	12.801
4.75	235.18	11.759	24.56
2.36	328.3	16.415	40.975
1.18	262.71	13.1355	54.1105
0.6	335.51	16.7755	70.886
0.3	189.61	9.4805	80.3665
0.15	183.05	9.1525	89.519
0.088	173.26	8.663	98.182
<0.088	15.98	0.799	98.981

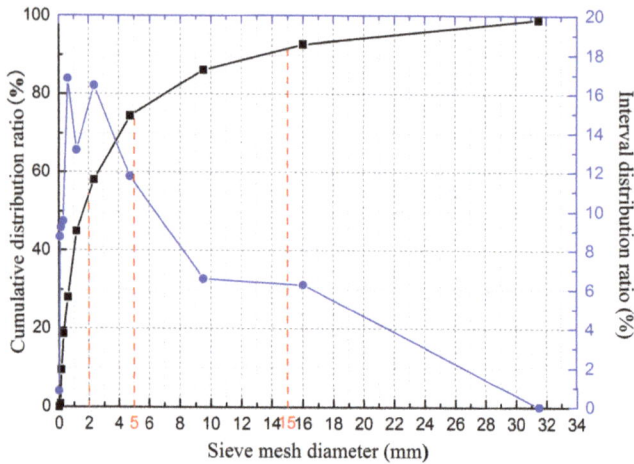

Figure 4. Cumulative and interval distribution curves of loose coal sample particle sizes.

Results of the initial screening show that coal particles >15 mm with very low distribution were not considered. Sieves with mesh diameters of 2, 5, and 15 mm were used for re-screening to obtain three particle size ranges of 0–2, 2–5, and 5–15 mm, which are similar to the particle size of three types of tectonically deformed coal, mylonitic coal (<1 mm), granulated coal (1–3 mm), and tectonically crushed coal (3–10 mm) after being broken by pressure [31,32], as shown in Figure 5. The above three size classes from the screening and raw coal (mixed particle sizes: 0–2 mm, 50%; 2–5 mm, 30%; 5–15 mm, 20%) were used in the porosity characterization and influencing factors tests.

Figure 5. Screened coal particles of three different particle size classes.

2.2. Methods

2.2.1. Porosity in the Particle Matrix

Porosity in the matrix of particles was characterized using the combined methods of mercury intrusion porosimetry (MIP) and low temperature nitrogen adsorption.

The MIP experiments were carried out using an Autopore 9510 Mercury Porosimeter (Micromeritics Instrument Corp., Norcross, GA, USA). This instrument can use mercury intrusion pressures from atmospheric pressure to 60,000 psi (414 MPa) and measure pore sizes in the 0.003–6 μm range with a ±0.10% transducer accuracy of full scale. The samples used was about 5–15 mm in size and dried at 110 °C for 24 h before the experiment.

The mercury porosimetry analysis technologies are based on the intrusion of mercury into aporous structure under stringently controlled pressures. They can capture numerous sample pore properties such as pore size distributions, total pore volume, or porosity. The porosity calculation formula is

$$n_p = V_1/V_2 \qquad (1)$$

where n_p is the particle matrix porosity; V_1 is the pore volume (mercury volume intruded at the maximum experimental pressure attained, 414 MPa) and V_2 is the sample volume which can be derived by Equation (2)

$$V_2 = \{w_4 - [w_3 - (w_2 - w_1)]\}/\rho_{Hg} = (w_4 - w_3 + w_2 - w_1)/\rho_{Hg} \qquad (2)$$

where ρ_{Hg} is the density of mercury at a certain temperature; w_1 is the mass of empty cell; w_2 is the mass of empty cell and sample; w_3 is the mass of mercury, empty cell and sample and after filling with mercury under a certain pressure; w_4 is the mass of mercury and empty cell after empty cell filled with mercury.

Due to compression of mercury, of the sample cell, and of the rest of the instrument components, but also of the sample itself, the mercury intrusion method will have a certain error, especially at higher mercury pressures [33,34]. Subtracting an empty cell run from an actual sample analysis is an effective means to correct compression of mercury and system components error (blank correction). For the problem of compressibility of coal under high pressure of mercury intrusion, low temperature nitrogen adsorption was used to instead of MIP to calculate the pore volume which was originally tested with mercury intrusion at high pressure.

The low temperature nitrogen adsorption experiments were carried out using a Micrometrics ASAP 2460 (Micrometrics Instrument Corp., Norcross, GA, USA) gas adsorption instrument. Samples were crushed to 250 μm and dried at 100 °C for 24 h to be used for low pressure isotherm analysis. After outgassing at 150 °C for 8 h, the N_2 gas isotherm adsorption measurements were carried out at 77 K (-196 °C). The pore volume and pore size distribution can be obtained from the low temperature nitrogen adsorption experiments.

2.2.2. Total Porosity of Loose Coal

According to the China National Standard [35], the extended formula for calculating total porosity while under overburden pressure is

$$n_t = (1 - \rho_b/\rho) \times 100\% \qquad (3)$$

where n_t is the total porosity of coal or rock (%); ρ_b is the bulk density of coal or rock, i.e., the ratio of mass to total volume of coal or rock (including the volume of skeleton and all pore spaces, g/cm^3) and ρ is the true density or mineral particle skeletal density of coal or rock (the ratio of mass to skeleton volume, not including the pore spaces, g/cm^3).

The test of true density (ρ) was done with the true density tester. The testing instrument was an AccuPyc 1330 Pycnometer (Micrometrics Instrument Corp., Norcross, GA, USA). The instrument uses the Archimedes principle of helium gas expansion. The molecular diameter of He is 0.26 nm, thus it can penetrate the smallest pores and irregular hollows in the surface of a sample. As a result, it is possible to obtain a skeleton volume very close to the true skeleton volume value, and the calculated true density is therefore a true reflection of the true density value of any given sample.

To determine bulk density (ρ_b) of coal samples at different pressures, the UTM5504 universal testing machine was used. The experimental equipment used and a schematic diagram of the method were shown in Figure 6.

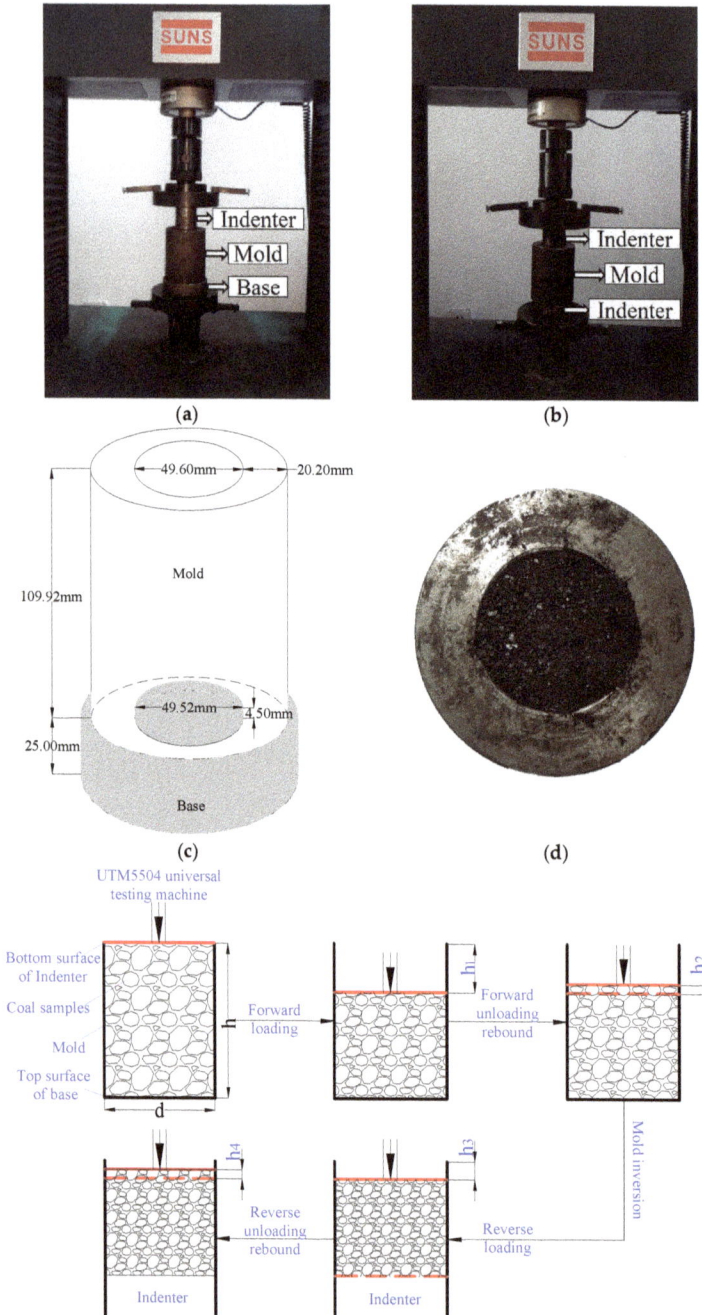

Figure 6. (**a**) Forward loading photograph; (**b**) reverse loading photograph; (**c**) mold schematic diagram; (**d**) schematic diagram of initial experimental coal samples; and (**e**) schematic diagram of the test apparatus and methodology for determining bulk density of coal at different pressures.

To accurately simulate the in situ state of pressure, the experiment was run in four stages:

(1) Forward loading stage: A loading speed of 2 mm/min was used, and after the specified pressure (Determined according to the in situ stresses—5, 10, 15, 20, and 25 MPa—was selected in this test because the in situ stress of Guobei Coal Mine 8105-1 working face is 20 MPa) level was reached, the force applied was kept constant for 30 min. The loading time was referenced by some scholars' briquette compressing schemes [24,25,36,37]. The forward decreased displacement of indenter h_1 was automatically recorded by the testing machine. At each initial experiment, the coal sample should be filled with the mold, that is, the initial position of the bottom surface of the indenter was flush with the top of the coal briquette mold, as shown in Figure 6d. The purpose of the operation is to conveniently calculate the height of the coal sample after compressing. After this stage, $h_{\text{coal}} = h - h_1$. h is the height of the briquette mold excluding the base embedded in the mold.

(2) Forward unloading rebound stage: The unloading speeds were set to 15 N/s corresponding to experimental pressures of 5 MPa, 30 N/s→10 MPa, 45 N/s→15 MPa, 60 N/s→20 MPa, and 75 N/s→25 MPa. The purpose is to keep the consistent unloading time at each pressure. When the force returned to zero, the compressed coal sample rebounded and the forward rebound amount of the coal sample h_2 was recorded.

(3) Reverse loading stage: The briquette mold was reversed, and the pressure was applied continuously. The loading speed was set to 2 mm/min and after the specified pressure level (5, 10, 15, 20, and 25 MPa) was reached the applied force was kept constant for 30 min. The reverse loading parameters were exactly the same as the parameters in the forward loading stage. The reverse decreased displacement of indenter h_3 was recorded.

(4) Reverse unloading rebound stage: The unloading speeds were set to 15 N/s corresponding to experimental pressures of 5 MPa, 30 N/s→10 MPa, 45 N/s→15 MPa, 60 N/s→20 MPa, and 75 N/s→25 MPa. When the reverse force returned to zero, the compressed coal sample rebounded and the reverse rebound amount h_4 was recorded.

(5) The mold was again turned upside down and the next level of pressure was applied. The previous four stages were repeated to complete the next pressure level tests.

The bulk density of coal under different pressures was then calculated as shown in Equation (4).

$$\rho_b = \frac{m}{\pi(d/2\,)^2[h - (h_1 + h_3 - h_2)]} \tag{4}$$

where ρ_b is the bulk density of coal under a certain overburden pressure (g/cm^3); m is the mass in grams of the coal sample in the briquette mold (the coal sample filled the mold at the beginning of the test); and d is the inner diameter of the briquette mold (cm). The inner diameter used in this experiment was 4.960 cm.

The various heights and displacements are as follows: h is the height of the briquette mold (10.542 cm) excluding the base embedded in the mold; h_1 is the forward decreased displacement (cm) of the indenter under a certain overburden pressure; h_2 is the forward rebound amount (cm) of the coal sample when the overburden pressure returns to zero; and h_3 is the reverse decreased displacement (cm) of the indenter under a certain overburden pressure. The result of $h - (h_1 + h_3 - h_2)$ is the height of the compressed coal samples under each pressure.

It should be noted that Equation (4) applies to the calculation of the bulk density of the independently filling and compressing coal samples under each pressure level, but if the coal sample compression of a certain pressure level continues to circulate on the basis of the previous pressure level, when calculating the height of the compressed coal sample, it is necessary to subtract the influence of the reverse rebound amount h_4 of the previous pressure level. For example, in the experiment scheme of this article, when the pressure is 5 MPa, the bulk density can be calculated directly using Equation (4), but when 10 MPa, the reverse rebound amount h_4 of 5 MPa needs to be subtracted when calculating the height of the compressed coal sample. Because the 10 MPa forward loading stage was followed by 5 MPa reverse unloading rebound stage.

In the bulk density experiment, the mass of the coal sample was measured using a precision balance (Yuyao Jiming Weighing Calibration Equipment Co., Ltd., Ningbo, China, JM-A20002, MAX = 2000 g, d = 0.01 g) and the height and inner diameter of the mold were measured using a Vernier caliper (Hangzhou Huafeng Big Arrow Tools Co., Ltd., Hangzhou, China, HF-8631215, MAX = 150 mm, d = 0.01 mm). The decreased displacement and rebound amount of the indenter were automatically recorded by the testing machine with a 0.04 μm displacement resolution. The maximum test force of the machine is 50 KN (resolution 1/500,000).

3. Results

3.1. Characterization of Total Porosity of Loose Coal

Total porosity refers to the ratio of the sum of the volume of all pore spaces in a rock sample to the volume of the sample. For loose coal, the total porosity consists of three parts: inter-particle voids, pores, and micro-fractures in the particle matrix.

The porosity of the loose coal matrix was obtained by mercury intrusion porosimetry and low temperature nitrogen adsorption. After testing the true density of the loose coal sample and the bulk density subjected to overburden pressure, the total porosity of the loose coal samples was determined from the functional interrelationships of total porosity, true density, and bulk density. Thus, the objective of accurately characterizing the porosity of loose coal which is multi-scale pore spaces (millimeter voids, micro-nano fractures and pores) was achieved.

3.1.1. True Density

The true densities of minerals in coal are much greater than the true densities of organic matter within coal. For example, the true densities of quartz, clay and pyrite are 2.15, 2.40, and 5.00 g/cm^3, respectively. The average true density of minerals in coal is approximately 3.00 g/cm^3, and the higher the inorganic mineral content, the higher the true density of the coal [38]. The associated mineral compositions of the coal samples were analyzed by the D8 ADVANCE X-ray diffractometer (Bruker AG, Karlsruhe, Germany), as shown in Figure 7.

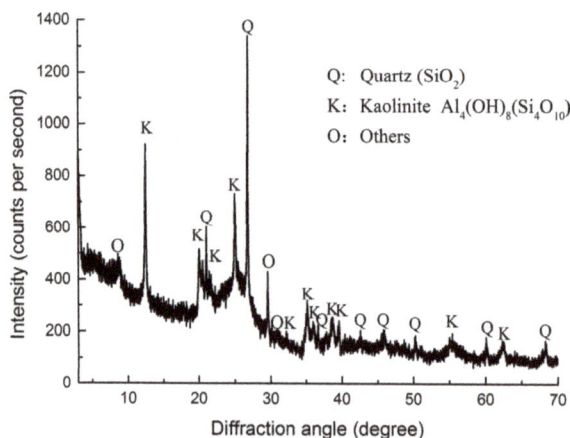

Figure 7. X-ray diffraction pattern of 8$_1$ coal sample.

Figure 7 shows that the coal sample collected in the experiment contained quartz, kaolinite, and other minerals, and the mineral contents were shown in Table 2.

Table 2. Results of quantitative analysis of 8_1 coal sample minerals.

Minerals	Quartz	Kaolinite	Others
Weight percentage	12.04%	9.47%	3.79%

As a result, the true densities measured for different particle size of coal were different, as shown in Table 3. The difference in true density values of different coal particle sizes may be due to the different mineral contents.

Table 3. True densities of coal measured for different size classes of coal particles.

Size	0–2 mm	2–5 mm	5–15 mm	Raw Coal
True density	1.562 g/cm^3	1.632 g/cm^3	1.827 g/cm^3	1.636 g/cm^3

3.1.2. Porosity in the Particle Matrix

The matrix porosity of 5–15 mm samples from coal seam 8_1 determined from the combined test of MIP and low temperature nitrogen adsorption was 4.23%. With 50 nm as the demarcation point of MIP and low temperature nitrogen adsorption [39,40], volume data for pores (diameter > 50 nm) used MIP data, and pores (diameter < 50 nm) used low temperature nitrogen adsorption data. The comparison between combined methods and separate MIP was shown in Figure 8a. In the figure, in order to show the comparison and correction more intuitively, the pore diameter data of low temperature nitrogen adsorption was converted into mercury intrusion pressure. Cumulative porosity and porosity distribution curves after correction were shown in Figure 8b. The matrix porosity of this sample is mainly distributed as transitional pores (reference to Hodot pore grading [41]), with pore throat diameters in the range of a 20–40 nm, and 90–347 μm diameter fractures and pores (visible to naked eye) which have an order of magnitude size difference compared to the transitional pores are also evident. There are also few abrupt peaks on the porosity distribution curve which may be due to the randomness of the samples.

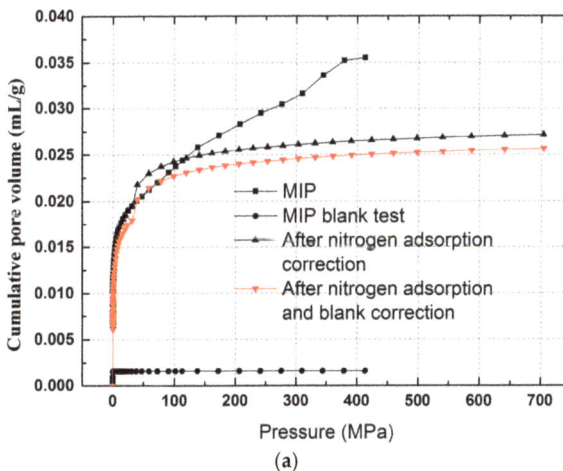

(a)

Figure 8. *Cont.*

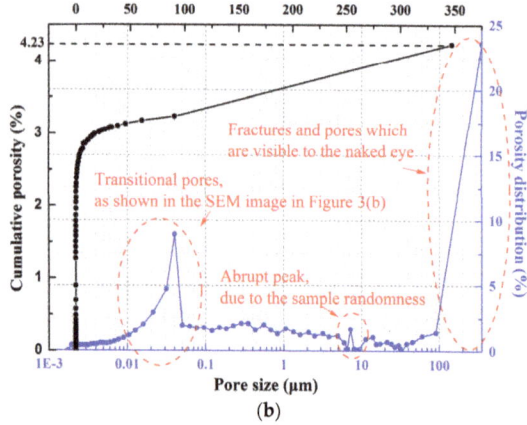

Figure 8. Results of pore space characterization using MIP and low temperature nitrogen adsorption: (**a**) pore volume and (**b**) porosity in the coal matrix.

3.1.3. Total Porosity of Loose Coal

The total porosity value was calculated by substituting the bulk density results and true density results into Equation (3). Results were shown in Figure 9. The actual depth (*H*) of the experimental sample collection site was about 800 m, based on the estimation of in situ stress ($\sigma = \gamma H$) where a value of 2.5 kN/m^3 was used for γ and the in situ stress at the sampling site was $\approx$20 MPa. The experimental results indicate a total porosity value of 10.22% for the sample from the 8_1 coal seam at 20 MPa.

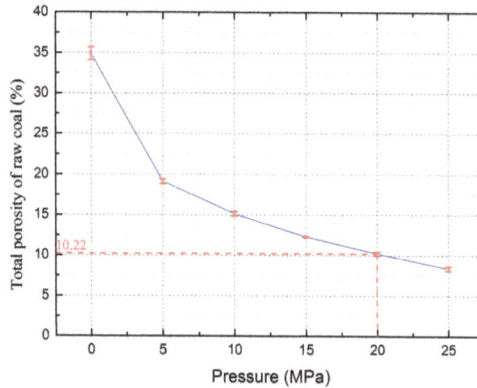

Figure 9. Curve of total porosity for a sample of coal from the 8_1 coal seam.

According to the MIP and low-temperature nitrogen adsorption data, the porosity in the matrix of particles from the 8_1 coal seam was approximately 4.23%. The remaining porosity of 5.99% for samples subjected to 20 MPa of overburden pressure was obtained from the inter-particle voids and the smaller pores (diameter < 1.77 nm) not measurable with the low temperature nitrogen adsorption experiment. In this article, the single coal sample of 5–15 mm was used in the MIP experiment, so the porosity result from MIP experiment does not include inter-particle voids. It can also be seen from the mercury intrusion curve that there is no obvious inflection point at low pressure.

3.2. Effects of Pressure and Particle Size on Total Porosity

The research object loose coal (tectonically deformed coal) has different burial depths and degrees of fragmentation. Different overburden pressures applied to loose, raw coal particles will result in deformation, breaking, rearrangement, and other changes that will cause the total porosity to change significantly. In addition, under the same overburden pressure, the arrangement and staggered relationship of coals with different particle sizes may also result in different porosity. To test the effects of pressure and particle size on total porosity, the characteristics of in situ stress and particle sizes of raw coal have been combined, and total porosity has been determined for different pressure gradients (0, 5, 10, 15, 20, and 25 MPa) and different particle size classes (0–2, 2–5, and 5–15 mm). The experimental results obtained are described in the following three subsections.

3.2.1. Pressure Effects on Total Porosity

Changes in total porosity with increasing pressure for the specified particle size classes have been constructed as curves, with data points fitted and analyzed (Figure 10 and Table 4). These data show that as pressure increases, the pores between the particles gradually close and the total porosity decreases for each particle size class. When the pressure is low, the degree of decrease is greatest, and as pressure increases the degree of decrease is gradually reduced.

Additionally, the total porosity and pressure obey an attenuated exponential function (all curves have an R-square of $\approx 98\%$). That is, the total porosity decreases exponentially with increasing pressure. Regression analysis of the experimental results show that the relationship can be defined as

$$n_t = ae^{-b\sigma} + c \tag{5}$$

where n_t is the total porosity of loose coal at a given pressure (%), σ is the pressure (MPa), the sum of a and c represents the total porosity of loose coal when the pressure is 0 MPa (%), and b represents the compression coefficient (MPa^{-1}).

Finally, the decreases in total porosity for each particle size class are similar as pressure increases, and the curves steepen. A mixed particle sizes (such as raw coal) will reduce the rate of total porosity decrease as pressure increase, evident in the gentle slope of the curve. The raw coal has a relatively low porosity at low pressure. For example, compared with 5–15 mm particles, the porosity value of the raw coal under 0 MPa is 15.31% less, and the value under 5 MPa is 3% less. This is because the staggered arrangement of large and small coal particles in raw coal and small particles filled large voids.

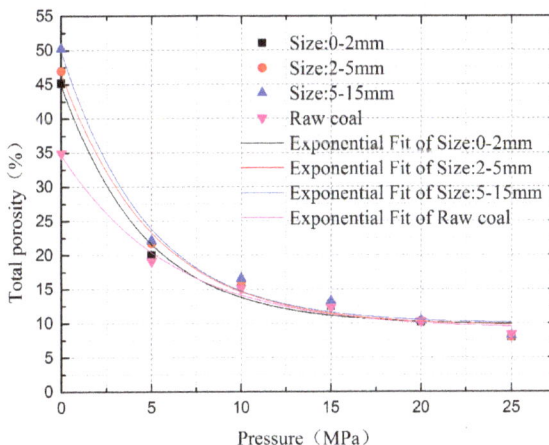

Figure 10. Curves of total porosity change with increasing pressure for specified particle size classes.

Table 4. Fitting formulae of total porosity for each particle size class.

Size	Fitting Formula	R-Square
0–2 mm	$n_t = 0.098 + 0.351e^{-0.217\sigma}$	0.97848
2–5 mm	$n_t = 0.095 + 0.371e^{-0.199\sigma}$	0.98336
5–15 mm	$n_t = 0.099 + 0.400e^{-0.211\sigma}$	0.98217
Raw coal	$n_t = 0.091 + 0.255e^{-0.162\sigma}$	0.98215

3.2.2. Particle Size Effects on Total Porosity

The study of loose coal particle size effects on total porosity includes two aspects: the initial particle size and the particle size of the compacted samples. The study of initial particle size has a reference value for the study of the influence of porosity on the physical and mechanical properties of coal by compressing coal samples with different porosity in the laboratory. The particle size of the compacted samples are the true particle size after laboratory or on-site compression, which truly reflects the relationship between the porosity and on-site tectonically deformed coal of different degrees of fragmentation. Using the cumulative PSD curve (Figure 4), the initial particle size D_{50} was selected as the average particle size of the segment. This is the particle size corresponding to 50% of the cumulative PSD in each segment of the three size classes. The fitting curve in Figure 4 shows that the D_{50} results for each segment are 0.54, 3.15, and 8.57 mm, respectively. From these data, it was possible to construct curves of total porosity change vs. initial particle size for each pressure interval, as shown in the line graph in Figure 11. The coal samples after each stage of loading screening was performed. Using the obtained particle size after loading, the curves of total porosity change vs. the particle size of the compacted samples were plotted, as shown in the histogram in Figure 11.

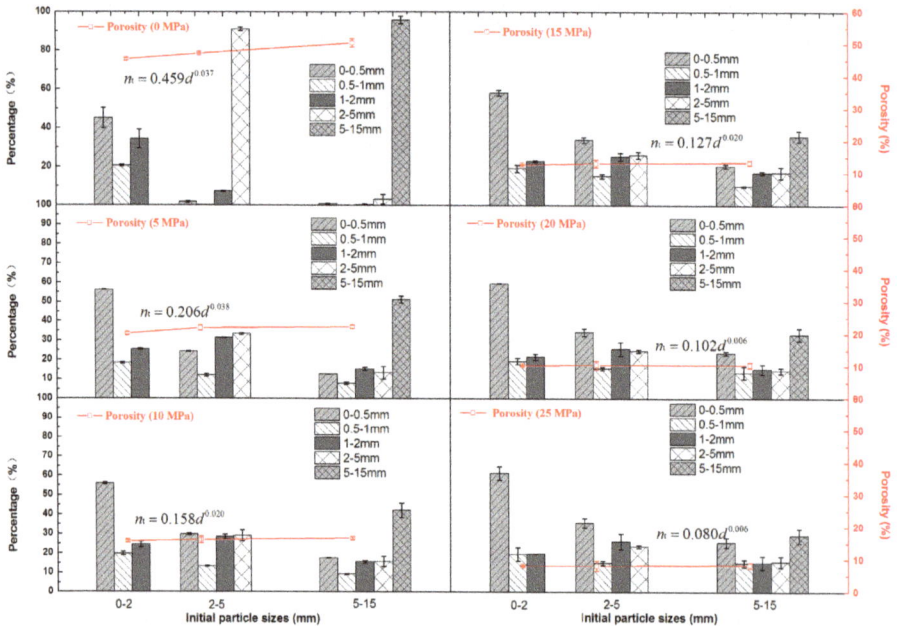

Figure 11. Curves of total porosity change vs. initial particle size and the particle size of the compacted samples.

The line graph curves show that, at each pressure, the total porosity increases with increasing initial particle size and, by fitting of the data points, the relationship between total porosity and initial particle size is shown to obey the power law

$$n_t = Ad^B \tag{6}$$

where n_t represents the total porosity; d represents the initial particle size and A and B are fitting constants.

The histograms show that the largest particles are gradually crushed and decreased as the pressure increase, and the degree of fragmentation is irregular. There is still a clear difference in particle size with the initial coal particles of 5–15, 2–5, and 0–2 mm subjected to the same pressure. Large initial particles correspond to low degree of on-site coal fragmentation.

Additionally, as pressure and initial particle size increases, the increased degree of total porosity undergoes a gradual decrease. When the pressure exceeds 20 MPa, the curve of initial particle size vs. total porosity approximates a straight line, indicating that the total porosity does not change with a change in initial particle size.

From the relationship between porosity and particle size of the compacted samples, it can be seen that the porosity of coal samples with low degree of fragmentation is larger than the coal samples with high degree of fragmentation, but the difference is small. As the overburden pressure increases, the difference gradually decreases. After 20 MPa, the difference tends to be zero. At this time, the porosity is considered to be independent of the degree of coal fragmentation.

3.2.3. Sensitivity of Total Porosity to Pressure and Particle Size

Evaluation of the sensitivity of total porosity to pressure and initial particle size is based on the use of total porosity reduction rate as the sensitivity parameter and a reference point with total porosity measured in coal in the 5–15 mm particle size range at a pressure of 0 MPa. The rate of change of total porosity with respect to the reference point at each pressure and initial particle size was defined as the total porosity reduction rate (Δn_t), as expressed in Equation (7). The greater the value of change in Δn_t as some other factor changes, the higher the sensitivity of total porosity to that factor, as

$$\Delta n_t = \frac{(n_0 - n_{ij})}{n_0} \times 100\% \tag{7}$$

where n_0 is the total porosity of coal samples in the 5–15 mm initial particle size range measured at 0 MPa pressure and n_{ij} is the total porosity measured for any particle size and pressure.

The reduction rate of total porosity for each initial particle size class and pressure was calculated according to Equation (7), and the results were shown in Table 5.

Table 5. Reduction rates of total porosity (Δn_t) for each particle size class and pressure.

Size \ Pressure	5–15 mm	2–5 mm	0–2 mm
0 MPa	0	6.54%	10.03%
5 MPa	55.96%	56.78%	60.18%
10 MPa	66.97%	67.96%	68.76%
15 MPa	73.64%	74.19%	75.04%
20 MPa	79.41%	79.55%	79.73%
25 MPa	83.74%	84.06%	84.06%

Table 5 shows that the total porosity of the loose coal is greatly reduced when the pressure is low (shallow burial conditions), and the sensitivity is high with respect to increasing pressure. Total porosity is less sensitive to pressure at higher stress levels (deep burial conditions). The change in total porosity reduction rate is not significant in terms of initial particle size changes, regardless of whether the pressure conditions are low or high, and the sensitivity is low. Generally, the sensitivity of total porosity to pressure is significantly higher than sensitivity to initial particle size (corresponding to degree of on-site coal fragmentation).

4. Discussions

4.1. Comparison with the Method of Compressing Briquette and Testing Porosity

There are two main problems with the method of compressing briquette and then testing porosity: one problem is that the difficulties of forming and taking out the briquette at low pressure, and another problem is that the briquette has a large amount of rebound at high pressure. The rebound amount can be reflected by the time–displacement curves measured from the UTM5504 universal testing machine, as shown in Figure 12a. The curves show that there is rebound after loading and unloading at all pressure levels. As the pressure level increases, the amount of rebound gradually increases. Thus, the error related to the porosity testing method by briquette mentioned in the introduction under overburden pressure also increases. The comparison of porosity differences caused by errors were shown in Figure 12b. The total porosity is calculated by Equation (3). When calculating, the method of compressing briquette accounts for the reverse rebound amount (h_4), but the method proposed in this paper eliminates the impact of the reverse rebound amount, not counting.

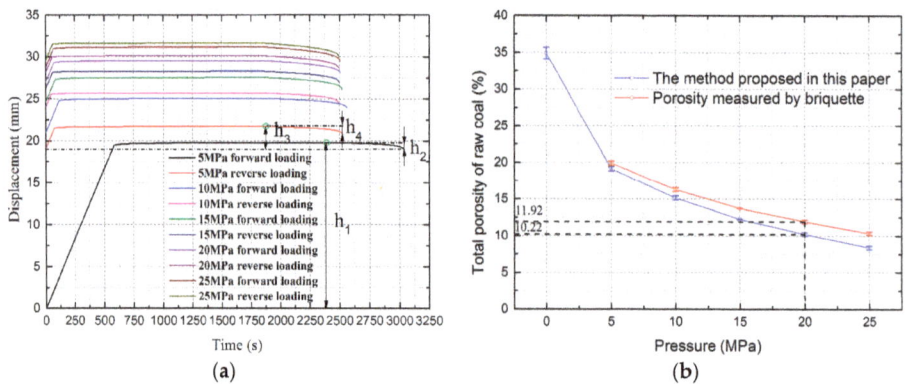

Figure 12. (a) Time–displacement curves determined during loading and unloading; and (b) comparison of porosity differences between the method of compressing briquette and then testing porosity and the method proposed in this paper.

4.2. Quality of Measurement Results

The true total porosity of on-site loose coal is difficult to measure or learn from relevant references, and the accuracy of this experiment cannot be quantitatively expressed. In these cases, uncertainty of measurement was usually used to evaluate the quality of measurement results [42,43]. It is the parameter, associated with the result of a measurement, that characterizes the dispersion of the values that could reasonably be attributed to the measurand. In most cases, a measurand is not measured directly, but is determined from other quantities through a functional relationship, where the uncertainty of output estimate or measurement result can be determined from the estimated standard deviation associated with each input estimate x_i, as shown in Equation (8)

$$u_c^2(nt) = \left[\frac{\partial f}{\partial \rho}\right]^2 u^2(\rho) + \left[\frac{\partial f}{\partial m}\right]^2 u^2(m) + \left[\frac{\partial f}{\partial d}\right]^2 u^2(d) + \left[\frac{\partial f}{\partial h}\right]^2 u^2(h) + \left[\frac{\partial f}{\partial h1}\right]^2 u^2(h1) + \left[\frac{\partial f}{\partial h3}\right]^2 u^2(h3) + \left[\frac{\partial f}{\partial h2}\right]^2 u^2(h2) + u^2(s) \quad (8)$$

where $u_c(n_t)$ is the combined standard uncertainty of output estimate n_t; $\partial f/\partial x_i$ is the partial derivative (sensitive coefficient) with respect to input quantities X_i ($\rho, m, d, h, h_1, h_3, h_2$) of functional relationship f between measurand n_t and input quantities X_i on which n_t depends, $u(x_i)$ is the standard uncertainty of input estimated x_i that estimates input quantity X_i, $u(s)$ is the standard uncertainty of the measurement repeatability of output estimate n_t, equal to the experimental standard deviation of the mean n_t.

Detailed calculation steps and precautions, see "Guide to the Expression of Uncertainty in Measurement" [42]. The uncertainty of the total porosity of loose raw coal under each pressure was calculated by Equation (8). The results under 20 MPa were shown in Table 6.

Table 6. Summary of standard uncertainty components.

Standard Uncertainty Component $u(x_i)$	Source of Uncertainty	Value of Standard Uncertainty $u(x_i)$	Sensitive Coefficient: $c_i = \partial f/\partial x_i$	Component of $u_c(n_t)$: $u_i(n_t) =$ $\lvert c_i \rvert u(x_i)$ (%)
$u(\rho)$	Instrument uncertainty ($\pm 0.03\%$ of the indication)	2.83×10^{-4} g/cm^3	0.14654	0.00438
	Measurement repeatability ($\pm 0.01\%$ of the indication)	4.15×10^{-5} g/cm^3		
$u(m)$	Instrument uncertainty (± 0.02 g)	0.0115 g	0.00111	0.00128
$u(d)$	Instrument uncertainty (± 0.002 cm)	0.00115 cm	0.09667	0.0869
	Measurement repeatability	0.00892 cm		
$u(h)$	Instrument uncertainty (± 0.002 cm)	0.00115 cm	0.00314	0.00286
	Measurement repeatability	0.00904 cm		
$u(h_1)$ $u(h_3)$ $u(h_2)$	Instrument uncertainty ($\pm 0.5\%$ of the indication)	0.00730 cm 0.00154 cm 0.000427 cm	0.00314	0.00229 0.000484 0.000134
$u(s)$	-	-	-	0.140

$$u_c{}^2(n_t) = \sum u_i{}^2(n_t) + u^2(s) = 2.74 \times 10^{-5}; u_c(n_t) = 0.17\%.$$

Although u_c can be universally used to express the uncertainty of a measurement result, it only corresponds to the standard deviation, and the measurement result $y \pm u_c$ represented by it has a low level of confidence. The additional measure of uncertainty that meets the requirement of providing an interval of the high level of confidence is termed expanded uncertainty and is denoted by U. The expanded uncertainty U is obtained by multiplying the combined standard uncertainty $u_c(y)$ by a coverage factor k: $U = ku_c(y)$.

Take $k = 2$, the corresponding level of confidence p is 95%, $U = ku_c(n_t) = 2 \times 0.17\% = 0.34\%$.

The uncertainty of the total porosity measurement of loose raw coal under 20 MPa is evaluated by using the expanded uncertainty. The measurement result is

$$n_t = 10.22\% \pm 0.34\%, p = 95\%$$

The expanded uncertainty of the total porosity measurement of loose raw coal under 5, 10, 15, and 25 MPa is 0.40%, 0.38%, 0.18%, and 0.36%.

Table 5 shows that the standard uncertainty of the measurement repeatability of output estimate $u(s)$ accounts for the vast majority (82%) in the combined standard uncertainty $u_c(n_t)$. The main sources of $u(s)$ are the randomness of the coal samples, and the measurement repeatability of the coal sample mass. The value of this part of the uncertainty can be reduced by increasing the number of repetitions of the measurements.

4.3. Engineering Significance, Novelty, Applicability, and Scalability of the Method

Using this experimental method, the value and variation of the total porosity of loose coal with different depths and different degrees of fragmentation can be estimated to provide data for field engineering.

The main novelty of this method is the following aspects: the extension of relationship between total porosity, true density, and bulk density to the total porosity test of the loose media under overburden pressure, the application of mechanical press to measure bulk density, the reverse loading and the record of rebound amounts during the bulk density measurement process.

This experimental method is mainly applicable to loose and broken deposition, especially for shallow loose deposits like sand and soil. The in situ stress that can be simulated in this method depends mainly on the maximum test force of the mechanical press.

Next, the accuracy of results can be improved by higher device performance and more pressure and displacement sensors. The relationship between temperature and the total porosity can be studied by using a thermally conductive mold. In addition, some devices can be added to study the pore distribution and permeability of loose media under overburden pressure.

5. Conclusions

The porosity of coal is very important in coal mining, such as for evaluating coal reservoirs and selection of grouting materials for reinforcement of coal roadways. This work has presented an experimental study on porosity characterization of loose coal media subjected to controlled overburden pressure. A new experimental method for such has been proposed, and the correlations between porosity and pressure and porosity and particle size have been examined. The main conclusions of this work are summarized below.

(1)   A new method for characterizing total porosity in loose media subjected to overburden pressure is proposed. It is based on the functional relationship between total porosity, true density, and bulk density.

(2)   After testing, the porosity of loose coal from the Guobei Coal Mine at 20 MPa in situ stress is found to be $\approx$ 10.22%. The total porosity experiences a downward trend as pressure increases for a fixed particle size, and the total porosity and pressure obey an attenuated exponential function. The decrease in total porosity with initial single particle sizes (0–2, 2–5, and 5–15 mm) is similar to that with increasing pressure, with steep curves of total porosity vs. pressure evident. There is reduction in the rate of total porosity decrease with increasing pressure with a mixed particle sizes.

(3)   At each selected pressure, the total porosity increases with increasing initial particle size (large initial particle size correspond to low degree of on-site coal fragmentation), and the total porosity and initial particle size obey a power function. The rate of total porosity increase becomes gradually reduced as particle size increases at higher stress levels. The curve of initial particle size vs. total porosity approximates a horizontal line when the pressure exceeds 20 MPa, and can thus be considered indicative of total porosity being insensitive to changes in initial particle size or the degree of on-site coal fragmentation.

(4)   When pressures are low (e.g., burial conditions are shallow), it is found that total porosity is greatly reduced and is highly sensitive to the increase in pressure. However, total porosity is less sensitive to pressure at higher stress levels (e.g., burial conditions are deep). The effect of particle size on the total porosity reduction rate in the loose coal is not significant irrespective of the pressure conditions (e.g., low or high). In general, the sensitivity of the total porosity to pressure is found to be significantly higher than sensitivity to particle size.

Author Contributions: All of the authors contributed extensively to the work. C.Z., N.Z., and D.Q. proposed key ideas. C.Z., N.Z., and D.Q. conceived and designed the experiment schemes. C.Z., D.Q., D.P., and Y.A. conducted the experiment. C.Z., D.Q., and D.P. analyzed the data. C.Z. wrote the paper. D.Q., D.P., Y.A., Y.Y., Z.X., and Y.W. modified the manuscript.

Funding: This work was financially supported by the National Natural Science Foundation of China (grant nos. 51704277 and 51674244), the Fundamental Research Funds for the Central Universities (grant nos. 2018QNA27, 2017CXNL01, and 2017CXTD02), the project funded by China Postdoctoral Science Foundation (grant no.

2017M621874), and a project Funded by the Priority Academic Program Development of Jiangsu Higher Education Institutions.

Acknowledgments: We are grateful to the staff at the Guobei coal mine for their assistance during the on-site sampling. We thank Warwick Hastie, from Liwen Bianji, Edanz Group China (www.liwenbianji.cn/ac), for editing the English text of a draft of this manuscript.

Conflicts of Interest: The authors declare no conflict of interest.

References

1. Zhao, P.N. *Loose Medium Mechanics*; Seismological Press: Beijing, China, 1995.

2. Fan, J.J.; Ju, Y.W.; Hou, Q.L.; Tan, J.Q.; Wei, M.M. Pore structure characteristics of different metamorphic-deformed coal reservoirs and its restriction on recovery of coalbed methane. *Earth Sci. Front.* **2010**, *17*, 325–335.

3. Curtis, M.; Ambrose, R.; Sondergeld, C. Structural Characterization of Gas Shales on the Micro and Nano-Scales. In Proceedings of the Canadian Unconventional Resources and International Petroleum Conference, Calgary, AB, Canada, 19–21 October 2010.

4. Sammartino, S.; Siitarikauppi, M.; Meunier, A.; Sardini, P.; Bouchet, A.; Tevissen, E. An Imaging Method for the Porosity of Sedimentary Rocks: Adjustment of the PMMA Method—Example of a Characterization of a Calcareous Shale. *J. Sediment. Res.* **2002**, *72*, 937–943. [CrossRef]

5. Hübner, W. Studying the pore space of cuttings by NMR and μCT. *J. Appl. Geophys.* **2014**, *104*, 97–105. [CrossRef]

6. Krzyżak, A.T.; Kaczmarek, A. Comparison of the Efficiency of 1H NMR and μCT for Determining the Porosity of the Selected Rock Cores. In Proceedings of the International Multidisciplinary Scientific Geoconference Green Sgem, Vienna, Austria, 11 November 2016.

7. Yao, Y.B.; Liu, D.M.; Cai, Y.D.; Li, J.Q. Advanced characterization of pores and fractures in coals by nuclear magnetic resonance and X-ray computed tomography. *Sci. Sin. Terrae* **2010**, *40*, 1598–1607. [CrossRef]

8. Andrew, M. Reservoir Condition Pore Scale Imaging of Multiphase Flow Using X-ray Microtomography. Ph.D. Thesis, Imperial College London, London, UK, 2014.

9. Menke, H.P.; Andrew, M.G.; Blunt, M.J.; Bijeljic, B. Reservoir condition imaging of reactive transport in heterogeneous carbonates using fast synchrotron tomography—Effect of initial pore structure and flow conditions. *Chem. Geol.* **2016**, *428*, 15–26. [CrossRef]

10. Andrew, M.; Menke, H.; Blunt, M.J.; Bijeljic, B. The imaging of dynamic multiphase fluid flow using synchrotron-based X-ray microtomography at reservoir conditions. *Transp. Porous Med.* **2015**, *110*, 1–24. [CrossRef]

11. Hu, X.; Liang, W.; Hou, S.J.; Zhu, X.G.; Huang, W.Q. Experimental study of effect of temperature and stress on permeability characteristics of raw coal and shaped coal. *Chin. J. Rock Mech. Eng.* **2012**, *31*, 1222–1229. [CrossRef]

12. Liu, X.J.; Gao, H.; Liang, L.X. Study of temperature and confining pressure effects on porosity and permeability in low permeability sandstone. *Chin. J. Rock Mech. Eng.* **2011**, *30*, 3771–3778.

13. Meng, Y.; Li, Z.P. Experimental study on the porosity and permeability of coal in net confining stress and its stress sensitivity. *J. China Coal Soc.* **2015**, *40*, 154–159. [CrossRef]

14. Tian, H.; Zhang, S.C.; Liu, S.B.; Ma, X.S.; Zhang, H. Parameter optimization of tight reservoir porosity determination. *Pet. Geol. Exp.* **2012**, 334–339. [CrossRef]

15. Chao, Z.M.; Wang, H.L.; Xu, W.Y.; Yang, L.L.; Zhao, K. Variation of permeability and porosity of sandstones with different degrees of saturation under stresses. *Chin. J. Rock Mech. Eng.* **2017**, *36*, 665–680. [CrossRef]

16. Dong, J.J.; Hsu, J.Y.; Wu, W.J.; Shimamoto, T.; Hung, J.H.; Yeh, E.C.; Wu, Y.H.; Sone, H. Stress-dependence of the permeability and porosity of sandstone and shale from TCDP Hole-A. *Int. J. Rock Mech. Min. Sci.* **2010**, *47*, 1141–1157. [CrossRef]

17. Tan, X.H.; Li, X.P.; Liu, J.Y.; Zhang, L.H.; Fan, Z. Study of the effects of stress sensitivity on the permeability and porosity of fractal porous media. *Phys. Lett. A* **2015**, *379*, 2458–2465. [CrossRef]

18. Meng, Y.; Li, Z.; Lai, F. Experimental study on porosity and permeability of anthracite coal under different stresses. *J. Pet. Sci. Eng.* **2015**, *133*, 810–817. [CrossRef]

19. Kong, Q.; Wang, H.L.; Xu, W.Y. Experimental study on permeability and porosity evolution of sandstone under cyclic loading and unloading. *Chin. J. Geotech. Eng.* **2015**, *37*, 1893–1900.

20. Jia, C.J.; Xu, W.Y.; Wang, H.L.; Wang, R.B.; Yu, J.; Yan, L. Stress dependent permeability and porosity of low-permeability rock. *J. Cent. South Univ.* **2017**, *24*, 2396–2405. [CrossRef]

21. Zheng, J.; Zheng, L.; Liu, H.H.; Ju, Y. Relationships between permeability, porosity and effective stress for low-permeability sedimentary rock. *Int. J. Rock Mech. Min. Sci.* **2015**, *78*, 304–318. [CrossRef]

22. Liu, J.J.; Liu, X.G. The effect of effective pressure on porosity and permeability of low permeability porous media. *J. GeoMech.* **2001**, *7*, 41–44. [CrossRef]

23. Wei, J.P.; Wang, D.K.; Wei, L. Comparison of permeability between two kinds of loaded coal containing gas samples. *J. China Coal Soc.* **2013**, *38*, 93–99. [CrossRef]

24. Yue, J.W.; Wang, Z.F. Imbibition characteristics of remolded coal without gas. *J. China Coal Soc.* **2017**, *S2*, 377–384. [CrossRef]

25. Zhu, C.Q.; Xie, G.X.; Wang, L.; Wang, C.B. Experimental study on the influence of moisture content and porosity on soft coal strength characteristics. *J. Min. Saf. Eng.* **2017**, 601–607.

26. Xu, J.; Lu, Q.; Wu, X.; Liu, D. The fractal characteristics of the pore and development of briquettes with different coal particle sizes. *J. Chongqing Univ.* **2011**, *9*, 81–89. [CrossRef]

27. Su, L.J.; Zhang, Y.J.; Wang, T.X. Investigation on permeability of sands with different particle sizes. *Rock Soil Mech. Rock Soil Mech.* **2014**, *35*, 1289–1294.

28. Yu, M.G.; Chao, J.K.; Chu, T.X. Experimental study on permeability parameter evolution of pressure-bear broken coal. *J. China Coal Soc.* **2017**, 916–922. [CrossRef]

29. Yu, B.Y.; Chen, Z.Q.; Wu, Y.; Zhang, S.B; Yu, L.L. Experimental study on the seepage characteristics of cemented broken mudstone. *J. Min. Saf. Eng.* **2015**, 853–858. [CrossRef]

30. Yuan, C.F. Tectonically deformed coal and coal gas outburst. *Coal Sci. Technol.* **1986**, 32–33. [CrossRef]

31. Tang, Y.Y.; Tian, G.L.; Sun, S.Q.; Zhang, G.C. Improvement and perfect way for the classification of the shape and cause formation of coal body texture. *J. Jiaozuo Inst. Technol. Nat. Sci.* **2004**, 161–164. [CrossRef]

32. Zhang, H. One of Parameters Reflecting Coal Reservoir Permeability—Block Coal Rate. *Coal Geol. Explor.* **2001**, *6*, 21–22. [CrossRef]

33. Y Leon, C.A.L. New perspectives in mercury porosimetry. *Adv. Colloid Interface Sci.* **1998**, *76*, 341–372. [CrossRef]

34. Guo, X.; Yao, Y.; Liu, D. Characteristics of coal matrix compressibility: An investigation by mercury intrusion porosimetry. *Energy Fuel* **2014**, *28*, 3673–3678. [CrossRef]

35. *Methods for Determining the Physical and Mechanical Properties of Coal and Rock—Part 4: Methods for Calculating the Porosity of Coal and Rock*; GB/T 23561.4-2009; Standards Press of China: Beijing, China, 2009; p. 6.

36. Guo, D.Y.; Li, C.J.; Zhang, Y.Y. Contrast Study on Porosity and Permeability of Tectonically Deformed Coal and Indigenous Coal in Pingdingshan Mining Area, China. *Earth Sci. J. China Univ. Geosci.* **2014**, *39*, 1600–1606. [CrossRef]

37. Wang, C. Research on Characteristics and Applications of the Permeability of Loaded Coal Containing Gas. Master's Thesis, Henan Polytechnic University, Jiaozuo, China, 2014.

38. Yang, X.P. *Physical Mineral Processing*; Metallurgical Industry Press: Beijing, China, 2014.

39. Labani, M.M.; Rezaee, R.; Saeedi, A.; Al Hinai, A. Evaluation of pore size spectrum of gas shale reservoirs using low pressure nitrogen adsorption, gas expansion and mercury porosimetry: A case study from the Perth and Canning Basins, Western Australia. *J. Petrol. Sci. Eng.* **2013**, *112*, 7–16. [CrossRef]

40. Zhang, S.W.; Meng, Z.Y.; Guo, Z.F.; Zhang, M.Y.; Han, C.Y. Characteristics and Major Controlling Factors of Shale Reservoirs in The Longmaxi Fm, Fuling Area, Sichuan Basin. *Nat. Gas Ind.* **2014**, *34*, 16–24. [CrossRef]

41. Hodot, B.B. *Outburst of Coal and Coalbed Gas (Chinese Translation)*; China Coal Industry Press: Beijing, China, 1966; p. 318.

42. ISO (International Organization for Standardization). *Guide to the Expression of Uncertainty in Measurement*; ISO Tag4; ISO: Genève, Switzerland, 1995.

43. JCGM (Joint Committee for Guides in Metrology). *International Vocabulary of Metrology-Basic and General Concepts and Associated Terms (VIM)*; In BIPM: Sèvres, France, 2008.

energies

MDPI

Article

Characterization and Prediction of Complex Natural Fractures in the Tight Conglomerate Reservoirs: A Fractal Method

Lei Gong [1], Xiaofei Fu [1,*], Shuai Gao [1,*], Peiqiang Zhao [2], Qingyong Luo [3], Lianbo Zeng [3], Wenting Yue [4], Benjian Zhang [5] and Bo Liu [1]

[1] College of Geosciences, Northeast Petroleum University, Daqing 163318, China; kcgonglei@nepu.edu.cn (L.G.); liubo@nepu.edu.cn (B.L.)
[2] Institute of Geophysics and Geomatics, China University of Geosciences, Wuhan 430074, China; zhaopq@cug.edu.cn
[3] State Key Laboratory of Petroleum Resource and Prospecting in China University of Petroleum, Beijing 100083, China; qingyong.luo@cup.edu.cn (Q.L.); lianbo.zeng@gmail.com (L.Z.)
[4] Department of Overseas Strategy & Production Planning Research in CNPC International Research Center, Beijing 100083, China; yuewenting@petrochina.com.cn
[5] Northwest Oil and Gas Field of Southwest Oil & Gas field Company, PetroChina, Jiangyou 621709, China; benjianz398@gmail.com
* Correspondence: fuxiaofei2008@gmail.com (X.F.); NEPUgaoshuai@gmail.com (S.G.);
  Tel.: +86-459-6504-955 (X.F.); +86-459-6504-027 (S.G.)

Received: 28 July 2018; Accepted: 31 August 2018; Published: 2 September 2018

Abstract: Using the conventional fracture parameters is difficult to characterize and predict the complex natural fractures in the tight conglomerate reservoirs. In order to quantify the fracture behaviors, a fractal method was presented in this work. Firstly, the characteristics of fractures were depicted, then the fracture fractal dimensions were calculated using the box-counting method, and finally the geological significance of the fractal method was discussed. Three types of fractures were identified, including intra-gravel fractures, gravel edge fractures and trans-gravel fractures. The calculations show that the fracture fractal dimensions distribute between 1.20 and 1.50 with correlation coefficients being above 0.98. The fracture fractal dimension has exponential correlation with the fracture areal density, porosity and permeability and can therefore be used to quantify the fracture intensity. The apertures of micro-fractures are distributed between 10 μm and 100 μm, while the apertures of macro-fractures are distributed between 50 μm and 200 μm. The areal densities of fractures are distributed between 20.0 m·m^{-2} and 50.0 m·m^{-2}, with an average of 31.42 m·m^{-2}. The cumulative frequency distribution of both fracture apertures and areal densities follow power law distribution. The fracture parameters at different scales can be predicted by extrapolating these power law distributions.

Keywords: tight conglomerate; fracture characterization and prediction; fractal method

1. Introduction

Natural fractures are one of the key factors affecting the exploration and development of tight oil and gas [1–5]. Many oil and gas fields in the world have been identified as fractured reservoirs [1,4] and the presence of natural fractures significantly increases the porosity and permeability of tight reservoirs [6–10]. Meanwhile, the existence of natural fractures may also facilitate the development and utilization of resources such as fracture water and geothermal resources [11–13], cause caprock failures [14–17] or reservoir leaks in projects (CO$_2$ sequestration, gas storage, nuclear waste disposal, etc.) [18,19], or induce geological disasters [18]. Therefore, quantitative characterization and comprehensive evaluation of natural fractures are very important and urgently needed.

Many studies were devoted to the characteristics, formation mechanisms, characterization methods and evaluations of natural fractures in tight sandstones, carbonatites, volcanics and shales [3–5,20–25]. Most of the fractures in these rocks are tectonic fractures. These tectonic fractures have long extension wing and stable occurrence, following a uniform orientation, and are often donated as systematic fractures [26–31]. However, since great variances exist in gravel diameter, gravel compositions, and interstitial material compositions, the tight conglomerate reservoir often shows a stronger rock mechanical heterogeneity than the other tight rocks [32]. The fractures developed in the tight conglomerates are drastically different in occurrence and geometry, with poor regularity, short extension and extremely complex spatial distributions, which are called non-systematic fractures [33].

At present, the fracture spacing (the vertical distance between two fractures), the linear density (the total number of fractures per unit length), or the areal density (the total length of fractures per unit area) are normally used to evaluate the intensity of systematic fractures [25,34–38]. However, for the complex non-systematic fractures, the spacing and the linear density are difficult to fully describe their development degree and spatial distribution characteristics. The areal density reveals only the trend of the fracture distribution which cannot be used to evaluate the contribution of the fractures to the reservoir. Therefore, it is essential to find a new parameter that can fully characterize the distribution of these complex non-systematic fractures, including the fracture abundance, spatial distribution characteristics.

Fractal geometry provides an effective mathematical tool to describe the irregular, nonlinear, complex, and naturally occurring objects, which covers almost all fields of the earth science [39–45]. In recent years, some studies showed that the complex morphology and distribution of faults and fractures have clear fractal characteristics, allowing to study and evaluate the fractures using fractal method [8,46–52]. In this study, fractal characterization and prediction methods were applied to study the complex fracture system of the tight conglomerate reservoirs of the Lower Jurassic Zhenzhuchong Formation in the Jiulongshan gas field, China, and the geological significance of the fractal method in characterizing and predicting fractures was discussed.

2. Geological Setting

The Jiulongshan gas field is located in the northern part of the Western Sichuan Foreland Basin, China, where the Longmen Mountain thrust belt intersects with the Micang Mountain uplift (Figure 1). The Jiulongshan gas field is a northwest-orientated dome-shaped anticline. Several small reverse faults develop in the study area with throw between 10 m and 100 m [53,54]. The lower Jurassic Zhenzhuchong Formation in this area is a tight conglomerate gas reservoir with formation thickness between 130 m and 210 m and burial depth more than 3000 m. The sedimentary environments are alluvial fan and fan delta front deposits. Reservoir lithology is mainly conglomerate, followed by lithic sandstone. The composition of conglomerate gravels is mainly quartz sandstone (70% to 80%), followed by chert (20% to 30%). The main components of the interstitial material are clastic particles and clay matrix. The sizes of gravels are non-uniform. Even though the gravel sorting is very poor, the degree of roundness is good. The gravel diameters are mainly distributed between 2 mm and 50 mm with a maximum of 80 mm. The pore types of the conglomerate reservoirs are mainly dissolved pores, intergranular pores and fractures.

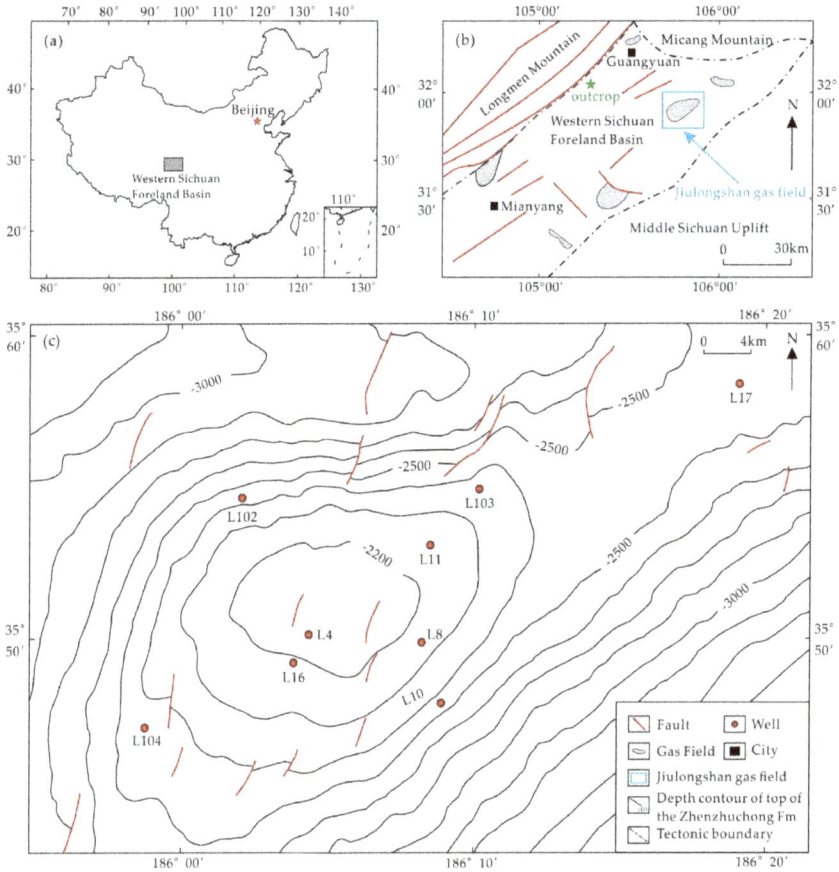

Figure 1. Location of the Jiulongshan gas field in the Western Sichuan Foreland Basin, China. (**a**) Location of the Western Sichuan Foreland Basin; (**b**) structure outline map of the northern part of Western Sichuan Foreland Basin; (**c**) depth contour and fault distribution of top of the Zhenzhuchong Formation, Jiulongshan oil field.

3. Methodology

Mandelbrot [39] firstly proposed the fractal theory, and defined fractals as shapes whose components are similar to the whole at different scales. The parameter that quantify the self-similarity is the fractal dimension, denoted as D. If the object distribution has fractal features, the number of objects and the measurement scale should follow a power law correlation, as Equation (1):

$$N(r) = C \cdot r^{-D}, \tag{1}$$

where $N(r)$ is the number of objects with the specified characteristics; r is the measure scale (cm); C is a constant; D is the fractal dimension.

Take the logarithm at the two sides of Equation (1), as Equation (2):

$$Log(N(r)) = Log(C) - D \cdot Log(r), \tag{2}$$

from Equation (2), $Log(N(r))$ is linear with $Log(r)$, and the slope D of the line is the fractal dimension.

There are many methods existing to calculate fractal dimension, such as the box-counting method, two-point correlation method and mass method, among which the box-counting method is most suitable for the fractal dimension measurement of the spatial distributions of natural fracture systems [42,55–58]. The generalized steps of the method include:

- Using a core scanner to obtain high-resolution 360° core images (Figure 2);
- Covering the image of the entire core with a mesh composed of square grids with side length of *r*; counting the number *N(r)* of boxes containing fractures;
- Gradually changing the side length *r* of the square grids, and repeatedly counting the corresponding *N(r)*;
- Taking *r* as the abscissa and *N(r)* as the ordinate, using the least-square method to perform regression analysis on the statistical data in the double logarithmic coordinate system (Figure 3).

If the fracture distribution on the core shows fractal features, the Log(*N(r)*) and Log(*r*) should follow the linear relationship in Equation (2), and the slope of the regression line is the fracture fractal dimension.

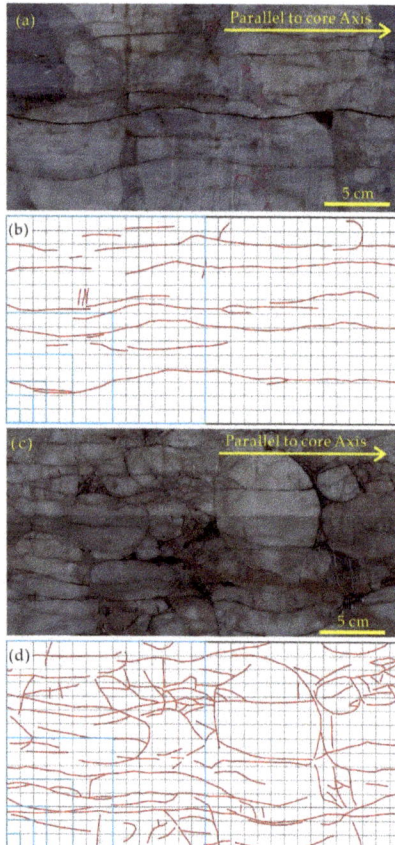

Figure 2. Core images, fracture traces and meshes. (**a**) Image of the surface of core A; (**b**) fracture traces of core A and the grids for box-counting; (**c**) image of the surface of core B; (**d**) fracture traces of core B and the grids for box-counting. The side length of the blue boxes is 1 cm, 2 cm, 3 cm, 5 cm, 8 cm, 10 cm and 15 cm, respectively.

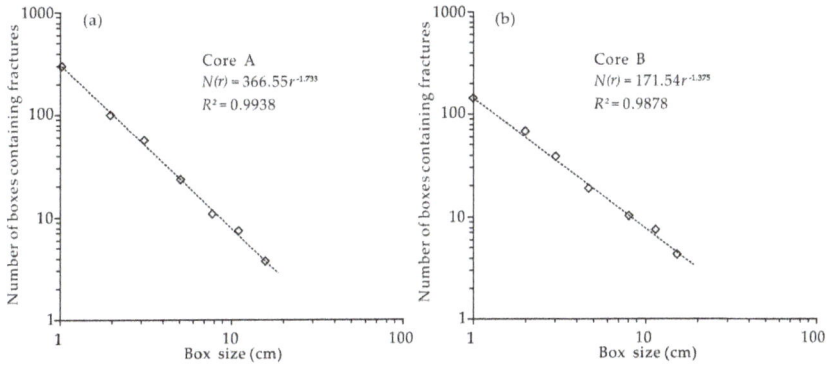

Figure 3. Schematic diagram for the calculation of fracture fractal dimension (modified from [59]). (a) Calculation of fracture fractal dimension for core A; (b) Calculation of fracture fractal dimension for core B.

To facilitate intensive analysis, some other parameters were also measured or calculated, such as the fracture aperture, the areal density, the porosity and the permeability. The feeler gauge is used to measure the fracture apertures directly since most of the fractures are opening-mode fractures that do not cut through the cores completely. The feeler gauge used in this study is composed of a set of thin steel sheets with different thicknesses. The thinnest steel sheet is 0.02 mm and the thickest steel sheet is 3 mm. The apertures of micro-fractures were measured with microscope. Fracture areal density is the total length of fractures per unit area, the fracture areal densities of both the cores and the thin sections were characterized. The porosities and permeabilities of 10 full-diameter cores (4 inches in diameter, containing fractures) and 10 core plugs (1 inch in diameter, containing no fracture) were derived using the automated test system for petrophysical parameters at Northeast Petroleum University. Based on the spacing, aperture, length and spatial distribution characteristics of fractures, the fracture porosity and permeability were calculated using the following empirical Equations [60]:

$$\varnothing_f = \frac{1}{S} \cdot \sum_{i=1}^{n} A_i \cdot L_i,$$ (3)

$$K_f = 5.66 \times 10^{-4} \cdot \overline{A}^2 \cdot \varnothing_f,$$ (4)

where $\varnothing_f$ is the fracture porosity (%); S is the core area (m^2); A_i is the aperture of the ith fracture (m); L_i is the length of the ith fracture (m); K_f is the fracture permeability (mD); and $\overline{A}$ is the average fracture aperture (m).

4. Fracture Characterization

4.1. Fracture Type and Characteristics

According to the spatial distribution characteristics and their relationship with gravels, the fractures in the conglomerate reservoirs can be divided into three types: Intra-gravel fractures (IGF), trans-gravel fractures (TGF) and gravel edge fractures (GEF) (Figure 4).

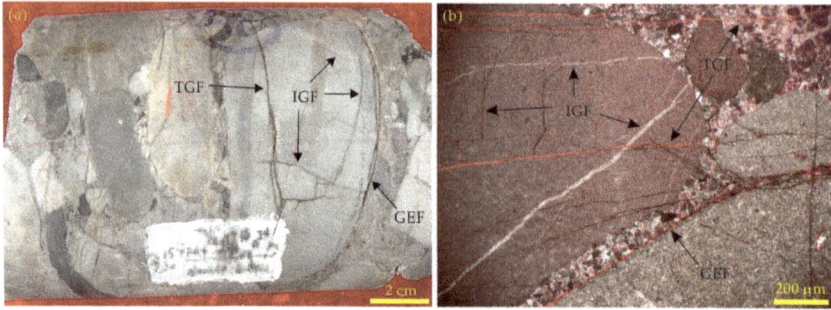

Figure 4. Fracture types in the tight conglomerates. (**a**) Well L4, 3164.32 m; (**b**) Well L102, 3198.35 m. IGF = intra-gravel fracture, TGF = trans-gravel fracture, GEF = Gravel edge fracture.

The intra-gravel fractures are mainly distributed inside the gravels, and they usually have a small extension length and do not cut through the edge of gravels (Figure 4). Such fractures are small in scale but high in density, and their apertures are generally less than 40 μm. The gravel edge fractures are mainly distributed along the edge of the gravels. Hence the surfaces of the gravel edge fractures are either spherical or ellipsoidal, and the fracture traces on cores or thin sections are curves (Figure 5a). This kind of fractures are also small in scale and short in extension, and their apertures are generally less than 20 μm.

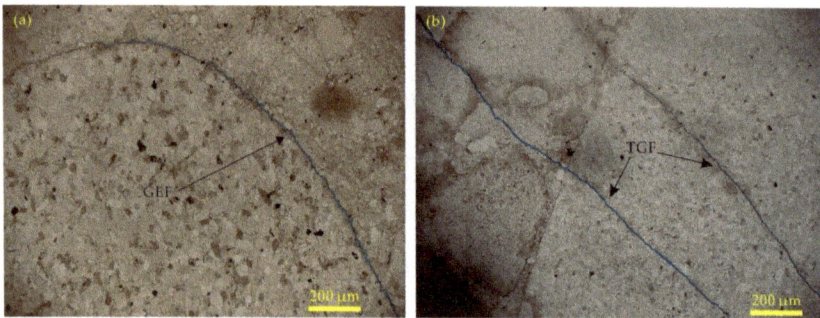

Figure 5. Gravel edge fracture (GEF) and trans-gravel fractures (TGF) in thin sections. (**a**) Gravel edge fracture (GEF), Well L104, 3175.05 m; (**b**) trans-gravel fractures (TGF), Well L16, 3159.34 m.

The trans-gravel fractures are the major fracture type in the study area (Figure 5b, Figure 6). Most of the trans-gravel fractures are tectonic shear fractures. Compared with the intra-gravel and gravel edge fractures, these trans-gravel fractures are relatively large in scale and long in length. They are not restricted by gravels and usually cut through two or more gravels. According to their dip angles, trans-gravel fractures can be subdivided into fractures with high dip angles and fractures with low dip angles (Figure 6). The high dip angle fractures have long extension on the core, and their height can be as large as 80 cm (Figure 6a). The low dip angle fractures are generally paralleled to each other, and their spacing is between 0.5 cm and 5.0 cm (Figure 6b). Due to the extremely high drilling-encounter ratio, it seems that they are the most important fractures in the cores (Figure 6b). Fractures with both high and low dip angles are very clear on outcrops, they usually appear as conjugate shear fractures (Figure 6c).

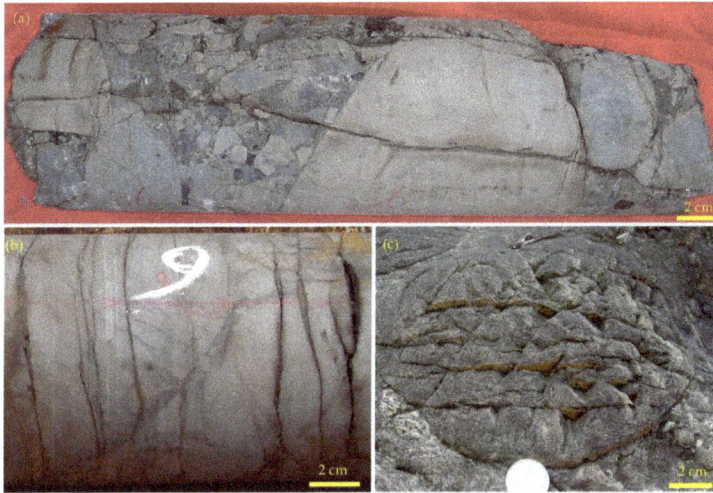

Figure 6. Trans-gravel fractures in cores and out crop. (**a**) Trans-gravel fractures with high dip angle in core, Well L102, 3185.75 m; (**b**) trans-gravel fractures with low dip angle in core, Well L104, 3174.65 m; (**c**) conjugate fractures in outcrop, see Figure 1 for outcrop location.

4.2. Fracture Parameters

Based on the high-resolution core scanning images, the fractures for 57 intervals from 4 wells were measured using grids with side length of 1 cm, 2 cm, 3 cm, 5 cm, 8 cm, 10 cm and 15 cm, respectively (Figure 2). The numbers of boxes containing fractures were counted and regressions were performed using the least squares method in a log-log coordinate system. Then the fracture fractal dimensions and their corresponding correlation coefficients were calculated. The correlation coefficients of all the cores are above 0.98, indicating that the spatial distribution of fractures in the tight conglomerate has good fractal characteristics (Table 1). The fractal dimensions of core fractures mainly distribute between 1.20 and 1.50, which is reasonable for the fractal range of the two-dimensional object (between 1 to 2) (Figure 7a).

Table 1. Fracture parameters (fractal dimension, areal density, porosity and permeability) calculated from cores.

Well Name	Interval		Fractal Dimension	Correlation Coefficient	Areal Density $(m \cdot m^{-2})$	Porosity (%)	Permeability (mD)
	Top (m)	Bottom (m)					
L4	3069.39	3069.55	1.38	0.9910	40.29	1.26	133.98
L4	3069.64	3069.72	1.22	0.9891	26.59	0.99	38.62
L4	3069.77	3069.78	1.22	0.9884	25.85	0.83	74.93
L4	3069.98	3070.11	1.28	0.9901	27.87	1.51	81.98
L4	3070.22	3070.29	1.31	0.9920	26.58	0.82	90.48
L4	3070.49	3070.57	1.40	0.9908	40.54	0.96	106.12
L4	3070.80	3070.93	1.43	0.9898	36.80	1.22	88.98
L4	3071.01	3071.12	1.10	0.9914	24.98	0.59	32.00
L4	3071.36	3071.56	1.24	0.9905	32.81	0.95	74.75
L4	3071.71	3071.80	1.58	0.9897	39.29	1.27	138.58
L4	3071.91	3072.12	1.17	0.9914	21.30	0.76	80.85
L4	3072.24	3072.41	1.01	0.9893	17.95	0.50	39.93
L10	3080.74	3080.91	1.15	0.9877	24.22	0.71	52.18
L10	3101.61	3101.80	1.28	0.9927	26.99	0.74	67.63
L10	3101.88	3102.07	1.03	0.9922	24.19	0.62	34.91
L10	3102.13	3102.34	1.65	0.9933	56.49	1.81	245.25
L10	3102.55	3102.71	1.24	0.9894	24.77	0.83	82.18

Table 1. *Cont.*

Well Name	Interval		Fractal Dimension	Correlation Coefficient	Areal Density (m·m⁻²)	Porosity (%)	Permeability (mD)
	Top (m)	Bottom (m)					
L10	3102.81	3102.95	1.31	0.9896	40.64	1.15	79.94
L10	3103.30	3103.42	1.24	0.9923	31.86	1.09	94.35
L10	3103.49	3103.53	1.33	0.9879	32.42	1.77	90.54
L10	3103.61	3103.79	1.36	0.9900	40.62	1.42	143.99
L10	3103.88	3103.93	1.40	0.9876	35.68	1.54	58.35
L10	3104.07	3104.18	1.37	0.9875	36.02	1.27	120.48
L10	3104.26	3104.37	1.36	0.9895	28.56	1.17	166.58
L102	3087.20	3087.34	1.46	0.9884	34.29	2.18	114.94
L102	3087.51	3087.57	1.35	0.9887	30.49	1.05	103.49
L102	3087.75	3087.85	1.50	0.9886	45.50	1.38	139.29
L102	3087.94	3088.08	1.55	0.9875	50.57	1.78	191.51
L102	3088.23	3088.33	1.04	0.9903	24.69	0.73	55.77
L102	3088.60	3088.70	1.05	0.9886	18.00	0.69	55.13
L102	3089.01	3089.17	1.51	0.9909	43.76	1.93	218.89
L102	3089.17	3089.28	1.10	0.9880	26.70	1.06	93.28
L102	3089.39	3089.55	1.08	0.9918	20.01	0.58	52.04
L103	3117.23	3117.38	1.24	0.9889	30.60	1.11	128.42
L103	3117.45	3117.54	1.43	0.9924	47.72	1.75	191.26
L103	3117.70	3117.84	1.44	0.9906	41.50	1.24	90.64
L103	3117.99	3118.05	1.28	0.9901	39.06	1.26	75.31
L103	3118.15	3118.26	1.39	0.9895	47.30	1.08	88.36
L103	3118.36	3118.58	1.19	0.9928	28.91	1.21	79.01
L103	3118.67	3118.82	1.22	0.9882	33.35	0.76	35.99
L103	3118.98	3119.03	1.41	0.9924	43.21	1.35	111.74
L103	3119.09	3119.21	1.19	0.9894	28.88	0.95	60.14
L103	3119.30	3119.40	1.56	0.9929	67.27	2.14	169.61
L103	3119.58	3119.72	1.15	0.9909	24.14	0.80	78.47
L103	3119.92	3120.16	1.31	0.9915	34.79	0.79	59.60
L103	3120.32	3120.50	1.44	0.9930	48.27	1.53	90.03
L103	3128.18	3128.37	1.52	0.9923	39.17	1.44	208.87
L103	3128.57	3128.71	1.24	0.9905	25.29	1.1	148.97

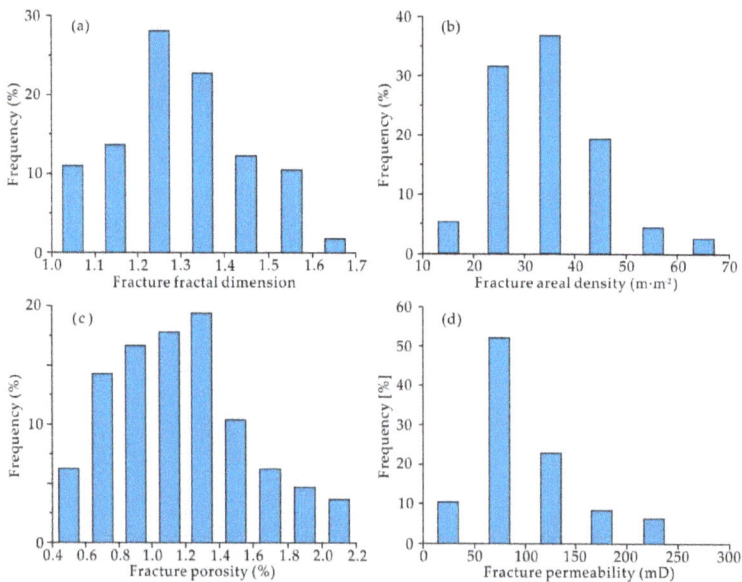

Figure 7. Distributions of fracture parameters. (**a**) Distribution of fracture fractal dimension; (**b**) distribution of fracture areal density; (**c**) distribution of fracture porosity; (**d**) distribution of fracture permeability.

The fracture parameters from cores and thin sections show that the apertures of micro-fractures are mainly distributed between 10 µm and 100 µm, while the apertures of macro-fractures are mainly distributed between 50 µm and 200 µm. The cumulative frequency distribution of both micro- and macro-fractures seems to follow a log-normal distribution (Figure 8). The areal densities of fractures are mainly distributed between 20.0 m·m^{-2} and 50.0 m·m^{-2}, with an average of 31.42 m·m^{-2} (Figure 7b). The fracture porosities are mainly distributed between 0.60% and 1.60%, with an average of 1.26% (Figure 7c), and the fracture permeabilities are mainly distributed between 50 mD and 150 mD (Figure 7d).

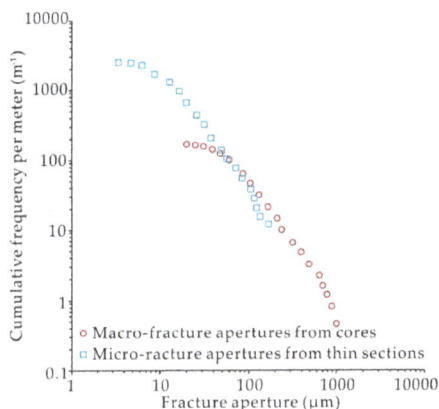

Figure 8. Cumulative frequency plots of micro-fracture apertures (blue squares) and macro-fracture apertures (red circles).

The core physical property tests show that (Table 2), the porosities of the core plugs are 0.29%–1.60% with an average of 0.97%, and the permeabilities are distributed between 0.0021 mD and 0.0191 mD with an average of 0.0089 mD. The porosity and permeability of the full-diameter cores are significantly larger than those of the core plugs. The porosities of full-diameter cores are distributed between 2.61% and 4.05% with an average of 3.52%. The vertical permeabilities of the full-diameter core are between 0.007 mD and 0.6730 mD with an average of 0.2586 mD, while the horizontal permeabilities are between 3.17 mD and 214.50 mD with an average of 77.7 mD.

Table 2. Porosities and Permeabilities of full-diameter cores and core plugs.

Number	Full Diameter Cores			Core Plugs		Φ_1/Φ_2	K_1/K_3
	Φ_1 (%)	K_1 (mD)	K_2 (mD)	Φ_2 (%)	K_3 (mD)		
1	3.90	214.50	0.0145	0.29	0.0021	13.45	102,142.86
2	3.74	201.69	0.0007	0.64	0.0084	5.84	24,010.71
3	3.80	166.57	0.0689	0.51	0.0056	7.45	29,744.64
4	3.12	5.75	0.0237	1.51	0.0105	2.07	547.62
5	2.61	83.58	0.0123	0.93	0.0096	2.81	8706.25
6	4.05	3.17	0.5917	1.60	0.0191	2.53	165.97
7	3.04	7.25	0.4628	1.26	0.0084	2.41	863.10
8	3.19	21.40	0.2450	1.09	0.0047	2.93	4553.19
9	3.06	32.30	0.4930	0.89	0.0079	3.44	4088.61
10	3.01	40.40	0.6730	0.95	0.0127	3.17	3181.10
Average	3.52	77.7	0.2586	0.97	0.0089	4.61	17,800.40

Note: Φ_1 and Φ_2 are the porosities of the full-diameter cores and the core plugs respectively; K_1 and K_2 are horizontal permeability and vertical permeability of the full-diameter cores respectively; K_3 is the permeability of the core plugs.

5. Discussion

5.1. Geological Significance of Fracture Fractal Dimension

The number of boxes containing fractures has a linear relationship (power law distribution) with the grid side length in double logarithmic coordinate system with correlation coefficients larger than 0.98, demonstrating that the development degree of the fractures in the reservoir follows good fractal features. Therefore, it is reasonable to quantify the development degree and distribution of reservoir fractures using the fractal D value. For example, the fracture fractal dimension D (D = 1.65) of core B is greater than the fracture fractal dimension (D = 1.37) of core A in Figure 2, which means that the fracture in core B is more complex than the fractures in core A. This conclusion is in accordance with the observation, indicating that the fractal dimension D is suitable in characterizing the development degree of these complex fractures. Figure 9 shows that the fracture areal density has an exponential relationship with fracture fractal dimension with a correlation coefficient of 0.8694, indicating that the fracture fractal dimension is a good indicator for the fracture areal density. In addition, the two cores with same fracture areal density have different fracture fractal dimension D. The fracture fractal dimension obtained by the box-counting method also reflects the centralization degree of the fracture distribution of the core specimens [40,42]. The denser the fracture distribution is, the greater the fracture fractal dimension is; vice versa. This indicates that the spatial geometry and complexity of fractures are affecting the fracture fractal dimension.

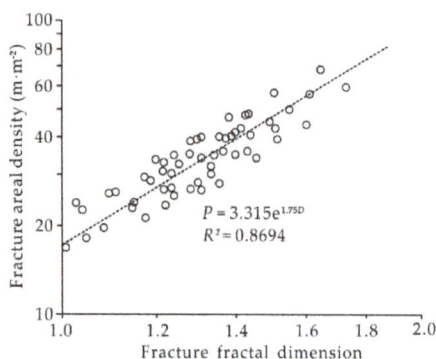

Figure 9. Relationship between fractal dimension and fracture areal density.

The fracture fractal dimension is also related to the porosity and permeability of fractures (Figures 10 and 11). Both the porosity and the permeability of fractures show exponential relationships with the fracture dimension with the correlation coefficients of 0.8457 and 0.7718, respectively in this case study. This result shows that the fracture fractal dimension is not limited to characterizing the fracture development degree and the spatial distribution complexity of fractures, but also a good indicator of the porosity and permeability of fractures, improving the understandings of the contributions of fractures to the tight reservoirs [59].

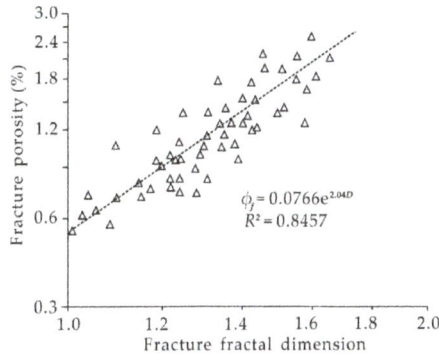

Figure 10. Relationship between fractal dimension and fracture porosity.

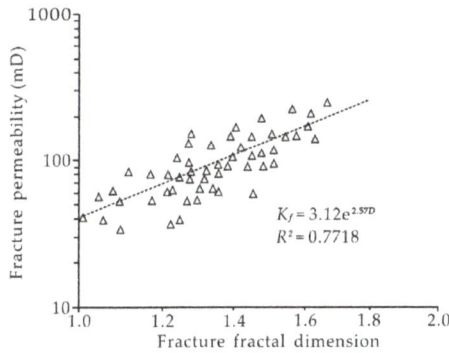

Figure 11. Relationship between fractal dimension and fracture permeability.

5.2. Power-Law Distribution of Fracture Parameters and Fracture Prediction

The cumulative frequency distribution of fracture parameters (e.g., aperture, areal density, etc.) shows fractal features, providing a theoretical foundation for fracture prediction at different scales [48,50,61–63]. Although the cumulative frequency distribution of fracture apertures seems to follow log-normal distribution at a single-observation scale (core scale or thin section scale), after integrating all the aperture data, they are subjected to a uniform power-law distribution, and each scale has almost the same slope (Figure 12). This phenomenon is caused by the truncation and censoring effects [61]. The truncation effect is defined as the phenomenon that the number of the small fractures is underestimated due to the limitation of the observation resolution. The truncation effect causes the upper part of the fracture cumulative frequency distribution curve deviate from the power law distribution. The fractures with large apertures usually cut through the cores making their apertures difficult to be accurately measured. Usually, the measured aperture of these large fractures is lower than the actual value. This is the censoring effect, and due to this effect, the lower part of the fracture cumulative frequency distribution curve will deviate from the power law distribution. Therefore, after eliminating these error data, the number or density of fractures at different scales can be predicted by fine extrapolation of the power law distribution. Using this method, the areal densities of micro-fractures were predicted by extrapolating the power law of cumulative areal density distribution of macro-fractures (Figure 13). Compared to the measured micro-fracture areal density, the errors are less than 5%, which indicates that the prediction results are reliable (Table 3).

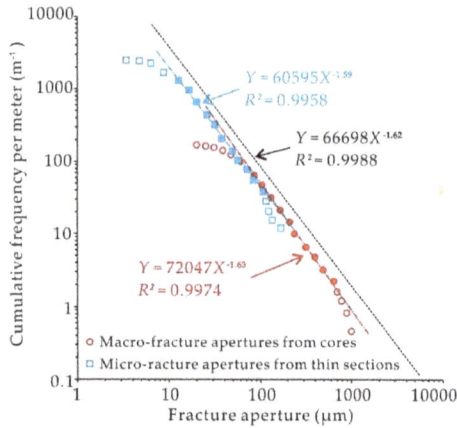

Figure 12. Cumulative plot of micro-fracture apertures (blue squares) and macro-fracture apertures (red circles). The power law distribution of micro-fractures (blue Equation) was obtained by fitting the solid square data, the power law distribution of macro-fractures (red Equation) was obtained by fitting the solid circle data, the power law distribution of all fractures (black Equation) was obtained by fitting the solid square and solid circle data and the best fitting line is slightly offset to the right for clarity of the plot.

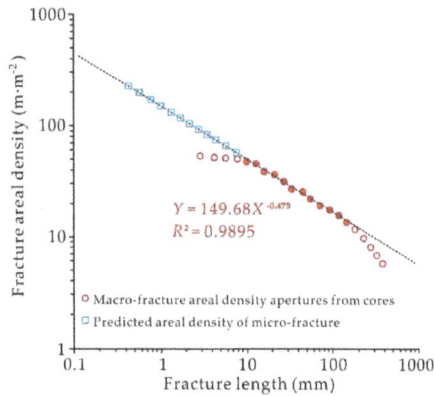

Figure 13. Cumulative plot of macro-fracture areal density (red circles) and prediction of micro-fracture areal density (blue square). The Equation was obtained by fitting the solid circle data, the opening squares are predicted areal density of micro-fractures by extrapolating the power law of cumulative areal density distribution of macro-fractures.

Table 3. Comparison of the measured and predicted areal density of micro-fractures.

Fracture Length (mm)	Fracture Areal Density		Absolute Error (m·m^{-2})	Relative Error (%)
	Measured (m·m^{-2})	Predicted (m·m^{-2})		
10	50.35	50.14	−0.21	0.42
9	51.89	52.71	0.82	1.58
8	56.05	55.74	−0.31	0.55
7	60.48	59.39	−1.09	1.80
6	62.36	63.91	1.55	2.48
5	67.07	69.69	2.62	3.90

5.3. Contribution of Fractures

Since the core plugs are drilled in a way to avoid fractures, while the full-diameter cores are usually drilled through natural fractures, the physical property test results of core plugs represent the physical property of the matrix, and the physical property test results of the full-diameter cores donate the total porosity and permeability of the matrix pores and the natural fractures. Their differences can be roughly regarded as the porosity and permeability of natural fractures. From Table 2, it can be estimated that the porosity of natural fractures is more than 2/3 of the total porosity, and the permeability of natural fractures is 2 to 5 orders of magnitude higher than that of the matrix pores. These observations indicate that natural fractures are the major contributor of the storage space and seepage channel of the tight conglomerate. In addition, the horizontal permeability of the full-diameter cores is much larger than the vertical permeability. This is attributed to the horizontal fractures being dominant in the full-diameter core samples. It also depicts that the natural fractures are the most important seepage channels in the tight conglomerate reservoirs.

The well productivity is also closely related to the development degree of fractures. From Figure 14, the daily gas production is exponentially related to the fracture fractal dimension. The well productivity increases as the fractures develop further. Therefore, the development degree of natural fractures is the most important factor controlling the natural gas enrichment and the capacity of the tight conglomerate reservoir in the Zhenzhuchong Formation.

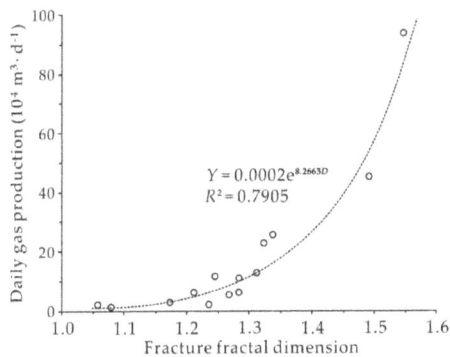

$$Y = 0.0002e^{8.2663D}$$
$$R^2 = 0.7905$$

Figure 14. Relationship between fracture fractal dimension and well productivity.

6. Conclusions

Three types of fractures (IGF, GEF and TGF) exist in the tight conglomerate reservoirs of Zhenzhuchong Formation of Jiulongshan gas field, China, forming a complex fracture network. The fractal dimensions of core fractures mainly distribute between 1.20 and 1.50, which is reasonable for the fractal range of the two-dimensional object (from 1 to 2). The correlation coefficients of all the fracture fractal dimensions are above 0.98, indicating that the spatial distribution of fractures in the tight conglomerate has fractal properties.

The areal densities of fractures are mainly distributed between 20.0 m·m^{-2} and 50.0 m·m^{-2}, with an average of 31.42 m·m^{-2}. The fracture porosities are mainly distributed between 0.60% and 1.60%, with an average of 1.26%, and the fracture permeabilities are mainly distributed between 50 mD and 150 mD. The fracture fractal dimension has an exponential correlation with the fracture areal density and can therefore be used to quantify the fracture intensity. A good exponential correlation also exists between the fracture fractal dimension and the fracture porosity and permeability, which can reflect their contributions to the physical properties of the tight reservoirs. Therefore, the fracture fractal dimension D is a good comprehensive index as a quantitative parameter to characterize the intensity of the complex fracture system and reflect the contributions of fractures to the tight reservoir.

The apertures of micro-fractures are mainly distributed between 10 μm and 100 μm, and the apertures of macro-fractures are mainly distributed between 50 μm and 200 μm. Although the cumulative frequency distribution of fracture apertures seems to follow log-normal distribution at a single-observation scale (core scale or thin section scale), after integrating all the aperture data, they are subjected to a uniform power-law distribution. The density of micro-fractures was predicted by fine extrapolation of the power law distribution. The errors of predicted results are less than 5%, which indicates that the prediction results are reliable.

Author Contributions: Conceptualization, X.F. and L.Z.; Data curation, P.Z. and B.Z.; Formal analysis, Q.L.; Investigation, B.L.; Methodology, L.G.; Writing—review & editing, L.G., S.G. and W.Y.

Funding: This research was funded by the National Natural Science Foundation of China (grant nos. 41502124 and U1562214), Natural Science Foundation of Heilongjiang Province (grant no. QC2018043), Young Innovative Talents Training Program for Universities in Heilongjiang Province (grant no. UNPYSCT-2018043), China Postdoctoral Science Foundation (grant nos. 2018M631908 and 2015M581424) and Cultivating Fund of Northeast Petroleum University (grant nos. SCXHB201705 and 2017PYQZL-14).

Acknowledgments: The authors thank Yuegang Li, Yong Duan, Hualing Ma, Senqi Pei and Yu Zhang at Northwest Oil and Gas Field of Southwest Oil & Gas field Company, PetroChina for their constructive help.

Conflicts of Interest: The authors declare no conflict of interest.

References

1. Zeng, L.; Su, H.; Tang, X.; Peng, Y.; Gong, L. Fractured tight sandstone oil and gas reservoirs: A new play type in the Dongpu depression, Bohai Bay Basin, China. *AAPG Bull.* **2013**, *97*, 363–377. [CrossRef]
2. Laubach, S.E.; Lamarche, J.; Gauthier, B.D.M.; Dunne, W.M.; Sanderson, D.J. Spatial arrangement of faults and opening-mode fractures. *J. Struct. Geol.* **2018**, *108*, 2–15. [CrossRef]
3. Gale, J.F.W.; Laubach, S.E.; Olson, J.E.; Eichhubl, P.; Fall, A. Natural fractures in shale: A review and new observations. *AAPG Bull.* **2014**, *98*, 2165–2216. [CrossRef]
4. Nelson, R.A. *Geologic analysis of naturally fractured reservoirs*; Elsevier: Amsterdam, Netherlands, 2001. Available online: https://www.sciencedirect.com/book/9780884153177/geologic-analysis-of-naturally-fractured-reservoirs (accessed on 17 August 2018).
5. Bisdom, K.; Gauthier, B.D.M.; Bertotti, G.; Hardebol, N.J. Calibrating discrete fracture-network models with a carbonate three-dimensional outcrop fracture network: Implications for naturally fractured reservoir modeling. *AAPG Bull.* **2014**, *98*, 1351–1376. [CrossRef]
6. Ghosh, K.; Mitra, S. Two-dimensional simulation of controls of fracture parameters on fracture connectivity. *AAPG Bull.* **2009**, *93*, 1517–1533. [CrossRef]
7. Wang, L.; Zhao, N.; Sima, L.Q.; Meng, F.; Guo, Y. Pore structure characterization of the tight reservoir: Systematic integration of mercury injection and nuclear magnetic resonance. *Energy Fuel.* **2018**, *32*, 7471–7484. [CrossRef]
8. Strijker, G.; Bertotti, G.; Luthi, S.M. Multi-scale fracture network analysis from an outcrop analogue: A case study from the Cambro-Ordovician clastic succession in Petra, Jordan. *Mar. Petrol. Geol.* **2012**, *38*, 104–116. [CrossRef]
9. Lei, Q.; Wang, X. Tectonic interpretation of the connectivity of a multiscale fracture system in limestone. *Geophys. Res. Lett.* **2016**, *43*, 1551–1558. [CrossRef]
10. Larsen, B.; Gudmundsson, A. Linking of fractures in layered rocks: Implications for permeability. *Tectonophysics* **2010**, *492*, 108–120. [CrossRef]
11. Magnusdottir, L.; Horne, R.N. Inversion of time-lapse electric potential data to estimate fracture connectivity in geothermal reservoirs. *Math. Geosci.* **2015**, *47*, 85–104. [CrossRef]
12. Roques, C.; Bour, O.; Aquilina, L.; Dewandel, B.; Leray, S.; Schroetter, J.; Longuevergne, L.; Le Borgne, T.; Hochreutener, R.; Labasque, T.; et al. Hydrological behavior of a deep sub-vertical fault in crystalline basement and relationships with surrounding reservoirs. *J. Hydrol.* **2014**, *509*, 42–54. [CrossRef]
13. Kong, L.; Ostadhassan, M.; Li, C.; Tamimi, N. Pore characterization of 3D-printed gypsum rocks: A comprehensive approach. *J. Mater. Sci.* **2018**, *53*, 5063–5078. [CrossRef]

14. Ogata, K.; Senger, K.; Braathen, A.; Tveranger, J. Fracture corridors as seal-bypass systems in siliciclastic reservoir-cap rock successions: Field-based insights from the Jurassic Entrada Formation (SE Utah, USA). *J. Struct. Geol.* **2014**, *66*, 162–187. [CrossRef]

15. Petrie, E.S.; Evans, J.P.; Bauer, S.J. Failure of cap-rock seals as determined from mechanical stratigraphy, stress history, and tensile-failure analysis of exhumed analogs. *AAPG Bull.* **2014**, *98*, 2365–2389. [CrossRef]

16. Ingram, G.M.; Urai, J.L. Top-seal leakage through faults and fractures: The role of mudrock properties. *Geol. Soc. Spec. Publ.* **1999**, *158*, 125–135. [CrossRef]

17. Jin, Z.J. A study on the distribution of oil and gas reservoirs controlled by source-cap rock assemblage in unmodified foreland region of Tarim Basin. *Oil Gas Geol.* **2014**, *35*, 763–770.

18. Smith, J.; Durucan, S.; Korre, A.; Shi, J. Carbon dioxide storage risk assessment: Analysis of caprock fracture network connectivity. *Int. J. Greenh. Gas Control* **2011**, *5*, 226–240. [CrossRef]

19. Alghalandis, Y.F.; Dowd, P.A.; Xu, C. Connectivity Field: a measure for characterising fracture networks. *Math. Geosci.* **2015**, *47*, 63–83. [CrossRef]

20. Gong, L.; Gao, S.; Fu, X.; Chen, S.; Lyu, B.; Yao, J. Fracture characteristics and their effects on hydrocarbon migration and accumulation in tight volcanic reservoirs: A case study of the Xujiaweizi fault depression, Songliao Basin, China. *Interpret. J. Sub.* **2017**, *5*, 57–70. [CrossRef]

21. Laubach, S.E.; Fall, A.; Copley, L.; Marrett, R.; Wilkins, S.J. Fracture porosity creation and persistence in a basement-involved Laramide fold, Upper Cretaceous Frontier Formation, Green River Basin, USA. *Geol. Mag.* **2016**, *153*, 887–910. [CrossRef]

22. Olson, J.E.; Laubach, S.E.; Eichhubl, P. Estimating natural fracture producibility in tight gas sandstones: coupling diagenesis with geomechanical modeling. *Lead. Edge* **2010**, *29*, 1494–1499. [CrossRef]

23. Peacock, D.C.P.; Sanderson, D.J.; Rotevatn, A. Relationships between fractures. *J. Struct. Geol.* **2018**, *106*, 41–53. [CrossRef]

24. Sanderson, D.J.; Nixon, C.W. The use of topology in fracture network characterization. *J. Struct. Geol.* **2015**, *72*, 55–66. [CrossRef]

25. Lyu, W.Y.; Zeng, L.; Zhang, B.; Miao, F.; Lyu, P.; Dong, S. Influence of natural fractures on gas accumulation in the Upper Triassic tight gas sandstones in the northwestern Sichuan Basin, China. *Mar. Petrol. Geol.* **2017**, *83*, 60–72. [CrossRef]

26. Gross, M.R.; Eyal, Y. Throughgoing fractures in layered carbonate rocks. *Geol. Soc. Am. Bull.* **2007**, *119*, 1387–1404. [CrossRef]

27. Laubach, S.E.; Olson, J.E.; Gross, M.R. Mechanical and fracture stratigraphy. *AAPG Bull.* **2009**, *93*, 1413–1426. [CrossRef]

28. Lyu, W.Y.; Zeng, L.B.; Liu, Z.Q.; Liu, G.P.; Zu, K.W. Fracture responses of conventional logs in tight-oil sandstones: A case study of the Upper Triassic Yanchang Formation in southwest Ordos Basin, China. *AAPG Bull.* **2016**, *100*, 1399–1417. [CrossRef]

29. Luo, Q.Y.; Gong, L.; Qu, Y.S.; Zahng, K.H.; Zhang, G.L.; Wang, S.Z. The tight oil potential of the Lucaogou Formation from the southern Junggar Basin, China. *Fuel* **2018**, *234*, 858–871. [CrossRef]

30. Zeng, L.B.; Tang, X.M.; Wang, T.C.; Gong, L. The influence of fracture cements in tight Paleogene saline lacustrine carbonate reservoirs, Western Qaidam Basin, Northwest China. *AAPG Bull.* **2012**, *96*, 2003–2017. [CrossRef]

31. Finn, M.D.; Gross, M.R.; Eyal, Y.; Draper, G. Kinematics of throughgoing fractures in jointed rocks. *Tectonophysics* **2003**, *376*, 151–166. [CrossRef]

32. Gong, L.; Zeng, L.B.; Zhang, B.J.; Zu, K.W.; Yin, H.; Ma, H.L. Control factors for fracture development in tight conglomerate reservoir of Jiulongshan structure. *J. China Univ. Petrol.* **2012**, *36*, 6–12.

33. Huang, N.; Jiang, Y.J.; Liu, R.C.; Li, B.; Zhang, Z.Y. A predictive model of permeability for fractal-based rough rock fractures during shear. *Fractals* **2017**, *25*, 1750051. [CrossRef]

34. Marrett, R.; Gale, J.F.W.; Gómez, L.A.; Laubach, S.E. Correlation analysis of fracture arrangement in space. *J. Struct. Geol.* **2018**, *108*, 16–33. [CrossRef]

35. Casini, G.; Hunt, D.W.; Monsen, E.; Bounaim, A. Fracture characterization and modeling from virtual outcrops. *AAPG Bull.* **2016**, *100*, 41–61. [CrossRef]

36. Santos, R.F.V.C.; Miranda, T.S.; Barbosa, J.A.; Gomes, I.F.; Matos, G.C.; Gale, J.F.W.; Neumann, V.H.L.M.; Guimaraes, L.J.N. Characterization of natural fracture systems: Analysis of uncertainty effects in linear scanline results. *AAPG Bull.* **2015**, *99*, 2203–2219. [CrossRef]

37. Zeeb, C.; Gomez-Rivas, E.; Bons, P.D.; Blum, P. Evaluation of sampling methods for fracture network characterization using outcrops. *AAPG Bull.* **2013**, *97*, 1545–1566. [CrossRef]
38. Procter, A.; Sanderson, D.J. Spatial and layer-controlled variability in fracture networks. *J. Struct. Geol.* **2018**, *108*, 52–65. [CrossRef]
39. Mandelbrot, B.B. *The fractal geometry of nature*; W.H. Freeman and Company: New York, NY, USA, 1982; Available online: https://us.macmillan.com/books/9780716711865 (accessed on 17 August 2018).
40. Cai, J.C.; Wei, W.; Hu, X.Y.; Liu, R.C.; Wang, J.J. Fractal characterization of dynamic fracture network extension in porous media. *Fractals* **2017**, *25*, 1750023. [CrossRef]
41. Cai, J.; Hu, X.; Xiao, B.; Zhou, Y.; Wei, W. Recent developments on fractal-based approaches to nanofluids and nanoparticle aggregation. *Int. J. Heat Mass Transf.* **2017**, *105*, 623–637. [CrossRef]
42. Liu, R.; Yu, L.; Jiang, Y.; Wang, Y.; Li, B. Recent developments on relationships between the equivalent permeability and fractal dimension of two-dimensional rock fracture networks. *J. Nat. Gas Sci. Eng.* **2017**, *45*, 771–785. [CrossRef]
43. Zhao, P.; Wang, L.; Sun, C.; Cai, C.; Wang, L. Investigation on the pore structure and multifractal characteristics of tight oil reservoirs using NMR measurements: Permian Lucaogou Formation in Jimusaer Sag, Junggar Basin. *Mar. Petrol. Geol.* **2017**, *86*, 1067–1081. [CrossRef]
44. Liu, K.Q.; Ostadhassan, M.; Zhou, J.; Gentzis, T.; Rezaee, R. Nanoscale pore structure characterization of the Bakken shale in the USA. *Fuel* **2017**, *209*, 567–578. [CrossRef]
45. Zhao, P.; Cai, J.; Huang, Z.; Ostadhassan, M.; Ran, F.Q. Estimating permeability of shale gas reservoirs from porosity and rock compositions. *Geophysics* **2018**, *83*, 1–36. [CrossRef]
46. Liu, R.; Li, B.; Jing, H.; Wei, W. Analytical solutions for water–gas flow through 3D rock fracture networks subjected to triaxial stresses. *Fractals* **2018**, *26*, 1850053. [CrossRef]
47. Jafari, A.; Babadagli, T. Estimation of equivalent fracture network permeability using fractal and statistical network properties. *J. Petrol. Sci. Eng.* **2012**, *92–93*, 110–123. [CrossRef]
48. Johri, M.; Zoback, M.D.; Hennings, P. A scaling law to characterize fault-damage zones at reservoir depths. *AAPG Bull.* **2014**, *98*, 2057–2079. [CrossRef]
49. Maerten, L.; Gillespie, P.; Daniel, J. Three-dimensional geomechanical modeling for constraint of subseismic fault simulation. *AAPG Bull.* **2006**, *90*, 1337–1358. [CrossRef]
50. Ortega, O.J.; Marrett, R.A.; Loubach, S.E. A scale-independent approach to fracture intensity and average spacing measurement. *AAPG Bull.* **2006**, *90*, 193–208. [CrossRef]
51. Zhu, J.; Cheng, Y. Effective permeability of fractal fracture rocks: Significance of turbulent flow and fractal scaling. *Int. J. Heat Mass Transf.* **2018**, *116*, 549–556. [CrossRef]
52. Huang, N.; Jiang, Y.J.; Liu, R.C.; Xia, Y.X. Size effect on the permeability and shear induced flow anisotropy of fractal rock fractures. *Fractals* **2018**, *26*, 1840001. [CrossRef]
53. Li, Y.; Gong, L.; Zeng, L.; Ma, H.; Yang, H.; Zhang, B.; Zu, K. Characteristics of fractures and their contribution to the deliverability of tight conglomerate reservoirs in the Jiulongshan Structure, Sichuan Basin. *Nat. Gas Ind.* **2012**, *32*, 22–26.
54. Pei, S.; Dai, H.; Yang, Y.; Li, Y.; Duan, Y. Evolutionary characteristics of T_3x^2 reservoir in Jiulongshan Structure, northwest Sichuan Basin. *Nat. Gas Ind.* **2008**, *28*, 51–53.
55. Babadagli, T.; Develi, K. On the application of methods used to calculate the fractal dimension of fracture surfaces. *Fractals* **2001**, *9*, 105–128. [CrossRef]
56. Klinkenberg, B. A review of methods used to determine the fractal dimension of linear features. *Math. Geol.* **1994**, *26*, 23–46. [CrossRef]
57. Walsh, J.J.; Watterson, J. Fractal analysis of fracture patterns using the standard box-counting technique: Valid and invalid methodologies. *J. Struct. Geol.* **1993**, *15*, 1509–1512. [CrossRef]
58. Roy, A.; Perfect, E.; Dunne, W.M.; McKay, L.D. Fractal characterization of fracture networks: An improved box-counting technique. *J. Geophys. Res.* **2007**, *112*, B12. [CrossRef]
59. Gong, L.; Zeng, L.; Miao, F.; Wang, Z.; Wei, Y.; Li, J.; Zu, K. Application of fractal geometry on the description of complex fracture systems. *J. Hunan Univ. Sci. Technol.* **2012**, *27*, 6–10.
60. Mu, L.; Zhao, G.; Tian, Z.; Yuan, R.; Xu, A. *Prediction of reservoir fractures*; Petroleum Industry Press: Beijing, China, 2009; pp. 111–121.
61. Odling, N.E. Scaling and connectivity of joint systems in sandstones from western Norway. *J. Struct. Geol.* **1997**, *19*, 1257–1271. [CrossRef]

62. Gauthier, B.D.M.; Lake, S.D. Probabilistic modeling of faults below the limit of seismic resolution in Pelican Field, North Sea, offshore United Kingdom. *AAPG Bull.* **1993**, *77*, 761–777.

63. Hooker, J.N.; Laubach, S.E.; Marrett, R. A universal power-law scaling exponent for fracture apertures in sandstones. *Geol. Soc. Am. Bull.* **2014**, *126*, 1340–1362. [CrossRef]

Article

Multifractal Characteristics and Classification of Tight Sandstone Reservoirs: A Case Study from the Triassic Yanchang Formation, Ordos Basin, China

Zhihao Jiang [1,2], Zhiqiang Mao [1,2,*], Yujiang Shi [3,4] and Daxing Wang [3,4]

[1]   State Key Laboratory of Petroleum Resources and Prospecting, China University of Petroleum, Beijing 102249, China; haojz1993@163.com

[2]   Beijing Key Laboratory of Earth Prospecting and Information Technology, China University of Petroleum, Beijing 102249, China

[3]   Research Institute of Exploration and Development, PetroChina Changqing Oilfield Company, Xi'an 710021, China; syj_cq@petrochina.com.cn (Y.S.); wdx1_cq@petrochina.com.cn (D.W.)

[4]   National Engineering Laboratory for Exploration and Development of Low-Permeability Oil & Gas Fields, Xi'an 710018, China

*   Correspondence: maozq@cup.edu.cn; Tel.: +86-010-8973-3318

Received: 26 July 2018; Accepted: 17 August 2018; Published: 27 August 2018

Abstract: Pore structure determines the ability of fluid storage and migration in rocks, expressed as porosity and permeability in the macroscopic aspects, and the pore throat radius in the microcosmic aspects. However, complex pore structure and strong heterogeneity make the accurate description of the tight sandstone reservoir of the Triassic Yanchang Formation, Ordos Basin, China still a problem. In this paper, mercury injection capillary pressure (MICP) parameters were applied to characterize the heterogeneity of pore structure, and three types of pore structure were divided, from high to low quality and defined as Type I, Type II and Type III, separately. Then, the multifractal analysis based on the MICP data was conducted to investigate the heterogeneity of the tight sandstone reservoir. The relationships among physical properties, MICP parameters and a series of multifractal parameters have been detailed analyzed. The results showed that four multifractal parameters, singularity exponent parameter (α_{min}), generalized dimension parameter (D_{max}), information dimension (D_1), and correlation dimension (D_2) were in good correlations with the porosity and permeability, which can well characterize the pore structure and reservoir heterogeneity of the study area, while the others didn't respond well. Meanwhile, there also were good relationships between these multifractal and MICP parameters.

Keywords: tight sandstone; pore structure; multifractal; classification; Ordos Basin

1. Introduction

Since advanced technologies of hydraulic fracture and horizontal wells were developed, tight sandstones have drawn much attention due to their considerable hydrocarbon productivity [1,2]. Different from conventional reservoirs, it is the pore structure that controls flow capacity, producible pore volumes and production capability of tight sandstone reservoirs, rather than total porosity [3,4]. Tight sandstone reservoirs often have strongly laterally and vertically heterogeneous pore structures, expressed as a big and fast change in porosity and permeability. Therefore, accurate description of the pore structure of tight sandstone is conducive to evaluating the productivity of reservoirs and searching for remaining oil [5–9]. The diagenetic transformation of tight sandstone reservoir is strong, and the types of diagenesis are complex and varied, mainly including mechanical compaction, cementation and dissolution. Varied sedimentation [7], diagenesis [8,9] and tectonism [10] will lead to complicated pore structures, among them compaction will further reduce the porosity, aggravating reservoir heterogeneity. The complex pore structure makes it hard to evaluate and type tight sandstone reservoirs [11,12]. Meanwhile, it is a problem to quantify the heterogeneity through conventional methods, such as log facies analysis, monofractal method.

With the decline in the oil-production of conventional reservoirs after years of production, tight sandstone is becoming more and more important, being regarded as the potential reservoir of the future. Lots of advanced equipment has been applied to investigate the pore structure and heterogeneity of tight sandstones, such as scanning electron microscopy (SEM), nano computed tomography (CT)-scan, mercury injection capillary pressure (MICP), nuclear magnetic resonance (NMR) and low-temperature nitrogen adsorption [13–19]. Among this, SEM and nano CT-scan equipment can be used to qualitatively describe the pore types and sizes, directly. MICP, NMR and nitrogen adsorption are indirectly indicated the pore structure in rocks by some media. NMR measurement is an important and effective method for studying the pore size distributions, and also, NMR data can be measured both underground and in the laboratory. Many researchers have described the pore size distribution of tight sandstone through NMR data [20–22]. However, from previous studies, different types of pore fluid will affect the morphology of NMR T2 spectrum, even when the reservoir is oil-wetted, the NMR T2 spectrum will not truly reflect pore structure [23].

The MICP method assumes that the inner space of the porous material is cylindrical, each pore can be extended to the outer surface of the sample to contact with the mercury directly, and the contact angle is about 140 [24,25]. Under a certain pressure, the method assumes mercury can only penetrate into the pores of the corresponding size, and the amount of mercury injected represents the volume of the internal pores. Increasing the pressure and calculating the amount of mercury entry, the pore volume distribution of porous materials can be measured. Some scholars have also tried to construct MICP data using conventional logs curves or NMR logs [26,27].

The fractal theory was firstly proposed by Mandelbrot [28] in 1982. It is a popular method to describe the self-similar characteristics of irregular geometric figures [29,30]. However, the shapes of most natural objects show the nonuniform and multifractal features. A constant fractal dimension cannot accurately describe an inhomogeneous object. Then, the multifractal theory, a method can provide more information about the pore properties was proposed, it transforms the measurement of self-similarity into multifractal function sets, which are characterized by the multifractal spectrum $\alpha \sim f(\alpha)$ and the generalized dimension spectrum $q \sim D_q$ [31,32].

Recently, a lot of research investigating the fractal characteristics of rocks has been done, Cai et al. [33] have shown that the multifractal features can be used to describe pore surfaces in hard rocks. Many researchers and scholars have investigated the multifractal characteristics of tight sandstone reservoirs using the NMR T2 spectra of samples at brine-saturated state [4,16,34]. Some researchers have studied the pore structures of tight sandstones using thin section images based on multifractal theory [20,35]. These studies all suggested that the multifractal theory is an effective and convenient method to study the pore structure in unconventional reservoirs.

In this study, firstly MICP data of some tight sandstone core samples from the Chang6 Formation from the later Triassic (Ordos Basin, China) were applied to characterize the heterogeneity of pore structure, and three types of pore structure were divided, from high to low quality, as Type I, Type II and Type III, separately. The pores have been divided into large pores, medium pores and small pores, and the relationships of the proportion of different pores in different types of pore structure were analyzed. Later, the multifractal parameters of MICP data were calculated. The relationships between petrophysical and multifractal parameters were investigated in detail. Through a series of comparisons, the most effective multifractal parameters for characterizing the pore structure of tight sandstone have been proposed.

2. Methodology

2.1. Geological Setting

The Ordos Basin, located in the midwest of China spanning five provinces, covers more than 370,000 square kilometers (Figure 1a,b). It is the second largest sedimentary basin in China [36,37]. The Ordos Basin is a large multicycle cratonic basin with a simple structure. The whole basin is flat, being higher in the east and lower in the west. Low porosity and low permeability are the most notable features of the Yanchang Formation sandstone reservoir. According to the standard of China's tight oil partition [38,39], the surface air permeability of reservoirs is less than 1 mD for tight oil. The main layers of oil exploration and development in the Ordos Basin, the Chang6 oil-bearing formation, Chang7 oil-bearing formation and Chang8 oil-bearing formation all belong to the tight oil category [10]. The lithostratigraphic section is shown in Figure 1c. In this study, 24 tight sandstone samples were collected by sealing core drilling from the Chang6 Triassic Yanchang Formation to evaluate the petrophysical properties and analyze multifractal features.

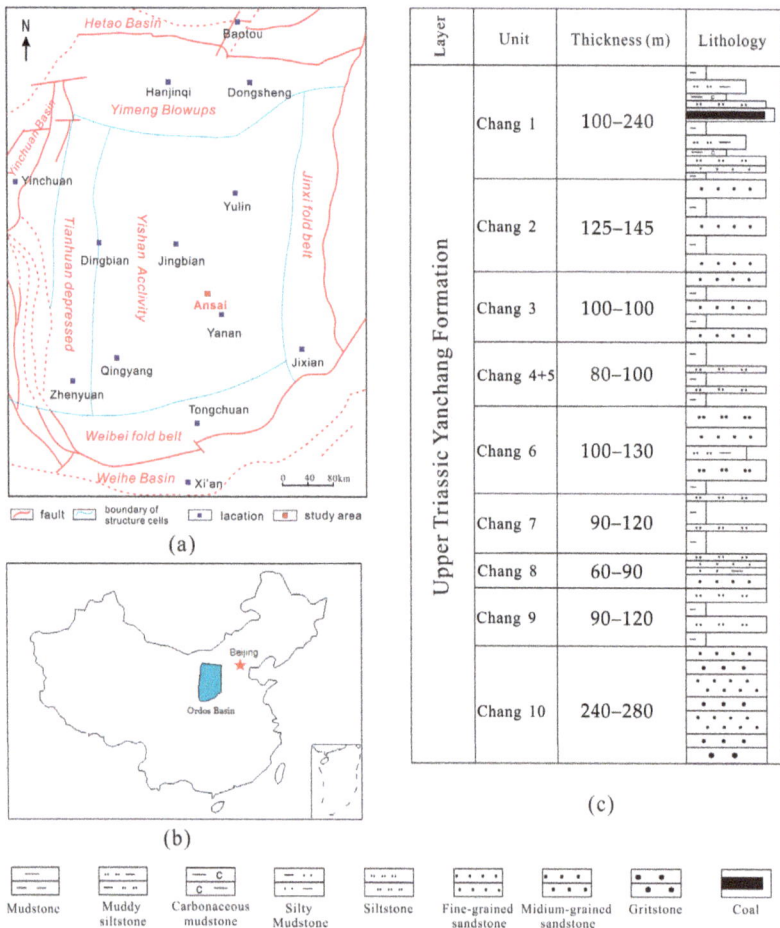

Figure 1. (**a**) Location of the study area, (**b**) location of the Ordos Basin, and (**c**) lithostratigraphic section of the Upper Triassic Yanchang Formation in the Ordos Basin [20,40].

2.2. Experiments

To better analyze the multifractal and petrophysical characteristics of the tight sandstone of the samples, a series of experiments were conducted, including routine petrophysical experiments, constant pressure MICP, thin section and SEM observations.

MICP measurements were performed to get the pore size distribution of the tight sandstone samples. The experiments were conducted using AutoPoreIII Mercury Injection equipment (Micromeritics, Norcross, GA, USA) under conditions of 18 °C temperature and relative air humidity (RH) of 55%, following the Chinese Oil and Gas Industry Standard SY/T 5346-2005. The maximum capillary pressure is 49.871 Mpa.

Before thin sections are made, the samples needed to be oil washed, then pumped under vacuum, and a colored epoxy resin injected into the pores. Finally, they were ground to 0.03 mm thickness. A polarizing microscope was used to observe the rock structure, mineral composition and pore characteristics.

The observation of the thin sections was conducted at a temperature of 24 °C and humidity of 35% RH at the China University of Petroleum (Beijing, China). During the course of observation, the magnification ranged from 60 to 2000 with an acceleration voltage of 20 kV.

2.3. Multifractal Analysis Theory

Compared with monofractal analysis, multifractal theory uses a continuous function, termed multifractal singularity spectrum, rather than a single fractal dimension, to characterize the multifractal features. Detailed multifractal theories have been introduced in some references [41,42]. In this paper, the box-counting algorithm multifractal method is adopted and briefly described here.

The MICP data can be distribution can be segmented in $N(\varepsilon)$ parts within scale ε: The multifractal theory is mainly used to study the normalized probability distribution of the target objects within the scale. Denoting n as the point number of the MICP data, n could equal to j power of 2 through interpolation method, i.e., $n = 2j, j = 1, 2, 3 \ldots$ The scale of MICP can be defined as $\varepsilon = n*2^{-k}$ ($k = 0, 1, 2, \ldots, j$). The MICP data can be segmented in $N(\varepsilon)$ part within scale ε:

$$N(\varepsilon) = \frac{L}{\varepsilon} = 2^k \tag{1}$$

For each box, the probability mass distribution of box densities can be expressed as:

$$P_i(\varepsilon) = \frac{v_i(\varepsilon)}{\sum_{i=1}^{N(\varepsilon)} v_i(\varepsilon)} \tag{2}$$

where $v_i(\varepsilon)$ is MICP data of the ith box, and $N(\varepsilon)$ is the total number of boxes for ε.

Defined the partition function as $\chi_q(\varepsilon)$, equal to the q square and weighted sum of the probability $P_i(\varepsilon)$. $\chi_q(\varepsilon)$ can be expressed as:

$$\chi_q(\varepsilon) = \sum_{i=1}^{N(\varepsilon)} P_i(\varepsilon)^q \propto \varepsilon^{\tau(q)} \tag{3}$$

where $\chi_q(\varepsilon)$ is a partition function of q with the scale of ε; $\tau(q)$ is the mass exponent.

The number of the boxes in multifractals is defined as $N_\alpha(\varepsilon)$, which is related to scale ε and can be expressed as [43,44]:

$$N_\alpha(\varepsilon) = \varepsilon^{-f(\alpha)} \tag{4}$$

Meanwhile, when taking the logarithm on both sides of Equation (3), $\tau(q)$ can be written as:

$$\tau(q) = \frac{Ln\chi_q(\varepsilon)}{Ln(\varepsilon)} \quad \varepsilon \to 0 \tag{5}$$

The mass exponent $\tau(q)$ can be taken as the slope of the curve $Ln\chi_q(\varepsilon) \sim Ln(\varepsilon)$. $\tau(q)$ can be used to calculate the generalized fractal dimension D_q. With the different values of q, the generalized fractal dimension has different definitions:

$$D_q = \begin{cases} \frac{\tau(q)}{q-1} = \frac{1}{q-1} \lim_{\varepsilon \to 0} \frac{\log \sum_{i=1}^{N(\varepsilon)} P_i(\varepsilon)^q}{\log \varepsilon}, & q \neq 1 \\ \frac{\sum_{i=1}^{N(\varepsilon)} P_i(\varepsilon) \log P_i(\varepsilon)}{\log \varepsilon}, & q = 1 \end{cases} \tag{6}$$

The singularity exponent α and multifractal spectrum $f(\alpha)$ can be transformed into [45,46]:

$$\alpha(q) = \frac{d\tau(q)}{dq} \tag{7}$$

and:

$$f(\alpha) = q\alpha(q) - \tau(q) \tag{8}$$

In fact, the multifractal spectrum $\alpha \sim f(\alpha)$ and the generalized dimension spectrum $q \sim D_q$ are two sets of different parameters describing the fractal characteristics [4,17,33].

2.4. Mercury Injection Capillary Pressure (MICP) Theory

The MICP method assumes that the inner space of the porous material is cylindrical, and each pore can be extended to the outer surface of the sample to contact with the mercury directly, and the contact angle is about 140° [47–49]. Under a certain pressure, the method assumes mercury can only penetrate into the pores of the corresponding size, and the amount of mercury injected represents the volume of the effective porosity, as the following Equation (9) (Young-Laplace law) shows [25]:

$$P_{Hg} = \frac{2\sigma_{Hg} \cos\theta_{Hg}}{r} \tag{9}$$

where P_{Hg} is capillary pressure, MPa; r is the pore throat radius, μm, when the capillary pressure is equal to P_{Hg}; σ_{Hg} is surface tension, mN/m, always taken as 480 mN/m, θ_{Hg} is contact angle, (°), about 140°. So, Equation (9) can be written as:

$$r = \frac{0.735}{P_{Hg}} \tag{10}$$

Increasing the pressure and calculating the amount of mercury entry, the pore volume distribution of porous materials can be measured [50,51].

In order to better analyze MICP data, several related parameters P_{50}, R_{50}, $\bar{r}$, P_d, S_{max} and W_e are introduced. P_{50} is median pressure, MPa, referring to the capillary pressure corresponding to the mercury saturation 50%. R_{50} is the median radius, μm, pore radius corresponding to median pressure P_{50}. $\bar{r}$ is the average throat radius, μm, indicating the pore structure distribution. The three front parameters, P_{50}, R_{50} and $\bar{r}$ all reflect the physical properties of rock pores. P_d, S_{max} and W_e are three MICP parameters, respectively, referring to displacement pressure (MPa), maximum mercury saturation (%) and efficiency of mercury withdrawal (%). The latter three parameters, P_d, S_{max} and W_e reflect the difficulty of mercury injection and mercury withdrawal, are also the embodiment of the complexity of the pore structure and clay content [52,53].

3. Results and Discussion

3.1. Porosity, Permeability and MICP Data

The porosity and permeability (K) values of the samples are listed in Table 1, together with a series of MICP parameters. The porosity range of the samples is from 5.7% to 15% with an average value of 12.10%. The permeability ranges from 0.108 mD to 9.987 mD with most of them less than 1 mD, and thus corresponding to tight sandstone. The relationship between the porosity and permeability is shown in Figure 2. The pore throat sizes calculated by Equation (10) are divided into three categories, large pores (pore throat sizes > 10 µm), medium pores (pore throat sizes range from 1–10 µm) and small pores (pore throat sizes < 1 µm). The proportions of different types of pores are listed in Table 1, too.

Table 1. The porosity, permeability and mercury injection capillary pressure (MICP) parameters of samples.

No.	Porosity (%)	K (mD)	P_{50} (MPa)	R_{50} (µm)	$\bar{r}$ (um)	P_d (MPa)	S_{max} (%)	W_e (%)	Large Pore (%)	Medium Pore (%)	Small Pore (%)
1	12.2	0.635	21.8635	0.0336	0.1362	1.6577	75.49	32.3168	0	36.47	39.02
2	14.1	9.987	4.5277	0.1624	1.4537	0.1341	72.59	37.2628	33.47	19.46	19.66
3	12.8	1.553	4.6343	0.1586	0.7140	0.3210	76.47	29.1339	25.96	27.38	23.13
4	13.3	1.318	7.9185	0.0928	0.5350	0.4078	73.78	30.1309	16.23	32.9	24.65
5	14.5	2.951	5.9248	0.1241	0.7622	0.2800	78.42	31.2342	27.2	24.23	26.99
6	13.4	1.285	9.3323	0.0788	0.5089	0.4364	74.11	30.1099	16.36	31.5	26.25
7	13.5	1.151	4.9514	0.1485	0.5617	0.3976	82.15	32.5816	22.36	30.91	28.88
8	15.0	5.365	5.8377	0.1259	0.9252	0.2202	74.77	32.7690	27.18	24.22	23.37
9	13.6	1.244	4.5437	0.1618	0.6162	0.4111	75.79	29.7635	23	30.77	22.02
10	13.6	1.515	4.0092	0.1834	0.6537	0.4114	76.13	30.5637	26.91	27.55	21.67
11	13.4	2.088	4.0521	0.1814	0.8023	0.3387	72.23	28.0463	30.56	23.35	18.32
12	5.7	0.108	29.7888	0.0247	0.0629	5.2180	64.38	31.5089	0	10.58	53.8
13	13.6	1.107	3.2311	0.2275	0.5251	0.3982	81.24	29.9167	15.25	42.66	23.33
14	13.6	1.107	3.9024	0.1884	0.3257	0.7812	79.67	33.3898	2.33	53.54	23.8
15	13.8	2.585	2.5664	0.2864	0.8658	0.1289	84.58	28.2272	30.89	28.38	25.31
16	14.0	0.486	6.0551	0.1214	0.1679	1.4097	84.15	31.8996	0	52.21	31.94
17	10.2	0.392	6.9278	0.1061	0.1420	1.9421	88.75	36.6435	0	50.19	38.56
18	9.7	0.156	13.7947	0.0533	0.0979	2.1185	75.9	20.6429	0	27.61	48.29
19	7.9	0.168	9.9355	0.0740	0.1976	0.9940	79.53	28.3013	0.91	43.11	35.51
20	10.8	0.271	10.0876	0.0729	0.1335	1.9989	81.85	31.1925	0	44.17	37.68
21	10.3	0.395	9.9605	0.0738	0.1609	1.5684	68.03	20.1933	0.43	40.5	27.1
22	11.7	0.800	4.5632	0.1611	0.3004	0.7287	80.32	29.9648	3.82	52.64	23.86
23	10.5	0.451	3.5488	0.2071	0.4010	0.4524	86.62	27.8590	11.59	49.73	25.3
24	10.5	0.279	6.7273	0.1093	0.1505	1.4602	88.1000	30.7172	0	50.65	37.45

Notes: Large pore: pore throat sizes > 10 µm; Medium pore: pore throat sizes range from 1–10 µm; Small pore: pore throat sizes < 1 µm.

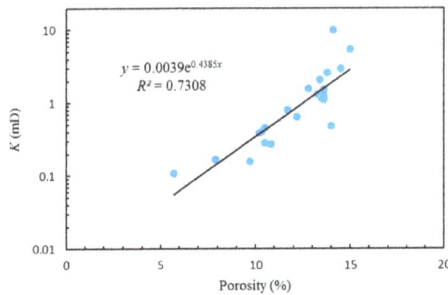

Figure 2. The relationship between the porosity and permeability.

The average pore radius $\bar{r}$ of the samples is distributed between 0.0629–5.2180 µm, with an average value of 0.4677 µm. The displacement pressure P_d ranges from 0.1289 MPa to 5.218 MPa, showing a good negative correlation with $\bar{r}$. The sensitivity to the permeability of different parameters listed in Table 1 is diverse. Among them, only the average pore radius $\bar{r}$ and displacement pressure P_d show a good correlation (or negative correlation) with permeability ($R^2 > 0.7$), while the correlations between permeability and the other petrophysical parameters are less the 0.25. $\bar{r}$ and P_d are good parameters to reflect the heterogeneity of tight sandstone in the study area. The maximum capillary pressure is only 49.871 Mpa, the maximum mercury saturation S_{max} of almost all samples reaches over 70%. This ensures that our experimental mercury injection curves can reflect most pore volume, some microporous information still will be omitted. When studying the reservoirs with lower permeability, the maximum capillary pressure should be higher than 100 MPa [54]. The mercury removal efficiency of the samples is about 30%. The efficiency of mercury withdrawal W_e is often influenced by physical properties, the type and content of clay. In general, the better the physical properties are, the greater mercury withdrawal efficiency is [55]. The values of P_d and W_e shows that the pore throats are fine and have poor connectivity and low percolation capacity.

3.2. Classification of the Pore Structure

Only partial thin sections and electron micrographs of samples were collected and observed. Through the observation and comparison of the thin sections, most of the samples belong to lithic feldspar sandstone and lithic feldspar sandstone (Table 2), and the rock particles are not well round, with a medium sorting, basically angular or sub angular. The primary intergranular pores as the most important pore space, are not developed. Plastic mica minerals are filled between particles, damaging the seepage capacity of rocks (Figure 3a,b). The surface of some particles is covered by padded chlorite, as Figure 3c shows. The development of chlorite always indicates the dominant reservoir [37]. Dissolution of feldspar particles has a constructive effect on pore structure (Figure 3a,d). In the study reservoir, the higher content of chlorite and feldspar represented the improvement of the porosity and permeability. Considering the accuracy of data from thin section observations, no further quantitative analysis of mineral content was made.

Table 2. The thin section observations of samples.

No.	Quartz (%)	Feldspar (%)	Mica (%)	Chlorite (%)	Iron Calcite (%)	Main Particle Size (μm)	Sorting	Grinding Roundness	Cementation Type
1	18	42	23	5	2	0.10–0.30	M	A	chlorite thin film
3	22	53	5	4	6	0.2–0.5	M	A	chlorite thin film
4	20	54	7	4	2	0.15–0.5	M	SA-A	chlorite thin film
7	21	55	3	4	7	0.15–0.4	M	A	pore-chlorite thin film
9	23	57	0	6	1	0.1–0.3	M-G	SA-A	chlorite thin film
11	22	55	5	6.5	1	0.1–0.32	M-G	SA-A	chlorite thin film
12	15	50	7	0	18	0.10–0.35	M	A	pore
13	20	60	3	6	2	0.10–0.30	M-G	A	chlorite thin film
15	21	56	5	5	3	0.10–0.35	M	A	pore-chlorite thin film
16	22	55	6	7	3	0.05–0.15	G	A	pore-chlorite thin film
20	30	39	8	2	1	0.05–0.20	M	SA	chlorite thin film-pore
21	23	53.5	3	3	1.5	0.2–0.5	M	SA	chlorite thin film-pore
22	22	48	8.5	3	2	0.12–0.38	M	A	pore-chlorite thin film
23	22	55	6	3	2	0.16–0.28	G	A	pore
24	22	52	6	3	0	0.12–0.24	G	A	pore-chlorite thin film

Notes: M: medium sorted; G: good sorted; A: angular; SA: sub-angular.

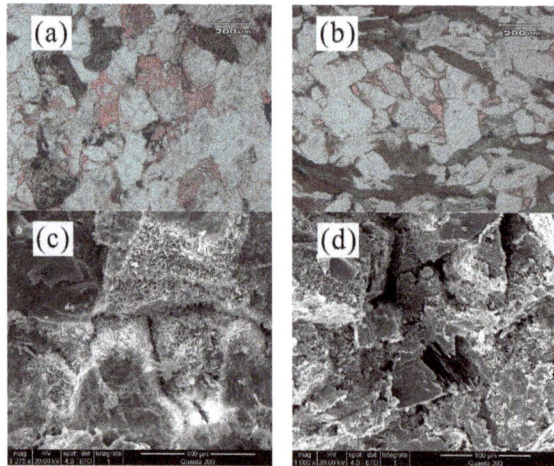

Figure 3. Thin section images and scanning electron microscopy (SEM) images of the tight sandstone in the study area: (**a**) intergranular pores and feldspar dissolution pore; (**b**) intergranular pores and enriched mica; (**c**) the surface of the rock particles covered with padded chlorite; (**d**) feldspar dissolved pore.

The samples with different pore structure have been divided into three types based on porosity, permeability, pore structure parameters, and the proportion of different pores. Pore structure of three types of reservoirs are represented by samples in Figure 4. The range and mean value of the parameters of different types are listed in Table 3.

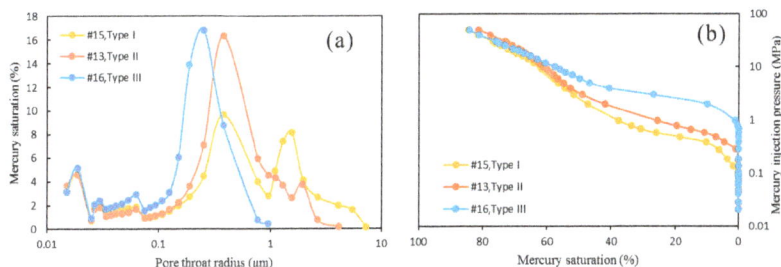

Figure 4. The mercury injection capillary pressure (MICP) curves of different types of reservoirs: (**a**) the pore throat size distributions of different types; (**b**) the mercury capillary pressure curves of different types.

Table 3. Classification of reservoirs according to physical properties and MICP parameters.

Type	Porosity (%)	K (Md)	P_{50} (MPa)	$\bar{r}$ (um)	P_d (MPa)	Large Pore (%)	Medium Pore (%)	Small Pore (%)
Type I	12.8–15	1.151–9.987	2.57–5.92	0.56–1.45	0.13–0.41	22.36–33.47	19.46–30.91	18.32–28.88
	13.81	3.160	4.56	0.82	0.29	27.50	26.25	23.26
Type II	10.5–13.6	0.451–1.318	3.23–9.33	0.30–0.54	0.40–0.78	2.33–16.36	31.5–53.54	23.33–26.25
	12.68	1.011	5.42	0.43	0.53	10.93	43.83	24.53
Type III	5.7–14	0.108–0.635	6.06–29.79	0.06–0.20	0.99–5.22	0–0.91	10.58–52.21	27.1–53.8
	10.3	0.369	11.97	0.15	1.91	0.52	40.81	37.32

3.3. Pore Characteristics of Reservoirs with Different Pore Structure Types

As Figures 4 and 5 show, different types of samples have different pore structures. Samples of Type I belong to the best reservoir in the study formation. The average porosity and permeability are the highest among the three types, are 13.81% and 3.160 mD, respectively. Large pores and medium pores are the main channels of Type I reservoir. The average proportion of large pore and medium pore are respectively 27.5% and 26.25%. Both of them are important indicators of permeability. With the porosity and permeability increasing in Type I reservoir, the proportion of medium pore decreases and the large pore proportion increases (Figure 6). The proportion of large pore has a good positive correlation with the permeability ($R^2 = 0.56$), while the proportion of medium pore has a better negative correlation with the permeability ($R^2 = -0.75$). Because the contribution of the large pores to the permeability is far greater than medium pores and small pores, and with the increasing proportion of the medium pores, the proportion of the large pores will decrease. Therefore, the proportion of the medium pores seems to be negatively correlated with the permeability. Meanwhile, as a non-major seepage channel, the change of small pore proportion has little effect on the permeability of Type I reservoirs.

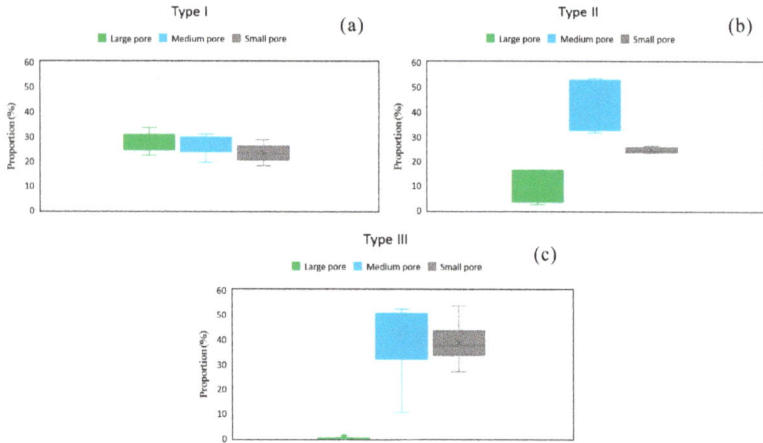

Figure 5. Different pore proportions of different reservoir types: (**a**) different pore proportion of Type I; (**b**) different pore proportion of Type II; (**c**) different pore proportion of Type III.

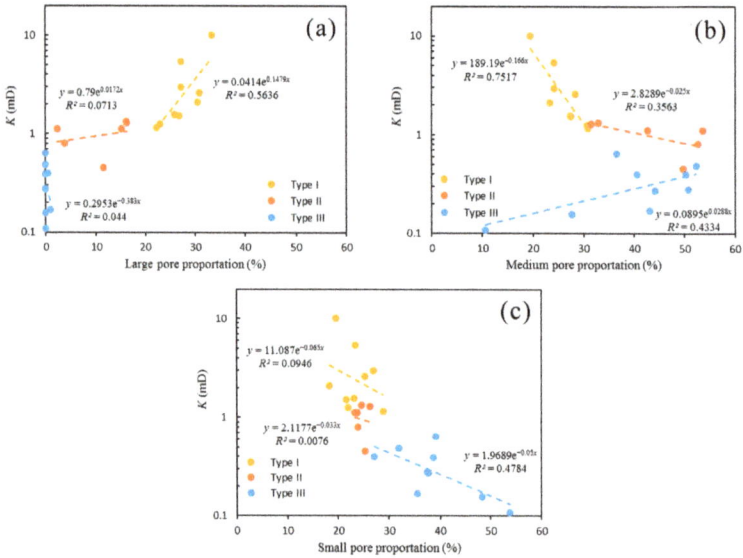

Figure 6. Relationships of the different pore proportion of different pore structure types: (**a**) relationships of large pore proportion of different pore structure types; (**b**) relationships of medium pore proportion of different pore structure types; (**c**) relationships of small pore proportion of different pore structure types.

The average porosity and permeability of Type II reservoir are 13.68% and 1.011 mD, tighter than Type I reservoir. In Type II reservoir, the proportion of medium pore ranges from 33.5% to 53.54%, with an average value of 43.83%. The medium pores, distributed from 1 μm to 10 μm seem to be the main seepage channels of the Type II reservoir, with the highest proportion. But with the increasing of the proportion of medium pore, permeability shows a reduced trend. In fact, the large pores of less proportion still greatly affected the permeability of samples. With the difference in diagenesis and cementation types, there is a bad correspondence between permeability and the proportion of different pores in Type II reservoir.

In reservoirs with Type III pore structure, the distribution of the pores is very different from the two above. The values of porosity and permeability are extremely low. There are little large pores; medium pores are the most major storage space and percolation path. With the increase of medium pore proportion, the permeability increases. In this kind of reservoirs, the proportion of medium pores and small pores both have good correlations with permeability. The Type III reservoir is the tightest type among them, with an average permeability of 0.369 mD.

3.4. Multifractal Spectrum Parameters

In this research, the values of q range from -10 to 10, with a step of 0.1. With the different values of q, the mass exponent $\tau(q)$, the slope of the curve $Ln\chi_q(\varepsilon) \sim Ln(\varepsilon)$, can be described by Figure 7. The mass exponent spectrum $q \sim \tau(q)$ is nonlinear, indicating the multifractal characteristics of MICP data. After the determination of $\tau(q)$, the multifractal spectrum $\alpha \sim f(\alpha)$ and the generalized dimension spectrum $q \sim D_q$ can be calculated by Equations (6) and (8), and the results are shown in Figure 7.

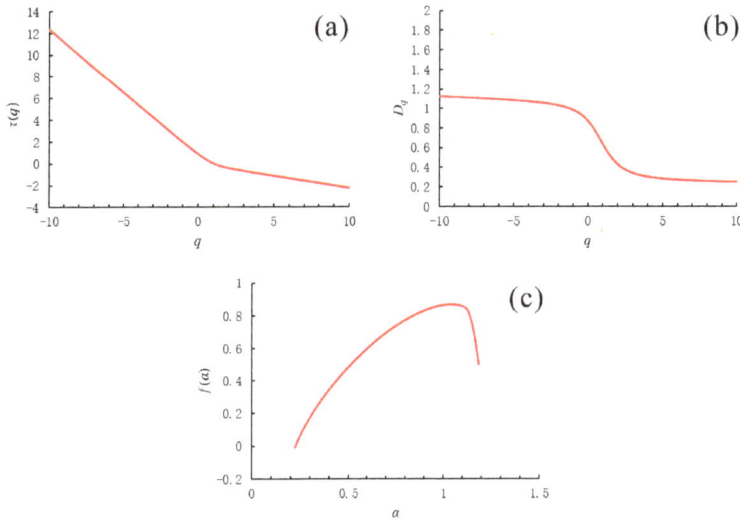

Figure 7. Multifractal analysis of the MICP data in the tight sandstone. (**a**) the mass exponent $\tau(q)$; (**b**) the generalized dimension spectrum $q \sim D_q$; (**c**) the multifractal spectrum $\alpha \sim f(\alpha)$.

As q increases from -10 to 10, D_q gradually decreased, eventually approaching a balance, where D_{-10} represents D_{min} and D_{10} represents D_{max}. D_0, D_1 and D_2 also are important dimension spectrum parameters, with the meanings of the capacity dimension parameter, information dimension parameter and correlation dimension parameter, respectively.

The multifractal spectrum $\alpha \sim f(\alpha)$ can also reveal the multifractal characteristics of the tight sandstones. α_{min} and α_{max} represent α_{-10} and α_{10}, respectively. At the same time, α_{min} is a subset of the maximum probability, while α_{max} is corresponding to a subset of the minimum probability. When $q < 0$, $f(\alpha)$ increases as α increasing. Contrary to when $q < 0$, $f(\alpha)$ decreases as α increases. The singularity strength range is defined as $\Delta\alpha = \alpha_{max} - \alpha_{min}$. $\Delta\alpha$ is used to describe the complexity and heterogeneity of multifractal objects. With the heterogeneity of the pore structure increasing, $\Delta\alpha$ increases. The parameter Δf is defined as the difference between the value of $f(\alpha_{max})$ and $f(\alpha_{min})$, which is equal to the ratio of the minimum value to the maximum value of multifractal singularity spectrum [4].

3.5. Multifractal Characteristics of Pore Structure in Tight Sandstones

Multifractal analysis was performed on all the MICP data of tight sandstone samples, and the relationship trends of mass exponent spectrum $q \sim \tau(q)$, the generalized dimension spectrum $q \sim D_q$ and the multifractal spectrum $\alpha \sim f(\alpha)$ were similar to Figure 7. However, the characteristics of the multifractal spectrum of samples with different pore structures are distinctly different. All three types of tight sandstones have strong multifractal characteristics and heterogeneity, as Figure 8 shows. The mass exponent spectrum $q \sim \tau(q)$ and the generalized dimension spectrum $q \sim D_q$ of different rock types are monotonically decreasing. Type I pore structure has the highest values of τ_{max} and D_{max}, while Type III pore structure always has the highest values of τ_{min} and D_{min}. In the multifractal spectrum $\alpha \sim f(\alpha)$, the highest points of $f(\alpha)$ are almost the same. With the changes of pore structure types, the left and right hook both change a lot, which indicates the heterogeneity of pore throat distribution of different types of reservoirs.

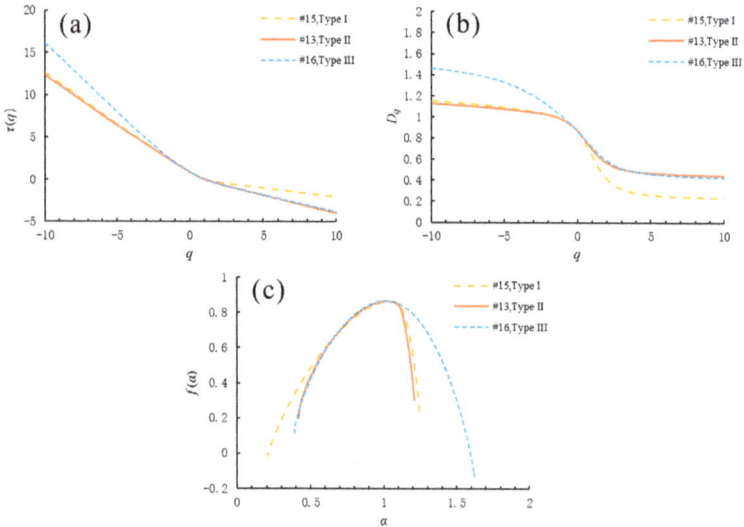

Figure 8. Multifractal spectrum of MICP data in tight sandstone samples with different types of reservoirs. (**a**) the mass exponent $\tau(q)$ of different types; (**b**) the generalized dimension spectrum $q \sim D_q$ of different types; (**c**) the multifractal spectrum $\alpha \sim f(\alpha)$ of different types.

The morphological parameters, such as α_{max}, α_{min}, D_{max}, D_{min}, D_1, D_2, f_{max}, f_{min}, $\Delta\alpha$ and Δf, were obtained from the analysis of the multifractal spectrum $\alpha \sim f(\alpha)$ and the generalized dimension spectrum $q \sim D_q$. Table 4 lists the multifractal parameters of three pore structure types. The average values of α_{max}, α_{min}, D_{max}, D_{min}, D_1, D_2 and f_{min} increase from Type I to Type III reservoir, while the average values of f_{max} and Δf decrease. In Table 4, among all the multifractal parameters, α_{min}, D_{max}, D_1, D_2 and f_{max} can be used to type the different pore structure types.

Table 4. Multifractal parameters of tight sandstone samples with different types of pore structure.

Type	α_{max}	α_{min}	D_{max}	D_{min}	D_1	D_2	f_{max}	f_{min}	$\triangle\alpha$	f
Type I	1.18–1.24	0.14–0.28	0.17–0.32	1.11–1.15	0.56–0.70	0.31–0.53	0.22–0.50	−0.01–0.00	0.90–1.06	0.22–0.50
	1.21	0.21	0.24	1.13	0.63	0.42	0.37	0.00	0.99	0.37
Type II	1.16–1.90	0.26–0.46	0.29–0.49	1.11–1.71	0.67–0.80	0.48–0.67	0.00–0.59	0.00–0.20	0.79–1.54	−0.01–0.55
	1.33	0.36	0.39	1.23	0.71	0.56	0.29	0.07	0.97	0.23
Type III	0.99–2.14	0.34–0.79	0.40–0.82	0.91–1.92	0.71–0.92	0.58–0.93	−0.28–0.19	−0.19–0.55	0.2–1.80	−0.30–0.45
	1.66	0.48	0.53	1.50	0.77	0.69	−0.07	0.08	1.17	0.05

3.6. The Relationship between Multifractal Parameters and the Porosity and Permeability

In Table 4, a series of multifractal parameters α_{max}, α_{min}, D_{max}, D_{min}, D_1, D_2, f_{max}, f_{min}, $\triangle\alpha$ and $\triangle f$ were used to classify the pore structure types. Figure 9 shows that the parameters α_{max}, α_{min}, D_{max}, D_{min}, D_1, D_2, and $\triangle\alpha$ are negatively correlated with porosity and permeability, whereas f_{max} and $\triangle f$ are negatively correlated with the porosity and permeability. As Figure 9 shows, some multifractal parameters, such as α_{min}, D_{max}, D_1, and D_2 are in strong correlations with porosity and permeability, with the absolute values of the correlation coefficients higher than 0.57.

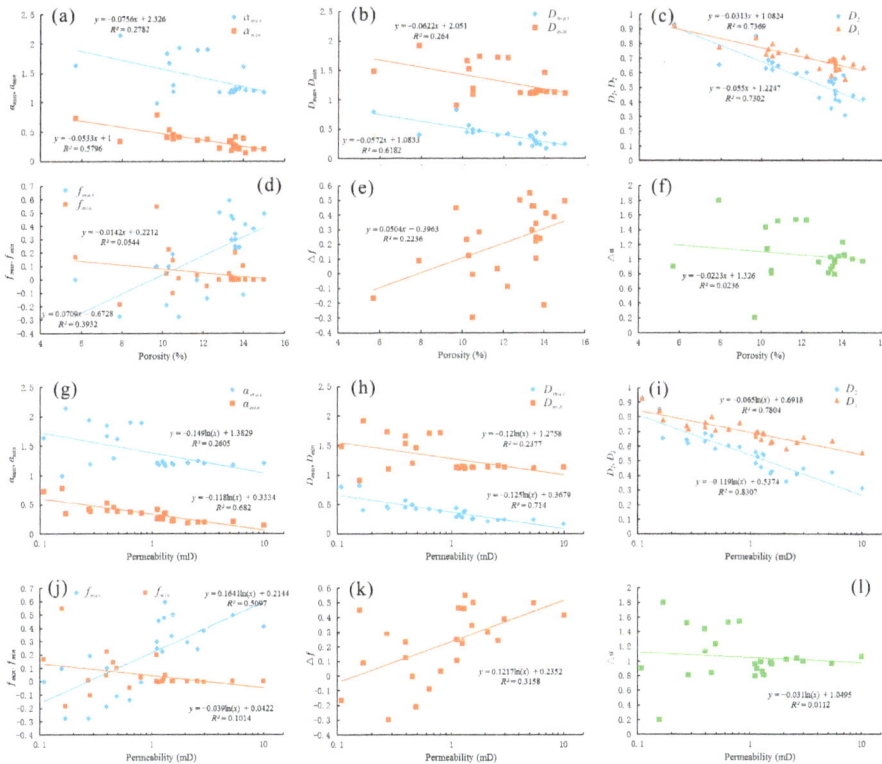

Figure 9. The cross plots between multifractal parameters and the porosity & permeability. (**a**) the cross plots between α_{max}, α_{min} and porosity; (**b**) the cross plots between D_{max}, D_{min} and porosity; (**c**) the cross plots between D_1, D_2 and porosity; (**d**) the cross plots between f_{max}, f_{min} and porosity; (**e**) the cross plot between $\triangle f$ and porosity; (**f**) the cross plot between $\triangle\alpha$ and porosity; (**g**) the cross plots between α_{max}, α_{min} and permeability; (**h**) the cross plots between D_{max}, D_{min} and permeability; (**i**) the cross plots between D_1, D_2 and permeability; (**j**) the cross plots between f_{max}, f_{min} and permeability; (**k**) the cross plot between $\triangle f$ and permeability; (**l**) the cross plot between $\triangle\alpha$ and permeability.

They can be used to effectively indicate the heterogeneity of pore structure. Meanwhile, the other multifractal parameters, α_{max}, D_{min}, f_{max}, f_{min}, $\Delta\alpha$ and $\triangle f$ display weak correlations with porosity and permeability, with correlation coefficients ranging from 0.0112 to 0.5097. Although these parameters are different in different types of reservoirs, the correlation coefficients are still not high enough, which might relate to the pore throat tortuosity and different cementation type. In conclusion, multifractal parameters, α_{min}, D_{max}, D_1, and D_2 are great indicators of the porosity and permeability in tight sandstone reservoirs.

3.7. The Relationship between Multifractal Parameters and Pore Structure Parameters

The cross plots between multifractal parameters and MICP parameters are shown in Figure 10, and the correlation coefficients are listed in Table 5. There are fine correlations between the pore structure parameters and multifractal parameters, too. P_{50} and P_d are positively correlated with α_{max}, α_{min}, D_{max}, D_{min}, D_1, D_2, f_{min} and $\Delta\alpha$, but negatively correlated with f_{max} and $\triangle f$, with logarithmic curve fitting. $\bar{r}$ is negatively correlated with α_{max}, α_{min}, D_{max}, D_{min}, D_1, D_2, f_{min} and $\Delta\alpha$, but positively correlated with f_{max} and $\triangle f$, with logarithmic curve fitting. The correlation coefficients between small pore proportion and large proportion to multifractal parameters are good but totally opposite, as their contributions to the permeability are different. Combined with the preceding conclusions, the average pore radius $\bar{r}$ and the displacement pressure P_d are good parameters to reflect the heterogeneity of tight sandstone in the study area; the multifractal parameters α_{min}, D_{max}, D_1 and D_2 display strong correlations with porosity and permeability. The multifractal parameters α_{min}, D_{max}, D_1 and D_2 are also in good correlation with both the average pore radius $\bar{r}$ and the displacement pressure P_d.

Table 5. The correlation coefficients between multifractal parameters and pore structure parameters.

	α_{max}	α_{min}	D_{max}	D_{min}	D_1	D_2	f_{max}	f_{min}	$\Delta\alpha$	f
Porosity	−0.278	−0.580	0.618	−0.264	−0.737	−0.730	0.393	−0.054	−0.024	0.224
K	−0.261	−0.682	−0.714	−0.238	−0.780	−0.831	0.510	−0.101	−0.011	0.316
P_{50}	0.192	0.443	0.476	0.188	0.439	0.514	−0.189	0.047	0.012	−0.062
$\bar{r}$	−0.328	−0.686	−0.718	−0.302	−0.712	−0.804	0.582	−0.114	−0.028	0.360
P_d	0.312	0.635	0.664	0.290	0.649	0.742	−0.523	0.099	0.028	−0.316
L-pore	−0.383	−0.512	−0.547	−0.359	−0.598	−0.660	0.572	−0.060	−0.069	0.373
M-pore	0.164	0.006	0.006	0.155	0.032	0.030	−0.225	−0.012	0.133	−0.205
S-pore	0.149	0.642	0.680	0.132	0.653	0.735	−0.363	0.118	0.000	−0.171

Notes: L-pore: Large pore proportion; M-pore: Medium pore proportion; S-pore: Small pore proportion.

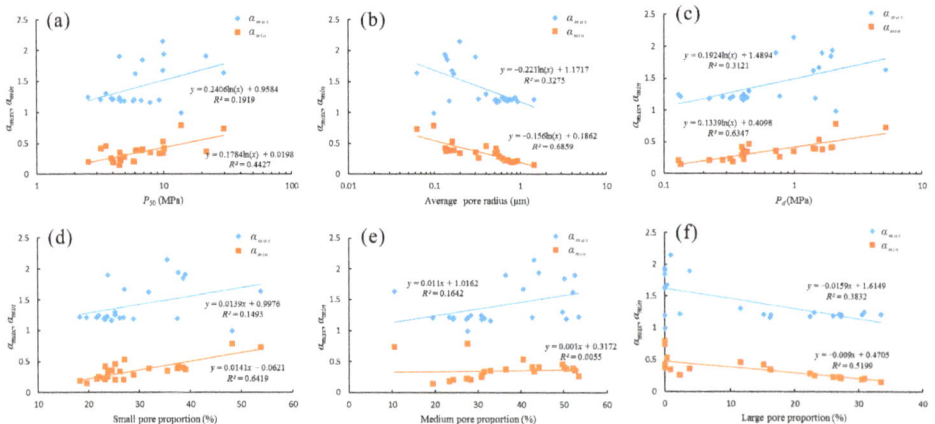

Figure 10. *Cont.*

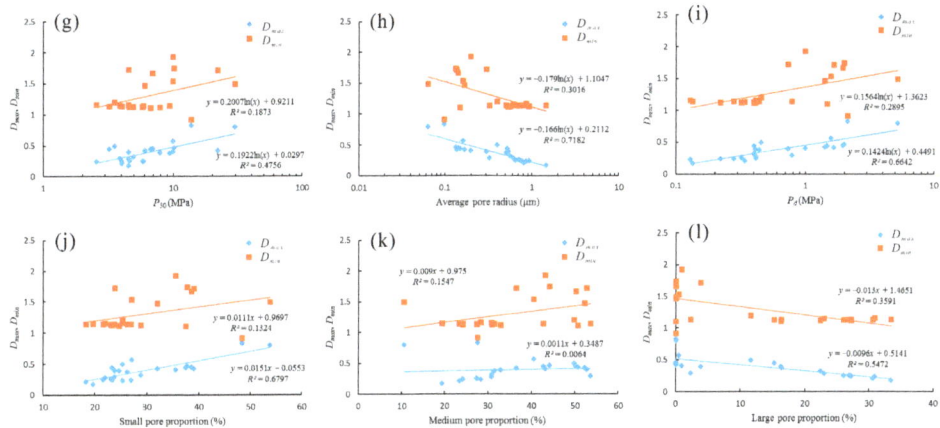

Figure 10. The cross plots between multifractal parameters and MICP parameters. (**a**) the cross plots between α_{max}, α_{min} and P_{50}; (**b**) the cross plots between α_{max}, α_{min} and average pore radius; (**c**) the cross plots between α_{max}, α_{min} and P_d; (**d**) the cross plots between α_{max}, α_{min} and small pore proportion; (**e**) the cross plots between α_{max}, α_{min} and medium pore proportion; (**f**) the cross plots between α_{max}, α_{min} and large pore proportion; (**g**) the cross plots between D_{max}, D_{min} and P_{50}; (**h**) the cross plots between D_{max}, D_{min} and average pore radius; (**i**) the cross plots between D_{max}, D_{min} and P_d; (**j**) the cross plots between D_{max}, D_{min} and small pore proportion; (**k**) the cross plots between D_{max}, D_{min} and medium pore proportion; (**l**) the cross plots between D_{max}, D_{min} and large pore proportion.

4. Conclusions

In this research, the pore structures of the Triassic Yanchang Formation in the Ordos Basin, China were studied. A series of experiments were conducted to collect more geophysical information about the tight sandstone samples. Different MICP parameters and multifractal parameters were used to characterize the heterogeneity of the tight sandstone and type the reservoirs. The following conclusions were obtained:

(1) The pore structures of the tight sandstone reservoir of the study area can be classified into three types. Type I reservoir indicates the most effective reservoir, and large pores with a pore throat bigger than 10 μm are the main seepage channel. Type III reservoir has little large pores, the main percolation channels of Type III are small pores. The permeability of Type II reservoir is between them.

(2) Different minerals have different effects on the physical properties of tight sandstone reservoirs. Plastic mica minerals filled between particles will negatively affect the seepage capacity of rocks. The higher content of chlorite and feldspar always results in an improvement of the porosity and permeability. The qualitative relationships between them need further research.

(3) The average pore radius $\bar{r}$ and the displacement pressure P_d are sensitive and effective parameters. They can be used to reflect the heterogeneity of tight sandstone in the study area.

(4) The multifractal parameters, α_{min}, D_{max}, D_1, and D_2 are great indicators of the heterogeneity of reservoirs, they are also in good correlation with the average pore radius $\bar{r}$ and the displacement pressure P_d.

Author Contributions: The main research idea was contributed by Z.J. and Z.M. The manuscript preparation and experiments parts were complicated by Z.J., with the assistance of Y.S. Z.M. and D.W. proposed several suggestions on geological knowledge and assisted on manuscript revising work. All authors revised and approved the publication of the paper.

Funding: This research was funded by the Major National Oil & Gas Specific Project of China (No. 2016ZX05050).

Acknowledgments: Research for this paper was supported by the Major National Oil & Gas Specific Project of China (No. 2016ZX05050). The authors also appreciate the help and support from State Key Laboratory of Petroleum Resources and Prospecting, and Beijing Key Laboratory of Earth Prospecting and Information Technology at China University of Petroleum, Beijing.

Conflicts of Interest: The authors declare no conflict of interest.

References

1. Guo, C.; Xu, J.; Wei, M.; Jiang, R. Experimental study and numerical simulation of hydraulic fracturing tight sandstone reservoirs. *Fuel* **2015**, *159*, 334–344. [CrossRef]
2. Zeng, F.; Guo, J.; Liu, H.; Yin, J. Optimization design and application of horizonal well-staged fracturing in tight gas reservoirs. *Acta Pet. Sin.* **2013**, *34*, 959–968.
3. Sakhaee-Pour, A.; Bryant, S.L. Effect of pore structure on the producibility of tight-gas sandstones. *AAPG Bull.* **2014**, *98*, 663–694. [CrossRef]
4. Zhao, P.; Wang, Z.; Sun, Z.; Cai, J.; Wang, L. Investigation on the pore structure and multifractal characteristics of tight oil reservoirs using NMR measurements: Permian Lucaogou Formation in Jimusaer Sag, Junggar Basin. *Mar. Pet. Geol.* **2017**, *86*, 1067–1081. [CrossRef]
5. Cai, J.; Wei, W.; Hu, X.; Wood, D.A. Electrical conductivity models in saturated porous media: A review. *Earth-Sci. Rev.* **2017**, *171*, 419–433. [CrossRef]
6. Anovitz, L.M.; Cole, D.R. Characterization and analysis of porosity and pore structures. *Rev. Mineral. Geochem.* **2015**, *80*, 61–164. [CrossRef]
7. Higgs, K.E.; Zwingmann, H.; Reyes, A.G.; Funnell, R.H. Diagenesis, porosity evolution, and petroleum emplacement in tight gas reservoirs, Taranaki basin, New Zealand. *J. Sediment. Res.* **2007**, *77*, 1003–1025. [CrossRef]
8. Hu, Q.; Ewing, R.P.; Dultz, S. Low pore connectivity in natural rock. *J. Contam. Hydrol.* **2012**, *133*, 76–83. [CrossRef] [PubMed]
9. Hansen, J.P.; Skjeltorp, A.T. Fractal pore space and rock permeability implications. *Phys. Rev. B* **1988**, *38*, 2635–2638. [CrossRef]
10. Fu, J.; Yu, J.; Xu, L.; Niu, X.; Feng, S.; Wang, X.; You, L.; Li, T. New progress in exploration and development of tight oil in Ordos basin and main controlling factors of large-scale enrichment and exploitable capacity. *China Pet. Explor.* **2015**, *20*, 9–19.
11. Gao, H.; Li, H. Pore structure characterization, permeability evaluation and enhanced gas recovery techniques of tight gas sandstones. *J. Nat. Gas Sci. Eng.* **2016**, *28*, 536–547. [CrossRef]
12. Lai, J.; Wang, G.; Cao, J.; Xiao, C.; Wang, S.; Pang, X.; Dai, Q.; He, Z.; Fan, X.; Yang, L.; et al. Investigation of pore structure and petrophysical property in tight sandstones. *Mar. Pet. Geol.* **2018**, *91*, 179–189. [CrossRef]
13. Zou, C.; Zhu, R.; Bai, B.; Yang, Z.; Wu, S.; Li, X. First discovery of nanopore throat in oil and gas reservoir in China and its scientific value. *Acta Pet. Sin.* **2011**, *27*, 1857–1864.
14. Shi, X.; Pan, J.; Hou, Q.; Jin, Y.; Wang, Z.; Niu, Q.; Li, M. Micrometer-scale fractures in coal related to coal rank based on micro-CT scanning and fractal theory. *Fuel* **2018**, *212*, 162–172. [CrossRef]
15. Pan, S.; Zou, C.; Yang, Z.; Dong, D.; Wang, Y.; Wang, S.; Wu, S.; Huang, J.; Liu, Q.; Wang, D.; et al. Methods for shale gas play assessment: A comparison between Silurian Longmaxi shale and Mississippian Barnett shale. *J. Earth Sci.* **2015**, *26*, 285–294. [CrossRef]
16. Lai, J.; Wang, G.; Fan, Z.; Chen, J.; Wang, S.; Zhou, Z.; Fan, X. Insight into the pore structure of tight sandstones using NMR and HPMI measurements. *Energy Fuels* **2016**, *30*, 10200–10214. [CrossRef]
17. Ge, X.; Fan, Y.; Zhu, X.; Chen, Y.; Li, R. Determination of nuclear magnetic resonance T2 cutoff value based on multifractal theory—An application in sandstone with complex pore structure. *Geophysics* **2014**, *80*, D11–D21. [CrossRef]
18. Cao, Z.; Liu, G.; Zhan, H.; Li, C.; You, Y.; Yang, C.; Jiang, H. Pore structure characterization of Chang-7 tight sandstone using MICP combined with N2GA techniques and its geological control factors. *Sci. Rep.* **2016**, *6*, 36919. [CrossRef] [PubMed]
19. Shao, X.; Pang, X.; Li, H.; Zhang, X. Fractal analysis of pore network in tight gas sandstones using NMR method: A case study from the Ordos basin, China. *Energy Fuels* **2017**, *31*, 10358–10368. [CrossRef]
20. Lai, J.; Wang, G.; Fan, Z.; Zhou, Z.; Chen, J.; Wang, S. Fractal analysis of tight shaly sandstones using nuclear magnetic resonance measurements. *AAPG Bull.* **2018**, *102*, 175–193. [CrossRef]

21. Mao, Z.; Zhang, C.; Xiao, L. A NMR-based porosity calculation method for low porosity and low permeability gas reservoir. *Oil Geophys. Prospect.* **2010**, *45*, 105–109.

22. Xiao, L.; Mao, Z.; Wang, Z.; Jin, Y. Application of NMR logs in tight gas reservoirs for formation evaluation: A case study of Sichuan basin in China. *J. Petrol. Sci. Eng.* **2012**, *81*, 182–195. [CrossRef]

23. Feng, C.; Shi, Y.; Hao, J.; Wang, Z.; Mao, Z.; Li, G.; Jiang, Z. Nuclear magnetic resonance features of low-permeability reservoirs with complex wettability. *Pet. Explor. Dev.* **2017**, *44*, 274–279. [CrossRef]

24. Xiao, L.; Liu, D.; Wang, H.; Li, J.; Lu, J.; Zou, C. The applicability analysis of models for permeability prediction using mercury injection capillary pressure (MICP) data. *J. Petrol. Sci. Eng.* **2017**, *156*, 589–593. [CrossRef]

25. Liu, Y.; Wang, Y.; Tang, H.; Wang, F.; Chen, L. Application of capillary pressure curves and fractal theory to reservoir classification. *Lithol. Reserv.* **2014**, *26*, 89–92.

26. Feng, C.; Shi, Y.; Li, J.; Chang, L.; Li, G.; Mao, Z. A new empirical method for constructing capillary pressure curves from conventional logs in low-permeability sandstones. *J. Earth Sci.* **2017**, *28*, 516–522. [CrossRef]

27. Xiao, L.; Zou, C.; Mao, Z.; Jin, Y.; Zhu, J. A new technique for synthetizing capillary pressure (Pc) curves using NMR logs in tight gas sandstone reservoirs. *J. Petrol. Sci. Eng.* **2016**, *145*, 493–501. [CrossRef]

28. Mandelbrot, B.B. *The Fractal Geometry of Nature*; W. H. Freeman: New York, NY, USA, 1982.

29. García-Gutiérrez, C.; Martínez, F.S.J.; Caniego, J. A protocol for fractal studies on porosity of porous media: High quality soil porosity images. *J. Earth Sci.* **2017**, *28*, 888–896. [CrossRef]

30. Yang, Z.Y.; Pourghasemi, H.R.; Lee, Y.H. Fractal analysis of rainfall-induced landslide and debris flow spread distribution in the Chenyulan Creek Basin, Taiwan. *J. Earth Sci.* **2016**, *27*, 151–159. [CrossRef]

31. Mandelbrot, B.B. Multifractal measures, especially for the geophysicist. *Pure Appl. Geophys.* **1989**, *131*, 5–42. [CrossRef]

32. Muller, J. Characterization of the North Sea chalk by multifractal analysis. *J. Geophys. Res. Solid Earth.* **1994**, *99*, 7275–7280. [CrossRef]

33. Cai, J.; Yu, B.; Zou, M.; Mei, M. Fractal analysis of surface roughness of particles in porous media. *Chin. Phys. Lett.* **2010**, *27*, 157–160.

34. Zhang, X.; Wu, C.; Li, T. Comparison analysis of fractal characteristics for tight sandstones using different calculation methods. *J. Geophys. Eng.* **2017**, *14*, 120–131. [CrossRef]

35. Peng, J.; Han, H.; Xia, Q.; Li, B. Evaluation of the pore structure of tight sandstone reservoirs based on multifractal analysis: A case study from the Kepingtage Formation in the Shuntuoguole uplift, Tarim Basin, NW China. *J. Geophys. Eng.* **2018**, *15*, 1122–1136. [CrossRef]

36. Dai, J.; Li, J.; Luo, X.; Zhang, W.; Hu, G.; Ma, C.; Guo, J.; Ge, S. Stable carbon isotope compositions and source rock geochemistry of the giant gas accumulations in the Ordos Basin, China. *Org. Geochem.* **2005**, *36*, 1617–1635. [CrossRef]

37. Shi, Y.; Liang, X.; Mao, Z.; Guo, H. An identification method for diagenetic facies with well logs and its geological significance in low-permeability sandstones: A case study on Chang 8 reservoirs in the Jiyuan Region, Ordos Basin. *Acta Pet. Sin.* **2011**, *32*, 820–828.

38. Zou, C.; Zhu, R.; Wu, S.; Yang, Z.; Tao, S.; Yuan, X.; Hou, L.; Yang, H.; Xu, C.; Li, D.; et al. Types, characteristics, genesis and prospects of conventional and unconventional hydrocarbon accumulations: taking tight oil and tight gas in China as an instance. *Acta Pet. Sin.* **2012**, *33*, 173–187.

39. Jia, C.; Zou, C.; Li, J.; Li, D.; Zheng, M. Assessment criteria, main types, basic features and resource prospects of the tight oil in China. *Acta Pet. Sin.* **2012**, *33*, 343–350.

40. Cai, J.; Wei, W.; Hu, X.; Liu, R.; Wang, J. Fractal characterization of dynamic fracture network extension in porous media. *Fractals* **2017**, *25*, 1750023. [CrossRef]

41. Dathe, A.; Tarquis, A.M.; Perrier, E. Multifractal analysis of the pore- and solid-phases in binary two-dimensional images of natural porous structures. *Geoderma* **2006**, *134*, 318–326. [CrossRef]

42. Xia, Y.; Cai, J.; Wei, W.; Hu, X.; Wang, X.I.N.; Ge, X. A new method for calculating fractal dimensions of porous media based on pore size distribution. *Fractals* **2018**, *26*, 1850006. [CrossRef]

43. Halsey, T.C.; Jensen, M.H.; Kadanoff, L.P.; Procaccia, I.I.; Shraiman, B.I. Fractal measures and their singularities: the characterization of strange sets. *Nucl. Phys. B* **1987**, *2*, 501–511. [CrossRef]

44. Li, K.; Zeng, F.; Cai, J.; Sheng, G.; Xia, P.; Zhang, K. Fractal characteristics of pores in Taiyuan formation shale from Hedong coal field, China. *Fractals* **2018**, *26*, 1840006. [CrossRef]

45. Chakraborty, B.; Haris, K.; Latha, G.; Maslov, N.; Menezes, A. Multifractal approach for seafloor characterization. *IEEE Geosci. Remote Sens. Lett.* **2014**, *11*, 54–58. [CrossRef]

46. Paz Ferreiro, J.; Vidal Vázquez, E. Multifractal analysis of Hg pore size distributions in soils with contrasting structural stability. *Geoderma* **2010**, *160*, 64–73. [CrossRef]

47. Nelson, P.H. Pore-throat sizes in sandstones, tight sandstones, and shales. *AAPG Bull.* **2009**, *93*, 329–340. [CrossRef]

48. Li, C.; Zhou, C.; Li, X.; Hu, F.; Zhang, L.; Wang, W. A novel model for assessing the pore structure of tight sands and its application. *Appl. Geophys.* **2010**, *7*, 283–291. [CrossRef]

49. Wang, L.; Zhao, N.; Sima, L.; Meng, F.; Guo, Y. Pore structure characterization of the tight reservoir: Systematic integration of mercury injection and nuclear magnetic resonance. *Energy Fuels* **2018**, *32*, 7471–7484. [CrossRef]

50. Zhao, P.; Cai, J.; Huang, Z.; Ostadhassan, M.; Ran, F. Estimating permeability of shale gas reservoirs from porosity and rock compositions. *Geophysics* **2018**, *83*, MR283–MR294. [CrossRef]

51. Zeng, D.; Li, S. Types and characteristics of low permeability sandstone reservoirs in China. *Acta Pet. Sin.* **1994**, *15*, 38–46.

52. Tang, W.; Tang, R. Fractal dimensions of mercury-ejection capillary pressure curves in Donghe-1 Oilfield. *Acta Pet. Sin.* **2005**, *26*, 90–93.

53. Xiao, D.; Jiang, S.; Thul, D.; Lu, S.; Zhang, L.; Li, B. Impacts of clay on pore structure, storage and percolation of tight sandstones from the Songliao Basin, China: Implications for genetic classification of tight sandstone reservoirs. *Fuel* **2018**, *211*, 390–404. [CrossRef]

54. Zhao, P.; Sun, Z.; Luo, X.; Wang, Z.; Mao, Z.; Wu, Y.; Xia, P. Study on the response mechanisms of nuclear magnetic resonance (NMR) log in tight oil reservoirs. *Chin. J. Geophys.* **2016**, *59*, 1927–1937.

55. Tang, R.; Ceng, Y. Study on several factors affecting the efficiency of mercury removal from rock. *Exp. Pet. Geol.* **1994**, *1*, 84–93.

Article

A Simple Fractal-Based Model for Soil-Water Characteristic Curves Incorporating Effects of Initial Void Ratios

Gaoliang Tao [1], Yin Chen [1], Lingwei Kong [2], Henglin Xiao [1], Qingsheng Chen [1,*] and Yuxuan Xia [3]

[1] Hubei Provincial Ecological Road Engineering Technology Research Center, Hubei University of Technology, Wuhan 430068, China; tgl1979@126.com (G.T.); agchen19930922@163.com (Y.C.); xiao-henglin@163.com (H.X.)

[2] State Key Laboratory of Geomechanics and Geotechnical Engineering, Institute of Rock and Soil Mechanics, Chinese Academy of Sciences, Wuhan 430071, China; lwkong@whrsm.ac.cn

[3] Hubei Subsurface Multi-Scale Imaging Key Laboratory, Institute of Geophysics and Geomatics, China University of Geosciences, Wuhan 430074, China; xiayx@cug.edu.cn

* Correspondence: chqsh2006@163.com

Received: 25 April 2018; Accepted: 29 May 2018; Published: 1 June 2018

Abstract: In this paper, a simple and efficient fractal-based approach is presented for capturing the effects of initial void ratio on the soil-water characteristic curve (SWCC) in a deformable unsaturated soil. In terms of testing results, the SWCCs (expressed by gravimetric water content) of the unsaturated soils at different initial void ratios were found to be mainly controlled by the air-entry value (ψ_a), while the fractal dimension (D) could be assumed to be constant. As a result, in contrast to the complexity of existing models, a simple and efficient model with only two parameters (i.e., D and ψ_a) was established for predicting the SWCC considering the effects of initial void ratio. The procedure for determining the model parameters with clear physical meaning were then elaborated. The applicability and accuracy of the proposed model were well demonstrated by comparing its predictions with four sets of independent experimental data from the tests conducted in current work, as well as the literature on a wide range of soils, including Wuhan Clay, Hefei and Guangxi expansive soil, Saskatchewan silt, and loess. Good agreements were obtained between the experimental data and the model predictions in all of the cases considered.

Keywords: soil-water characteristic curve; initial void ratio; air-entry value; fractal dimension; fractal model

1. Introduction

The soil-water characteristic curve (SWCC) of a soil reflects the relationship between volumetric water content and matric suction. It is a useful tool for indicating the hydraulic properties of the soil and has various applications in the field of unsaturated soil mechanics, such as strength theory, percolation theory, consolidation theory, and constitutive theory [1–5]. The SWCC is also commonly employed in estimating the shear strength, stress–strain relationships, and permeability of unsaturated soils [6–8].

Over the past decades, a large number of studies have been carried out on the SWCCs of unsaturated soils [6–10]. It has been recognized that SWCCs are affected by various factors, such as pore-size distribution (PSD), particle-size distribution, and dry density [11,12]. One specific factor that affects the SWCC is the porosity of the soil, which can change considerably for the soils with variation of stress and suction states, as well as the stress and suction history of the soil [13]. Zhou et al. [14] reported that the samples of a given soil with different void ratios could be regarded as entirely different soils. As also stated by Assouline [15], a change of soil porosity can result in a significant change of SWCC, and such a change in soil porosity is a common feature of natural soils.

In recent decades, the issue of the effect of soil porosity on the hydraulic properties of a soil has received great attentions of various researchers [16–24]. In the meanwhile, various models have been proposed to account for the effect of initial void ratio on SWCCs. For example, Van Genuchten [6] proposed an equation to empirically describe the SWCC data, in which there are three parameters and the parameter α is the main influence factor of air-entry value. Gallipoli et al. [17] modified the equation by assuming the parameter α to be related to the void ratio by a power law. Tarantino [25] presented a void ratio-dependent SWCC model. For high suction stage, it was assumed that the initial void ratio has insignificant influence on the SWCC data, with a result that the SWCC could be represented by a single equation. This model is very similar to the model proposed by Gallipoli et al. [17], but it incorporates one parameter less than Gallipoli's model. Masin [23] proposed a hydraulic model that is capable of capturing the dependence of the degree of saturation on the void ratio and suction using the effective stress principle. In this model, three material parameters were required. Most recently, Zhou et al. [14] proposed a method to take account of the effects of initial void ratio on SWCC through introducing one more additional parameter in the existing empirical SWCC equations. Gallipoli et al. [26] formulated an SWCC model considering the effects of hysteresis and void ratio of the soils with various degrees of saturation. However, there are seven parameters in this model. Although substantial contributions have been made in the above models for predicting the water retention behavior of unsaturated soils, the application of these models are still largely restricted due to their complexity. As a result, it is of immense necessity to propose a simple and efficient approach with fewer parameters for estimating the SWCC of unsaturated soils, so that the engineers can readily apply it in practice.

Recently, fractal theory has been increasingly recognized as a useful tool to analyze the physico-geometrical properties (e.g., particle-size distribution, pore volume, and pore surface) of porous media [27–31]. It was pointed out that the hydraulic characteristics of porous media are closely related to its pore characteristics [32], thus, fractal theory is also suitable to describe the hydraulic characteristics of porous media. Shi et al. [33] developed a fractal model for describing the spontaneous imbibition of wetting phase into the porous media. Most recently, some researchers have also devoted to fractal-based SWCC model [34–36]. It was reported that this approach is capable of combining the empirical models with the parameters with clear physical meaning, and it can also be readily applied in engineering practice [37]. Nevertheless, most of existing fractal-based SWCC models could not account for the effects of void ratios of soils. Most recently, Khalili et al. [38] proposed a fractal-based model considering volume change, and the fractal dimension of PSD in the model was taken to be void ratio dependent, however, this model has seven parameters, which are not easily determined. Meanwhile, it has been verified for clayey silty sand only.

To address the above issues, this paper presents a simple and efficient SWCC model incorporating the effects of initial void ratios by employing the fractal theory. On the basis of experimental results, the SWCCs of the soils at different initial void ratios were found to be mainly controlled by air-entry value, while fractal dimension can be seen as a constant. Combined with fractal theory, a new model for predicting the air-entry value was proposed. Thereafter, a simple and efficient SWCC model capable of predicting the SWCCs at deformed state considering the effects of initial void ratio was established. In contrast to the complexity of existing models, only two parameters are needed in the proposed model. The applicability and accuracy of this model were demonstrated by comparing its predictions with four sets of independent experimental data from the tests conducted in current work, as well as the literatures. Good agreement was obtained between the experimental data and the model predictions in all of the cases considered.

2. Proposed Model

2.1. Fractal Description of a Soil

Assuming continuous pore-size distribution (PSD) over a range of pore sizes (r) between r_{min} and r_{max}, in which r_{min} and r_{max} are the smallest and largest pores, respectively, the probability density function of pore size r is written as [39]

$$f(r) = cr^{-1-D} \tag{1}$$

where c is a constant, D is the fractal dimension, and r is the effective pore size of the connected pore channel. It is assumed that the minimum pore size r_{min} is near zero. Then, the total volume $V (\leq r)$ of pores having size less than or equal to r can be expressed as

$$V(\leq r) = \int_0^r cr^{-1-D} k_V r^3 dr = \frac{ck_V}{3-D} r^{3-D} \tag{2}$$

where k_V is a pore volume shape-related constant corresponding to the volume of the pores. Assuming that pores having a size less than or equal to r are fully filled with water, then the gravimetric water content in the pores of soil particles weighing 1 g is

$$w = \rho_w V(\leq r) = \frac{c\rho_w k_V}{3-D} r^{3-D} \tag{3}$$

where ρ_w is the mass density of water.

The soil sample is regarded as fully saturated soil when the largest pores with r_{max} are filled with water. Then, substituting r with r_{max} in Equation (3) yields the following expression

$$w_s = \frac{c\rho_w k_V}{3-D} r_{max}^{3-D} \tag{4}$$

where w_s is gravimetric water content of the soil in saturated condition.

2.2. Fractal-Based Model for Variation of Soil-Water Characteristic Curve (SWCC) with Initial Void Ratio

According to the Young-Laplace equation, the relationship between matrix suction ψ and effective pore size r can be described as

$$\psi = 2T_s \cos \alpha / r \tag{5}$$

where T_s is the surface tension and α is the contact angle. In the constant temperature condition, $2T_s \cos \alpha$ can be assumed as a constant. The matrix suction ψ corresponding to maximum pore size can be approximately regarded as the air-entry value ψ_a, which can be captured by substituting r with r_{max} in Equation (5)

$$\psi_a = 2T_s \cos \alpha / r_{max} \tag{6}$$

Then, substituting Equation (5) into Equation (3), and Equation (6) into Equation (4), respectively, yields the following expressions

$$w = \frac{c\rho_w k_V}{3-D} \left(\frac{2T_s \cos \alpha}{\psi}\right)^{3-D} \tag{7}$$

$$w_s = \frac{c\rho_w k_V}{3-D} \left(\frac{2T_s \cos \alpha}{\psi_a}\right)^{3-D} \tag{8}$$

Dividing Equation (8) by Equation (7) gives

$$\frac{w}{w_s} = \left(\frac{\psi_a}{\psi}\right)^{3-D} \tag{9}$$

Note that Equations (7) and (9) are only valid in the range of $\psi \geq \psi_a$. If matrix suction ψ is less than ψ_a, the soil sample is assumed to be fully saturated. Then, the fractal model for soil-water characteristic curve is written as

$$
\begin{cases}
\frac{w}{w_s} = \left(\frac{\psi_a}{\psi}\right)^{3-D} & \psi \geq \psi_a \\
w = w_s & \psi < \psi_a
\end{cases}
\tag{10}
$$

In the saturated condition, the water content of soil can be determined by

$$
w_s = \frac{e}{G_s}
\tag{11}
$$

where G_s is the relative density and e refers to the initial void ratio.

Then, the variation of SWCC with void ratio can be captured by substituting Equation (11) into Equation (10)

$$
\begin{cases}
w = \frac{e}{G_s}\left(\frac{\psi_a}{\psi}\right)^{3-D} & \psi \geq \psi_a \\
w = \frac{e}{G_s} & \psi < \psi_a
\end{cases}
\tag{12}
$$

2.3. Determination of Model Parameters

The main objective of this study is to predict the SWCCs at the deformed state considering the effects of initial void ratio. It should be noted that the SWCC at the reference state is corresponding to the experimental soil sample with the maximum initial void ratio e_0, while the SWCCs at the deformed state refer to the experimental soil samples with the other arbitrary initial void ratio e_1 ($e_1 < e_0$). As observed in Equation (12), the prediction of SWCCs at arbitrary initial void ratio is mainly governed by only two parameters: (i) fractal dimension D and (ii) air-entry value ψ_a at the deformed state. The procedure for determining the model parameters (i.e., D and ψ_a) are now elaborated as follows.

2.3.1. Fractal Dimension at Reference State

By taking the logarithm of both sides of Equation (7), it could be rewritten as

$$
\begin{aligned}
\ln w &= \ln \frac{c\rho_w k_v}{3-D}\left(\frac{2T_s \cos\alpha}{\psi}\right)^{3-D} \\
&= \ln \frac{c\rho_w k_v}{3-D} + (3-D)\ln \frac{2T_s \cos\alpha}{\psi} \\
&= (3-D)[\ln 2T_s \cos\alpha - \ln\psi] + \ln\frac{c\rho_w k_v}{3-D} \\
&= (3-D)(-\ln\psi) + \ln\left[\frac{c\rho_w k_v}{3-D}(2T_s \cos\alpha)^{3-D}\right]
\end{aligned}
$$

Then, it is easy to obtain Equation (13)

$$
\ln w \propto (3-D)(-\ln\psi)
\tag{13}
$$

By plotting the $\ln w$ against $(-\ln\psi)$, $(3-D)$ can be evaluated from the gradient k of the graph, then the fractal dimension can be determined as $D = 3 - k$.

2.3.2. Fractal Dimension at Deformed State

Based on the experimental data, it was found by Tao et al. [40] that SWCC expressed by gravimetric water content at different initial void ratios presents a type of "broom shape" distribution. More precisely, SWCCs at different initial void ratios are almost consistent when the high matrix suction is greater than air-entry value. As shown in Figure 1, SWCCs corresponding to initial void ratio e_0, e_1, and e_2 (i.e., a-b-c-d, e-b-c-d, and f-c-d) almost overlap at the tail. It should be noted that the SWCC is expressed by gravimetric water content herein. As a result, the fractal dimension D_1 for e_1 at deformed state is approximatively equal to the fractal dimension D_0 for e_0 at reference state. That is, the fractal dimension D of saturated soil at deformed state can be assumed as a constant (i.e., $D_1 = D_0$).

It should be highlighted that similar ideas have been presented by Bird et al. [34] and Russell and Buzzi [41].

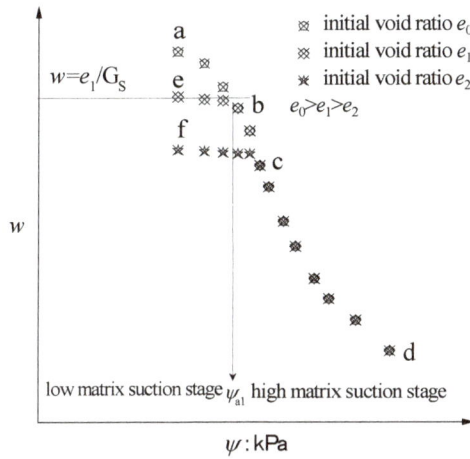

Figure 1. Schematic sketch of soil-water characteristic curves (SWCCs) in terms of gravimetric water content of unsaturated soils with different initial void ratios.

2.3.3. Air-Entry Value at Reference State

If the SWCC at reference state is measured, the corresponding fractal dimension D_0 can be calculated by following the procedure described above, and the air-entry value ψ_{a0} can be determined by best fitting Equation (12) to the experimental SWCCs. Then, the SWCC at reference state can be expressed as

$$\begin{cases} w = \frac{e_0}{G_s} \left(\frac{\psi_{a0}}{\psi}\right)^{3-D_0} & \psi \geq \psi_{a0} \\ w = \frac{e_0}{G_s} & \psi < \psi_{a0} \end{cases} \tag{14}$$

2.3.4. Air-Entry Value at Deformed State

Drawing a horizontal line $w = e_1/G_s$, the horizontal line would have an intersection with the SWCC at reference state. The abscissa of this intersection can be approximately regarded as ψ_{a1} corresponding to e_1, as shown in Figure 1. By substituting $w = e_1/G_s$ into the first formula in Equation (14), the following expression is obtained

$$\frac{e_1}{e_0} = \left(\frac{\psi_{a0}}{\psi}\right)^{3-D_0} \tag{15}$$

Equation (15) can be then simplified to

$$\psi = \frac{\psi_{a0}}{\left(\frac{e_1}{e_0}\right)^{1/(3-D_0)}} \tag{16}$$

The matrix suction expressed by Equation (16) is approximately considered as air-entry value ψ_{a1} for e_1.

3. Model Validation

To demonstrate the performance of the proposed fractal model (Equation (12)) for SWCCs of unsaturated soils considering effects of the initial void ratio, a series of simulations were carried out using a set of independent laboratory tests conducted in the current work, as well as by others in the literatures [42–44] on a wide range of soils, including Wuhan Clay, Hefei and Guangxi expansive soil, Saskatchewan silt, and loess. The comparisons and discussions are presented as follows.

3.1. Wuhan Clay

Pressure plate tests were conducted to determine the SWCCs for samples of clay collected from the bottom of a foundation ditch in Wuhan, China. The basic physical property index of soil is shown in Table 1. The experiment was completed at the Institute of Rock and Soil Mechanics, Wuhan Institute of Chinese Academy of Sciences. The pressure plate instrument used in the experiment, which consists of pressure cell, higher air-entry value (HAE) ceramic disc, pressure gage, and nitrogen source, is shown in Figure 2.

Table 1. Basic physical property index of Wuhan clay.

Natural Density (g/cm^3)	Relative Density	Natural Water Content (%)	Liquid Limit (%)	Plastic Limit (%)
2.03	2.75	21.9	38.9	20.4

Figure 2. Pressure plate apparatus instrument.

In the tests, seven samples were compacted using hydraulic jack to form different initial dry densities, ranging from 1.30 to 1.71 g/cm^3. The specific test procedure is as follows: (1) The sample with different dry densities, together with the HAE ceramic disc, was saturated. (2) The specimen, together with the stainless steel cutting rings, was placed on the HAE ceramic disc in the pressure cell. (3) The applied air pressure was imposed on the specimen when the pressure cell was sealed. (4) The water drained from the specimens was recorded during the whole process of the test. It was assumed to reach the equilibrium state at the current suction level when the water drainage of specimen was constant, then the next suction level would be imposed. (5) At the end of each suction level step, the drainage valve was closed and then the applied air pressure was released. Meanwhile, the weights of specimens needed to be measured. (6) The above-mentioned procedure was repeated until the whole test was completed.

The initial void ratio of the specimen at the loosest state (e_0 = 1.115) was regarded as the reference state. Following the procedure presented previously, the fractal dimension for the Wuhan Clay at reference state was determined as D_0 = 2.869, with a fitting correlation coefficient of up to 0.99, while the air-entry value at reference state was estimated to be $\psi_{a0} = 1.66$ kPa by fitting Equation (12) to the experimental SWCCs, as shown in Figures 3 and 4, respectively.

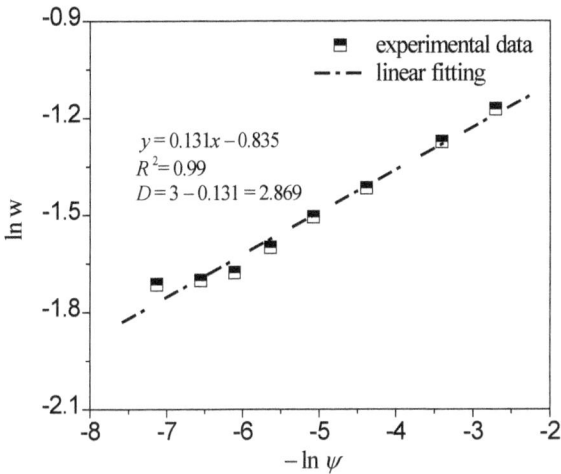

Figure 3. Determination of the values of fractal dimension at reference state through plotting experimental data of ln$w$ against $(-\ln\psi)$ for Wuhan Clay with $e_0 = 1.115$.

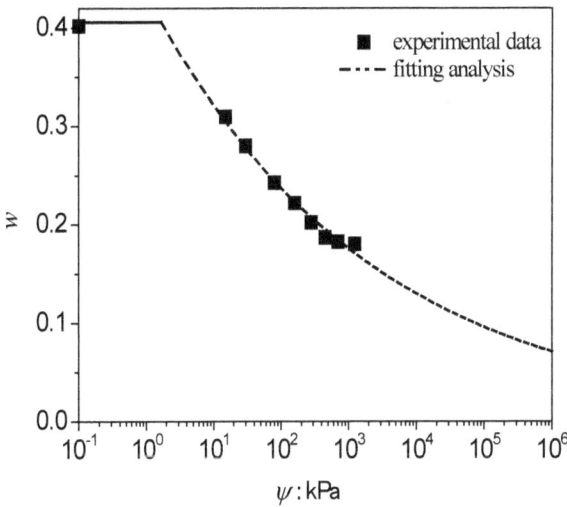

Figure 4. Determination of air-entry values at reference state through fitting Equation (12) to the experimental SWCCs for Wuhan Clay with $e_0 = 1.115$.

As a result, the air-entry values ψ_a of Wuhan Clay at deformed state were determined using Equation (16), as shown in Table 2.

Table 2. Predictions of fractal dimensions and air-entry values at deformed state.

Soil Type	Initial Void Ratio	Fractal Dimension	Air-Entry Value/kPa	Soil Type	Initial Void Ratio	Fractal Dimension	Air-Entry Value/kPa
Wuhan clay	1.037	2.869	2.89	Saskatchewan silt/T1	0.54	2.640	15.50
	0.964		5.04		0.528		16.50
	0.897		8.74		0.501		19.08
	0.833		15.37		0.483		21.12
	0.719		47.28		0.466		23.33
	0.613		159.74				
Hefei expansive soil	0.838	2.514	60.58	Saskatchewan silt/T2	0.513	2.604	18.77
	0.766		72.89		0.490		21.08
					0.474		22.92
					0.454		25.56
					0.426		30.02
Guangxi expansive soil	0.824	2.589	33.28	Xian Loess (5 °C)	0.88	2.825	3.73
	0.753		41.44		0.75		9.29
					0.72		11.73

Figure 5 demonstrates that the calculated SWCCs at deformed state using the proposed model (Equation (12)) agree well with the experimental data of Wuhan Clay at various void ratios (i.e., $e_0 = 1.037$, 0.964, 0.897, 0.833, 0.719, and 0.613, respectively).

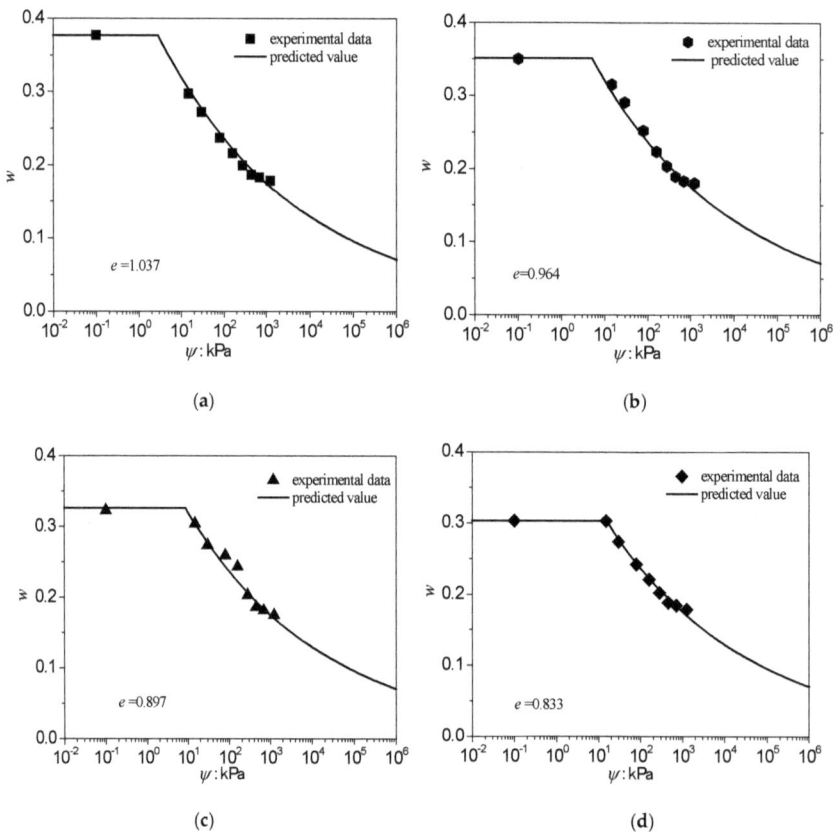

(a)

(b)

(c)

(d)

Figure 5. *Cont.*

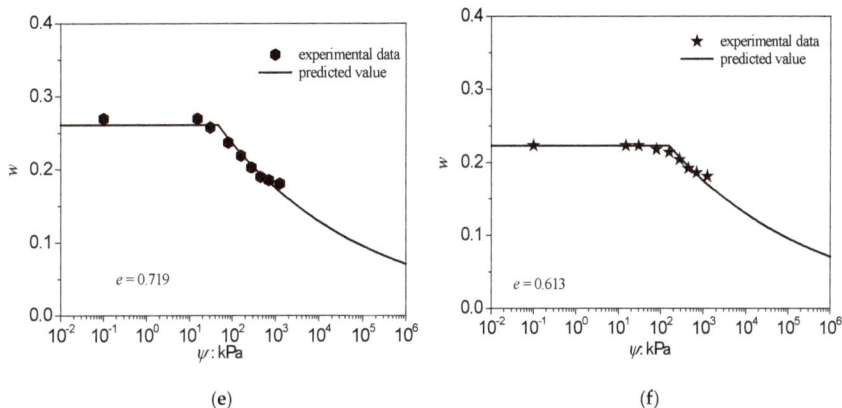

Figure 5. Measured and predicted SWCCs for Wuhan clay at deformed state at (a) $e = 1.037$; (b) $e = 0.964$; (c) $e = 0.897$; (d) $e = 0.833$; (e) $e = 0.719$; (f) $e = 0.613$.

3.2. Hefei and Guangxi Expansive Soils

Miao et al. [42] performed a series of pressure plate tests on Hefei and Guangxi expansive soil, where the specimens were compacted by the static compaction in steel rings at three dry densities, 1.42 g/cm^3, 1.48 g/cm^3, 1.54 g/cm^3, respectively. The physical properties of Hefei and Guangxi expansive soils are summarized in Tables 3 and 4, respectively. The void ratios of the specimens at the loosest state ($e_0 = 0.915$ for Hefei expansive soil, $e_0 = 0.901$ for Guangxi expansive soil) were regarded as the initial void ratios at reference state.

Similarly, the parameters at reference state for Hefei expansive soil ($D_0 = 2.514$, $\psi_{a0} = 50.56$ kPa) were obtained as shown in Figures 6 and 7, respectively, while the corresponding air-entry values ψ_a at deformed state were determined using Equation (16), as shown in Table 2. As can be seen in Figure 6, the fitting correlation coefficients for fractal dimension are up to 0.98, indicating that the SWCCs of Hefei specimens have obvious fractal features.

Table 3. Physical properties of Hefei expansive soils (data after Miao et al. [42]).

Relative Density	Liquid Limit (%)	Plastic Limit (%)	Plasticity Index (%)
2.72	58.6	26.4	32.2

Table 4. Physical properties of Guangxi expansive soils (data after Miao et al. [42]).

Relative Density	Liquid Limit (%)	Plastic Limit (%)	Plasticity Index (%)
2.70	61.4	30.3	31.1

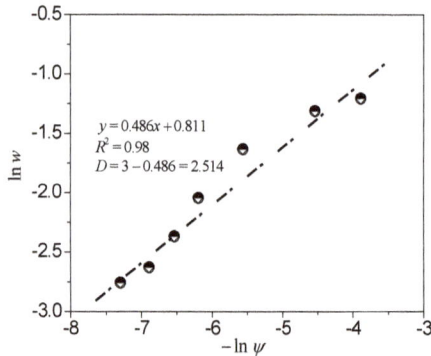

Figure 6. Determination of the values of fractal dimension at reference state through plotting experimental data of ln$w$ against ($-$lnψ) for Hefei expansive soil with $e_0 = 0.915$.

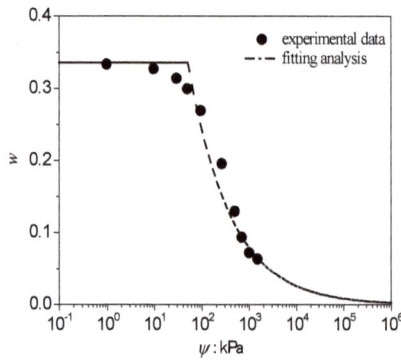

Figure 7. Determination of air-entry values at reference state through fitting Equation (12) to the experimental SWCCs for Hefei expansive soil with $e_0 = 0.915$.

Figure 8 indicates that the SWCCs computed by the current model compared well with the laboratory data of the unsaturated Hefei expansive soils at all initial void ratios.

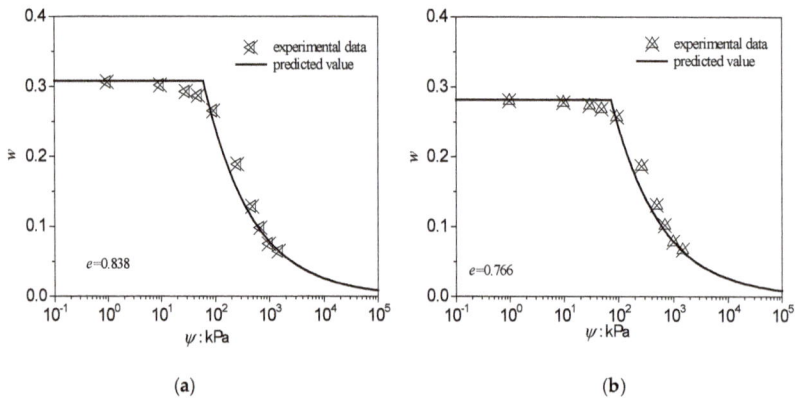

(a)

(b)

Figure 8. Measured and predicted SWCCs for Hefei expansive soils at deformed state at (a) $e = 0.838$; (b) $e = 0.766$ (data after Miao et al. [42]).

The parameters at reference state for Guangxi expansive soils (D_0 = 2.589, ψ_{a0} = 26.78 kPa) were obtained as shown in Figures 9 and 10, respectively, while the corresponding air-entry values ψ_a at deformed state were determined using Equation (16), as shown in Table 2. As can be seen in Figure 9, the fitting correlation coefficients for fractal dimension are up to 0.97, indicating that the SWCCs of Guangxi specimens have obvious fractal features.

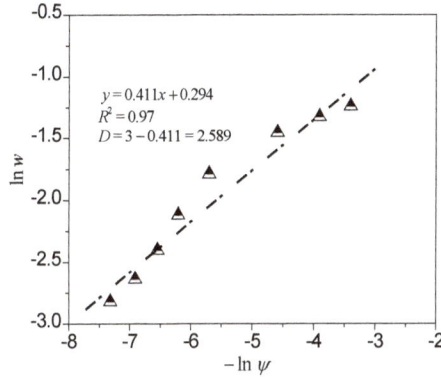

Figure 9. Determination of the values of fractal dimension at reference state through plotting experimental data of ln$w$ against ($-$lnψ) for Guangxi expansive soil with e_0 = 0.901.

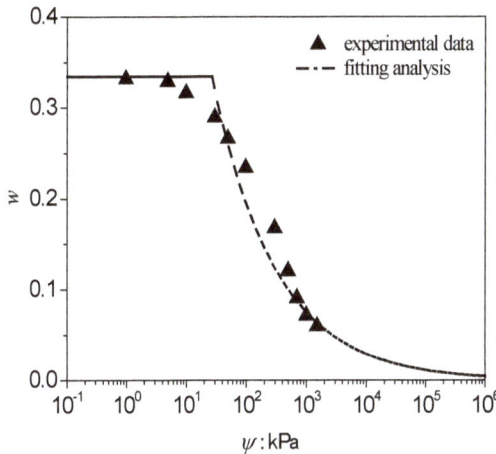

Figure 10. Determination of air-entry values at reference state through fitting Equation (12) to the experimental SWCCs for Guangxi expansive soil with e_0 = 0.901.

Figure 11 indicates that the SWCCs computed by the current model compared well with the laboratory data of the unsaturated Guangxi expansive soils at all initial void ratios.

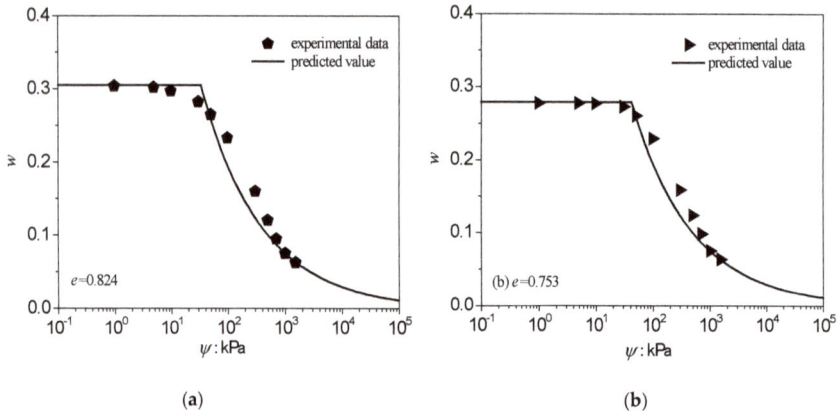

(a)   (b)

Figure 11. Measured and predicted SWCCs for Guangxi expansive soils at deformed state at (a) $e = 0.824$; (b) $e = 0.753$ (data after Miao et al. [42]).

3.3. Saskatchewan Silt

The change of the SWCCs with the variation of initial void ratios for pressure plate cells tests performed by Huang [43] on the Saskatchewan silt was simulated in this section for further model validation. The physical properties of Saskatchewan silt are shown in Table 5. The specimens were compressed in stainless sample rings by conventional odometer under the same consolidation pressures. Two groups of tests were conducted. In the first group of tests (T1), the initial void ratios of the specimens were 0.692, 0.540, 0.528, 0.501, 0.483, and 0.466, respectively, while the initial void ratios of specimens in the second group of tests (T2) were 0.525, 0.513, 0.490, 0.474, 0.454, and 0.426, respectively.

The void ratio of the specimen at the loosest state $e_0 = 0.692$ for T1 was regarded as the initial void ratio at reference state. The calibrated SWCC parameters of T1 at reference state for the first tests ($D_0 = 2.640$, $\psi_{a0} = 7.78$ kPa) were obtained as shown in Figures 12 and 13, respectively. The corresponding air-entry values at deformed state are presented in Table 2.

Table 5. Physical properties of Saskatchewan silt (data after Huang [43]).

Relative Density	Natural Water Content (%)	Liquid Limit (%)	Plastic Limit (%)	Plastic Index (%)
2.68	0.86	22.2	16.6	5.6

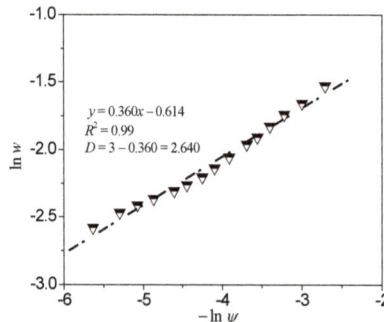

Figure 12. Determination of the values of fractal dimension at reference state through plotting experimental data of $\ln w$ against $(-\ln\psi)$ for Saskatchewan silt (T1) with $e_0 = 0.692$.

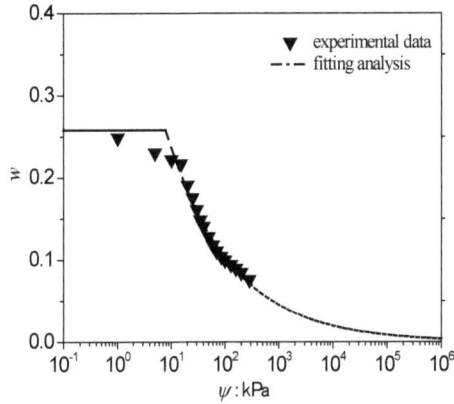

Figure 13. Determination of air-entry values at reference state through fitting Equation (12) to the experimental SWCCs for Saskatchewan silt (T1) with $e_0 = 0.692$.

Figure 14 presents the comparison of the experimental SWCCs and their predictions, demonstrating that the SWCCs of unsaturated Saskatchewan silt (T1) with the variation of initial void ratios were captured well.

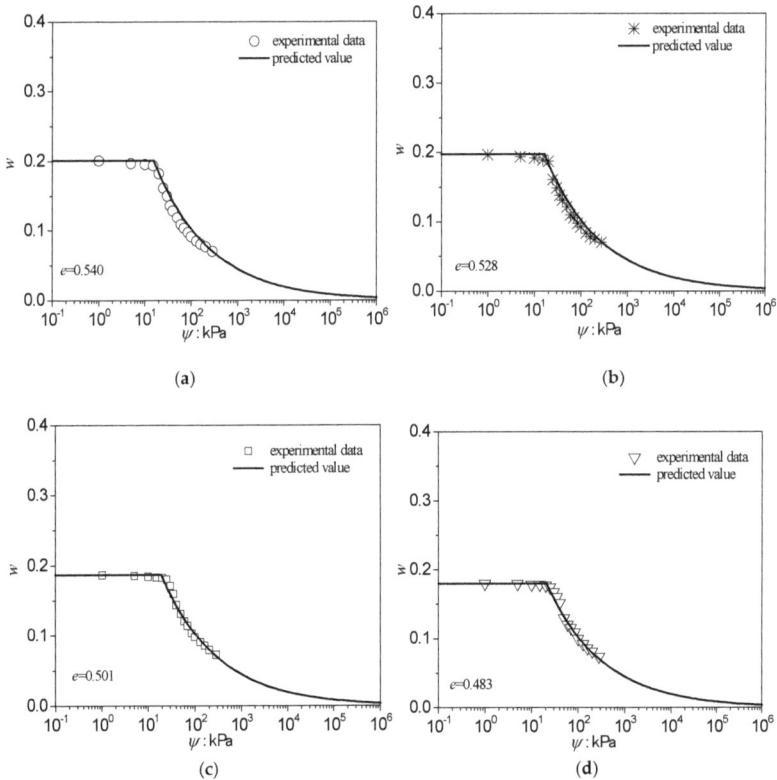

(a)

(b)

(c)

(d)

Figure 14. *Cont.*

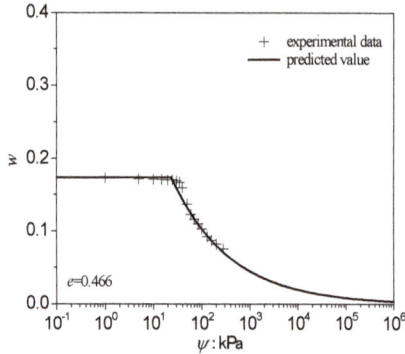

(e)

Figure 14. Measured and predicted SWCCs for Saskatchewan silt specimens (T1) at deformed state at (**a**) $e = 0.540$; (**b**) $e = 0.528$; (**c**) $e = 0.501$; (**d**) $e = 0.483$; (**e**) $e = 0.466$ (data after Huang [43]).

The void ratio of the specimen at the loosest state $e_0 = 0.525$ for T2 was regarded as the initial void ratio at reference state. The calibrated SWCC parameters of Saskatchewan silt of T2 at reference state for the second tests ($D_0 = 2.604$, $\psi_{a0} = 17.71$ kPa) were obtained as shown in Figures 15 and 16, respectively. The corresponding air-entry values at deformed state are presented in Table 2.

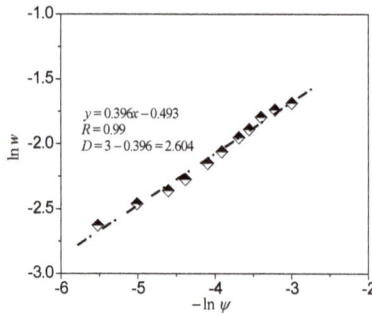

Figure 15. Determination of the values of fractal dimension at reference state through plotting experimental data of $\ln w$ against $(-\ln \psi)$ for Saskatchewan silt (T2) with $e_0 = 0.525$.

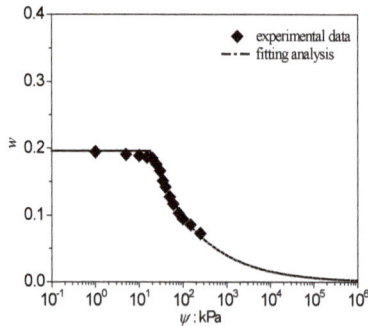

Figure 16. Determination of air-entry values at reference state through fitting Equation (12) to the experimental SWCCs for Saskatchewan silt (T2) with $e_0 = 0.525$.

Figure 17 presents the comparison of the experimental SWCCs and their predictions, demonstrating that the SWCCs of unsaturated Saskatchewan silt (T2) with the variation of initial void ratios were captured well.

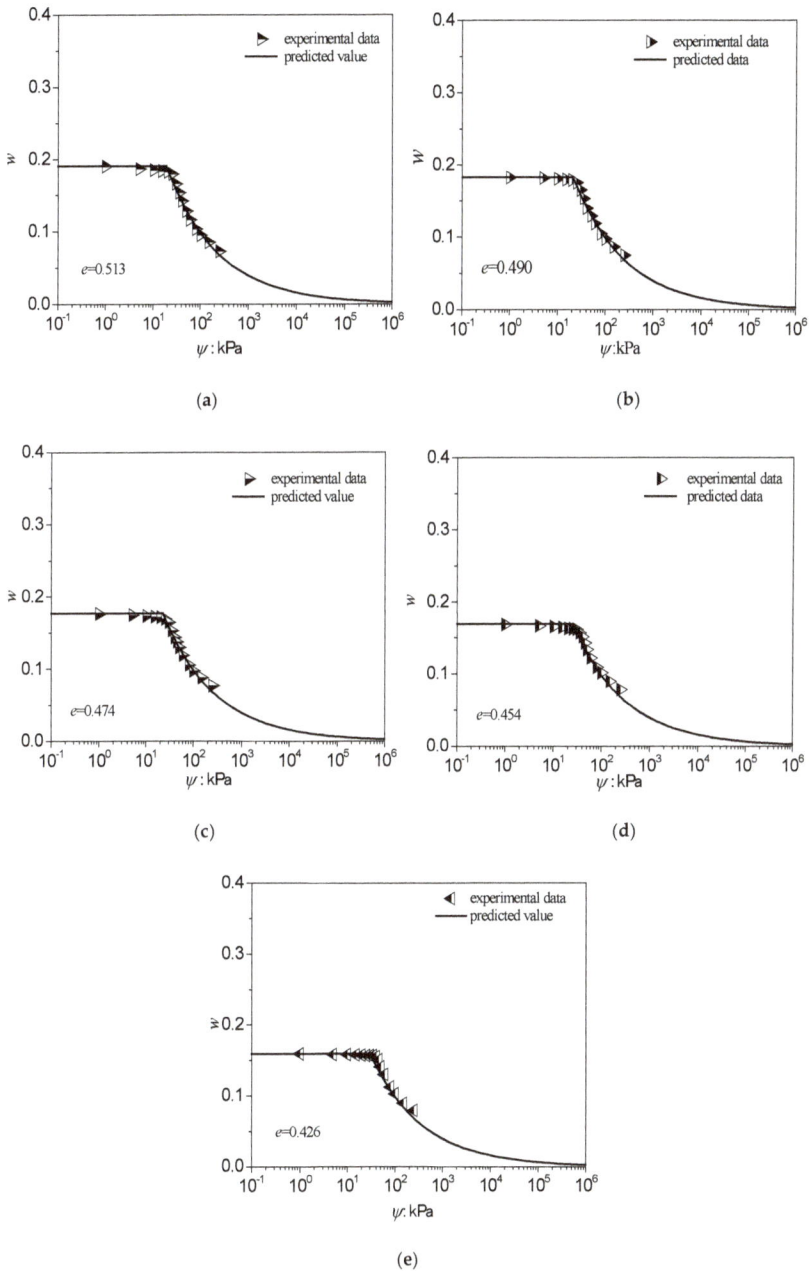

Figure 17. Measured and predicted SWCCs for Saskatchewan silt specimens (T2) at deformed state at (a) e = 0.513; (b) e = 0.490; (c) e = 0.474; (d) e = 0.454; (e) e = 0.426 (data after Huang [43]).

3.4. Loess

Wang et al. [44] conducted a series of laboratory tests to investigate the effect of dry density and temperature on the SWCC of Xi'an loess. The physical properties of Saskatchewan silt are shown in Table 6. The compacted samples with different densities of 1.2 g/cm^3, 1.4 g/cm^3, 1.5 g/cm^3, and 1.6 g/cm^3, respectively, were prepared and tested by the high-speed centrifuge method. The void ratio of a specimen at the loosest state (e_0 = 1.23) was regarded as the initial void ratio at reference state. The experimental temperatures were controlled at 5 °C, 15 °C, 25 °C, and 35 °C, respectively. For model validation, experimental data at 5 °C was employed. The parameters of the loess at reference state (D_0 = 2.825, ψ_{a0} = 0.55 kPa) were obtained as shown in Figures 18 and 19, respectively, while the corresponding air-entry values at deformed state with various initial void ratios (i.e., e_0 = 0.88, 0.75, and 0.72, respectively) are shown in Table 2. It is shown in Figure 18 that the fitting correlation coefficient is up to 1.00, which highlights that the SWCCs of loess specimens have significant fractal features.

Table 6. Physical properties of Xi'an loess (data after Wang et al. [44]).

Depth	Liquid Limit (%)	Plastic Limit (%)	Plastic Index (%)
2.68	30.7	18.4	12.3

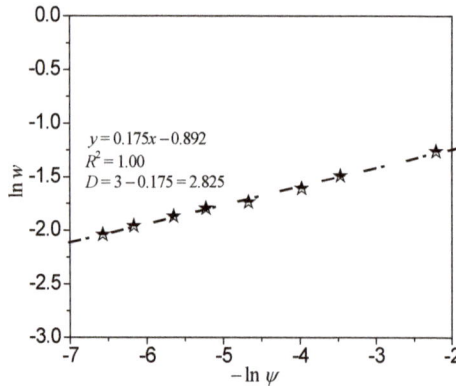

Figure 18. Determination of the values of fractal dimension at reference state through plotting experimental data of ln*w* against (−lnψ) for loess with e_0 = 1.230.

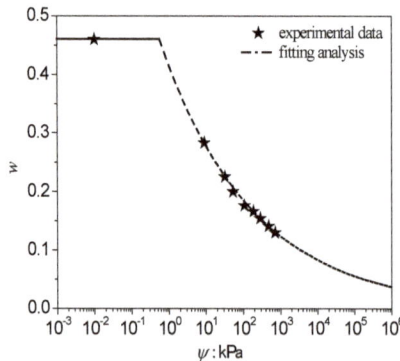

Figure 19. Determination of air-entry values at reference state through fitting Equation (12) to the experimental SWCCs for loess with e_0 = 1.230.

The model prediction in terms of variation of initial void ratio is shown in Figure 20. A very good agreement is obtained with the experimental data, which demonstrates the capability of the model in capturing the change of SWCCs with void ratio for unsaturated soils.

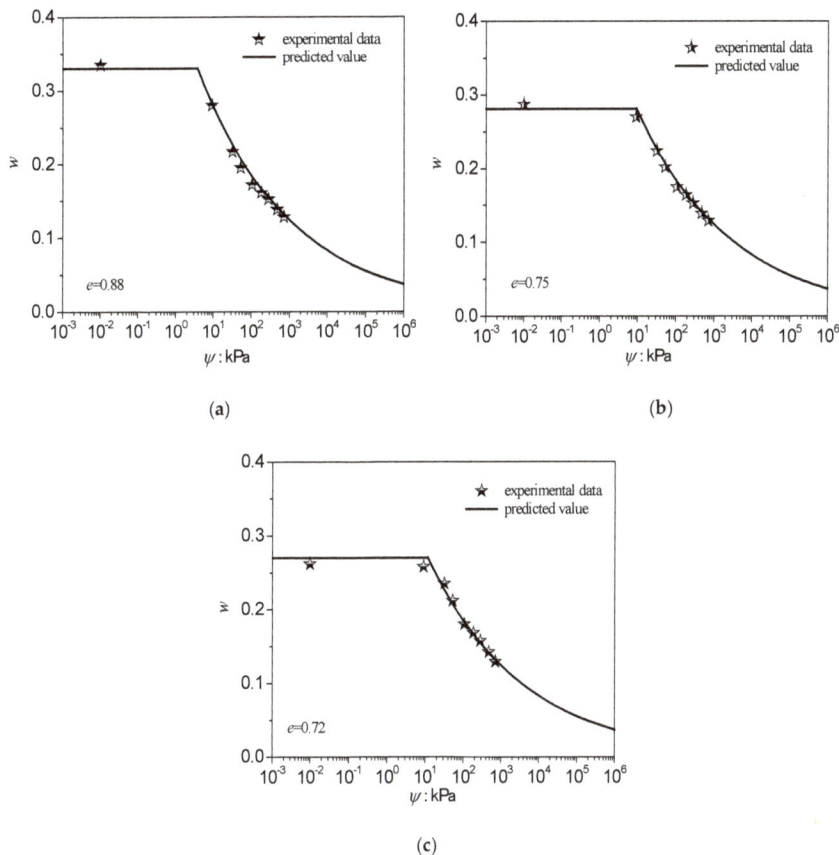

(a)

(b)

(c)

Figure 20. Measured and predicted SWCCs for Xi'an loess specimens at deformed state at (a) $e = 0.88$; (b) $e = 0.75$; (c) $e = 0.72$ (data after Wang et al. [44]).

4. Discussion

There are generally three forms of SWCC expression, including gravimetric water content–suction curve (w–ψ curve), volumetric water content–suction curve (θ–ψ curve), and degree of saturation–suction curve (S_r–ψ curve). In the current study, the theoretical principle of the proposed model is illustrated in Figure 1, where the SWCC presents a type of "broom shape" distribution in terms of the gravimetric water content. If it is necessary to obtain the SWCCs expressed by volumetric water content or degree of saturation, the gravimetric water content can be converted to the volumetric water content or degree of saturation. Respectively, the transformation formulas are expressed as

$$\theta = wG_S / (1 + e) \tag{17}$$

$$S_r = wG_S / e \tag{18}$$

In the drying process, water drains through large pores when the suction is low, which means the particles around large pores are gradually subjected to capillary pressure. As a result, the size of large pores decreases gradually. As the suction increases, small pores begin to drain water and the pore size is reduced correspondingly, leading to the shrinkage of the macroscopic volume of soils. The larger the initial void ratio, the more obvious this phenomenon. In the above process, it is only when the water is drained that the pore size can be reduced, and the drainage of water occurs firstly at large pores, and then at small pores. Hence, the radius of pores starting to drain water under a certain suction can be approximatively seen as its initial size. In contrast, in the wetting process, the drainage of water has been completed, and the volume shrinkage of large and small pores has also finished. So, the pore-size distribution characteristics at this stage have great difference from that in saturated state, which leads to the phenomenon of SWCC hysteresis. It should be stated that only the influence of initial void ratio on SWCC in drying path was investigated in this work, while the scenario for SWCC in wetting path will be studied in the future study.

5. Conclusions

A simple and efficient fractal-based approach capable of capturing effects of initial void ratio was presented for the SWCC of unsaturated soils. In terms of the experimental data, the SWCCs of the unsaturated soils at different initial void ratios were found to be mainly controlled by the air-entry value, while the fractal dimension could be assumed to be constant. In contrast to the complexity of existing models, only two parameters (i.e., fractal dimension D and air-entry value ψ_a) were employed in the current model. Determination of the model parameters with clear physical meaning were elaborated. The application of the model to a wide range of experimental data from the tests conducted in the current work, as well as the literatures, was examined. Good agreement was obtained between the experimental data and the model predictions in all of the cases considered.

Author Contributions: The research study was carried out successfully with contribution from all authors. The main research idea and manuscript preparation were contributed by G.T. and Q.C.; Y.C. contributed on the manuscript preparation and performed the correlative experiment. L.K. and H.X. gave several suggestions from the industrial perspectives. Y.X. assisted on finalizing research work and manuscript. All authors revised and approved the publication of the paper.

Acknowledgments: The financial support from the National Key R&D Program of China (No. 2016YFC0502208), National Natural Science Foundation of China (No. 51708189 and 51409097), and Research project of Hubei Provincial Education Department (No. D20161405) are gratefully acknowledged.

Conflicts of Interest: The authors declare no conflict of interest.

References

1. Fredlund, D.G.; Xing, A.; Fredlund, M.D.; Barbour, S.L. Relationship of the unsaturated soil shear strength to the soil-water characteristic curve. *Can. Geotech. J.* **1996**, *33*, 440–448. [CrossRef]
2. Wheeler, S.J. Inclusion of specific water volume within an elasto-plastic model for unsaturated soil. *Can. Geotech. J.* **1996**, *33*, 42–57. [CrossRef]
3. Assouline, S. A model for soil relative hydraulic conductivity based on the water retention characteristic curve. *Water Resour. Res.* **2001**, *37*, 265–271. [CrossRef]
4. Gallipoli, D.; Gens, A.; Sharma, R.; Vaunat, J. An elasto-plastic model for unsaturated soil incorporating the effects of suction and degree of saturation on mechanical behaviour. *Geotechnique* **2003**, *53*, 123–135. [CrossRef]
5. Fredlund, D.G.; Rahardjo, H.; Fredlund, M.D. *Unsaturated Soil Mechanics in Engineering Practice*; John Wiley & Sons: Hoboken, NJ, USA, 2012; Volume 132, pp. 286–321.
6. Van Genuchten, M.T. A closed-form equation for predicting the hydraulic conductivity of unsaturated soils. *Soil Sci. Soc. Am. J.* **1980**, *44*, 892–898. [CrossRef]
7. Fredlund, D.G.; Xing, A. Equations for the soil-water characteristic curve. *Can. Geotech. J.* **1994**, *31*, 521–532. [CrossRef]
8. Haj, K.M.A.A.; Standing, J.R. Soil water retention curves representing two tropical clay soils from Sudan. *Geotechnique* **2015**, *66*, 1–14.
9. Brooks, R.H. *Hydraulic Properties of Porous Media*; Colorado State University: Fort Collins, CO, USA, 1964; Volume 3, pp. 352–366.

10. Romero, E.; Vecchia, G.D.; Jommi, C. An insight into the water retention properties of compacted clayey soils. *Geotechnique* **2011**, *61*, 313–328. [CrossRef]

11. Simms, P.H.; Yanful, E.K. Predicting soil-water characteristic curves of compacted plastic soils from measured pore-size distributions. *Geotechnique* **2002**, *52*, 269–278. [CrossRef]

12. Chan, T.P.; Govindaraju, R.S. A new model for soil hydraulic properties based on a stochastic conceptualization of porous media. *Water Resour. Res.* **2003**, *39*, 1195. [CrossRef]

13. Salager, S.; Nuth, M.; Ferrari, A.; Laloui, L. Investigation into water retention behaviour of deformable soils. *Can. Geotech. J.* **2013**, *50*, 200–208. [CrossRef]

14. Zhou, A.-N.; Sheng, D.; Carter, J.P. Modelling the effect of initial density on soil-water characteristic curves. *Geotechnique* **2012**, *62*, 669–680. [CrossRef]

15. Assouline, S. Modeling the relationship between soil bulk density and the water retention curve. *Vadose Zone J.* **2006**, *5*, 554–562. [CrossRef]

16. Ng, C.W.W.; Pang, Y.W. Influence of stress state on soil-water characteristics and slope stability. *J. Geotech. Geoenviron. Eng.* **2000**, *126*, 157–166. [CrossRef]

17. Gallipoli, D.; Wheeler, S.J.; Karstunen, M. Modelling the variation of degree of saturation in a deformable unsaturated soil. *Geotechnique* **2003**, *53*, 105–112. [CrossRef]

18. Wheeler, S.J.; Sharma, R.S.; Buisson, M.S.R. Coupling of hydraulic hysteresis and stress–strain behaviour in unsaturated soils. *Geotechnique* **2003**, *53*, 41–54. [CrossRef]

19. Sun, D.A.; Sheng, D.; Cui, H.B.; Sloan, S.W. A density-dependent elastoplastic hydro-mechanical model for unsaturated compacted soils. *Int. J. Numer. Anal. Methods Geomech.* **2007**, *31*, 1257–1279. [CrossRef]

20. Khalili, N.; Habte, M.A.; Zargarbashi, S. A fully coupled flow deformation model for cyclic analysis of unsaturated soils including hydraulic and mechanical hystereses. *Comput. Geotech.* **2008**, *35*, 872–889. [CrossRef]

21. Miller, G.A.; Khoury, C.N.; Muraleetharan, K.K.; Liu, C.; Kibbey, T.C.G. Effects of soil skeleton deformations on hysteretic soil water characteristic curves: Experiments and simulations. *Water Resour. Res.* **2008**, *44*, 137–148. [CrossRef]

22. Nuth, M.; Laloui, L. Advances in modelling hysteretic water retention curve in deformable soils. *Comput. Geotech.* **2008**, *35*, 835–844. [CrossRef]

23. Masin, D. Predicting the dependency of a degree of saturation on void ratio and suction using effective stress principle for unsaturated soils. *Int. J. Numer. Anal. Methods Geomech.* **2010**, *34*, 73–90. [CrossRef]

24. Sheng, D.; Zhou, A.N. Coupling hydraulic with mechanical models for unsaturated soils. *Can. Geotech. J.* **2011**, *48*, 826–840. [CrossRef]

25. Tarantino, A. A water retention model for deformable soils. *Geotechnique* **2009**, *59*, 751–762. [CrossRef]

26. Gallipoli, D.; Bruno, A.W.; Onza, F.D.; Mancuso, C. A bounding surface hysteretic water retention model for deformable soils. *Geotechnique* **2015**, *65*, 793–804. [CrossRef]

27. Wei, W.; Cai, J.; Hu, X.; Han, Q.; Liu, S.; Zhou, Y. Fractal analysis of the effect of particle aggregation distribution on thermal conductivity of nanofluids. *Phys. Lett. A* **2016**, *380*, 2953–2956. [CrossRef]

28. Wei, W.; Cai, J.; Hu, X.; Han, Q. An electrical conductivity model for fractal porous media. *Geophys. Res. Lett.* **2015**, *42*, 4833–4840. [CrossRef]

29. Cai, J.; Hu, X.; Xiao, B.; Zhou, Y.; Wei, W. Recent developments on fractal-based approaches to nanofluids and nanoparticle aggregation. *Int. J. Heat Mass Transf.* **2017**, *105*, 623–637. [CrossRef]

30. Cai, J.; Wei, W.; Hu, X.; Liu, R.; Wang, J. Fractal characterization of dynamic fracture network extension in porous media. *Fractals* **2017**, *25*, 1750023. [CrossRef]

31. Cai, J.; Hu, X. *Fractal Theory in Porous Media and its Applications*; Science Press: Beijing, China, 2015.

32. Zolfaghari, A.; Dehghanpour, H.; Xu, M. Water sorption behaviour of gas shales: II. Pore size distribution. *Int. J. Coal. Geol.* **2017**, *179*, 187–195. [CrossRef]

33. Shi, Y.; Yassin, M.R.; Dehghanpour, H. A modified model for spontaneous imbibition of wetting phase into fractal porous media. *Colloids Surf. A Physicochem. Eng. Asp.* **2018**, *543*, 64–75. [CrossRef]

34. Bird, N.R.A.; Perrier, E.; Rieu, M. The water retention function for a model of soil structure with pore and solid fractal distributions. *Eur. J. Soil Sci.* **2000**, *51*, 55–63. [CrossRef]

35. Xu, Y.F.; Sun, D.A. A fractal model for soil pores and its application to determination of water permeability. *Physica A* **2002**, *316*, 56–64. [CrossRef]

36. Russell, A.R. How water retention in fractal soils depends on particle and pore sizes, shapes, volumes and surface areas. *Geotechnique* **2014**, *64*, 379–390. [CrossRef]

37. Huang, G.; Zhang, R. Evaluation of soil water retention curve with the pore–solid fractal model. *Geoderma* **2005**, *127*, 52–61. [CrossRef]

38. Khalili, N.; Khoshghalb, A.; Pasha, A.Y. A fractal model for volume change dependency of the water retention curve. *Geotechnique* **2015**, *65*, 141–146.

39. Tao, G.L.; Zhang, J.R. Two categories of fractal models of rock and soil expressing volume and size-distribution of pores and grains. *Sci. Bull.* **2009**, *54*, 4458–4467. [CrossRef]

40. Tao, G.L.; Zhang, J.R.; Zhuang, X.S.; Yang, L. Influence of compression deformation on the soil-water characteristic curve and its simplified representation method. *J. Hydraul. Eng.* **2014**, *45*, 1239–1245.

41. Russell, A.R.; Buzzi, O. A fractal basis for soil-water characteristics curves with hydraulic hysteresis. *Geotechnique* **2012**, *62*, 269–274. [CrossRef]

42. Miao, L.; Jing, F.; Houston, S.L. Soil-water characteristic curve of remolded expansive soils. In Proceedings of the Fourth International Conference on Unsaturated Soils, Carefree, Arizona, 2–6 April 2006; pp. 997–1004.

43. Huang, S. Evaluation and Laboratory Measurement of the Coefficient of Permeability in Deformable. Ph.D. Thesis, University of Saskatchewan, Saskatoon, SK, Canada, 1994.

44. Wang, T.H.; Lu, J.; Yue, C.K. Soil-water characteristic curve for unsaturated loess considering temperature and density effect. *Rock Soil Mech.* **2008**, *29*, 1–5.

energies

MDPI

Article

Engineering Simulation Tests on Multiphase Flow in Middle- and High-Yield Slanted Well Bores

Dan Qi [1,2,3,*]**, Honglan Zou** [3]**, Yunhong Ding** [3]**, Wei Luo** [4] **and Junzheng Yang** [3]

[1] School of Engineering Science, University of Chinese Academy of Sciences, Beijing 100049, China
[2] Institute of Porous Flow and Fluid Mechanics, Chinese Academy of Sciences, Langfang 065007, China
[3] PetroChina Research Institute of Petroleum Exploration & Development, Beijing 100083, China;
zouhl69@petrochina.com.cn (H.Z.); dyhong@petrochina.com.cn (Y.D.);
yangjunzheng69@petrochina.com.cn (J.Y.)
[4] The Branch of Key Laboratory of CNPC for Oil and Gas Production, Yangtze University,
Wuhan 430115, China; luoruichang@163.com
* Correspondence: qidan69@petrochina.com.cn; Tel.: +86-010-8359-6155

Received: 8 August 2018; Accepted: 26 September 2018; Published: 28 September 2018

Abstract: Previous multiphase pipe flow tests have mainly been conducted in horizontal and vertical pipes, with few tests conducted on multiphase pipe flow under different inclined angles. In this study, in light of mid–high yield and highly deviated wells in the Middle East and on the basis of existent multiphase flow pressure research on well bores, multiphase pipe flow tests were conducted under different inclined angles, liquid rates, and gas rates. A pressure prediction model based on Mukherjee model, but with new coefficients and higher accuracy for well bores in the study block, was obtained. It was verified that the newly built pressure drawdown prediction model tallies better with experimental data, with an error of only 11.3%. The effect of inclination, output, and gas rate on the flow pattern, liquid holdup, and friction in the course of multiphase flow were analyzed comprehensively, and six kinds of classical flow regime maps were verified with this model. The results showed that for annular and slug flow, the Mukherjee flow pattern map had a higher accuracy of 100% and 80–100%, respectively. For transition flow, Duns and Ros flow pattern map had a higher accuracy of 46–66%.

Keywords: wellbore multiphase flow; inclined angle; liquid rate; gas rate; pressure drawdown model with new coefficients

1. Introduction

Multiphase flow of oil, gas, and water universally exist in the course of oil and gas exploitation as the multiphase flow in well bores have a great impact on the selection of rational oil production method, the adjustment of production parameters, improvement in lift efficiency, and achievement of optimal profits [1–3]. Therefore, it is very important to study the multiphase flow features of well bores, with the flow pattern and pressure loss especially crucial for oil recovery engineering. In petroleum engineering, the multiphase flow is mostly studied as per the following procedure: Firstly, lab simulation test is conducted. Then, based on the test phenomena and data analysis, the flow parameters in the course of multiphase flow are figured out to build a pressure drawdown prediction model. Finally, the field production data is used to verify the model. Although research on multiphase pipe flow in well bores has been going on for more than half a century, a general method to discriminate the flow pattern and predict the pressure variation has still not been developed.

The main task of multiphase flow research is to analyze the flow behaviors of each phase and the mixed phase and build their correlating relations. As the flow pattern variation is the most intuitive and fundamental phenomenon, analysis on forming mechanisms and features of flow pattern is the foundation for studying the features of multiphase flow. However, there is no uniform standard for the classification of flow patterns yet, and therefore most are classified according to the variation of flow shapes. This method often results in big errors as it is restricted by objective conditions, such as observation instrument, and affected by subjective factors, such as the observer. At present, the flow patterns mostly used for classification include bubble flow, slug flow, churn flow, annular flow, laminar flow, and wave flow, with the latter two mainly occurring in horizontal or slightly inclined strings.

The pressure variation of oil and gas well system is not only a foundation for judging the natural flow of oil and gas wells and studying the production effects of various lift modes, but it is also an important parameter considered in the optimum design of oil and gas wells to guarantee high and stable yield. Currently, research on wellbore pressure drawdown mostly regard oil, gas, and water as gas and liquid phases, taking the flow features of each flow pattern as the basis to figure out the pressure distribution along the well bore based on momentum conservation and energy conservation. There are very few research projects that regard oil, gas, and water as three phases due to the complicated oil–water mixture property and an unclear understanding of the flow pattern and formation mechanism.

After the multiphase flow technology was first applied in petroleum engineering in the 1950s, researchers like Govier, Griffith, Fancher, and Brown presented some related expressions to calculate pressure loss based on experiments [4–8]. However, these related expressions were mostly derived based on small pipe diameter and vertical pipe and were not introduced in practical applications. Starting from the 1960s, researchers like Duns and Ros, Hagedorn and Brown, Orkiszawski, Govier, and Aziz [8–11] were devoted to study more applicable equations and obtained good results. The related expressions they presented had higher accuracy when the flow conditions matched; however, they did not take the effect of inclined angle into account [3,11]. Beggs and Brill (1973) [12] derived multiphase-flow-related expression suitable for any inclined angle in the lab, but the experimental flow media only included air and water, and the maximal experimental pipe diameter was only 1.5 inch, meaning its scope of application was restricted. Barnea et al (1986) [13] believed that laminar flow would not occur when the inclined angle was larger than 10°, but Shoham (1990) [14] believed that 20° inclined angle was the limit for the occurrence of laminar flow.

In the following decades, based on the previous research results, many researchers have committed to improving the accuracy of parameters related to multiphase flow and studying the temperature field of multiphase flow. On the basis of two-phase flow pattern classification method of Taitel and Barnea, Chen (1992) [15] proposed a comprehensive mechanical model for the flow characteristic of gas–liquid two-phase flow in vertical tubes. Barnea (1993) [16] proposed a model that can be applied to calculate the slug length of the slug flow at any desired position along the pipe. Liu (1995) [17] put forward a calculation method for multiphase flow in wellbore that is suitable for different reservoir types, including black oil, volatile oil, condensate gas, moisture, dry gas, etc. Liao (1998) [18] divided the multiphase flow into bubble flow, slug flow, turbulence flow, and annular flow and established the calculation formula of the split pressure gradient with higher accuracy compared to other correlations under the condition of high gas to oil ratio. Hibiki (1999) [19] held that the basic structure of the two-phase flow can be characterized by two fundamental geometrical parameters—known as void fraction and interfacial area concentration—and that the mechanism of the interfacial area transport depended on the bubble mixing length, turbulence intensity, and so on. Zhang (2001) [20] developed a new correlation for two-phase friction pressure drop in small diameter tubes by modifying the previous correlation. Hou (2004) [21] believed that the relevant parameters in the multiphase flow analysis of the wellbore are closely related to the well depth. On account of this, assuming that the calculation parameters in the microsegment are unchanged, a correlation model of the pressure gradient was established. Chen (2006) [22] investigated the flow boiling flow patterns in small diameters and sketched a new flow map that showed the transition boundaries of slug to churn and churn to annular flow depended strongly on the diameter. Yu (2008) [23] established a temperature field model for multiphase flow in a vertical wellbore based on multiphase flow dynamics and heat transfer theory. Cheng (2008) [24] researched the two-phase flow patterns in adiabatic and diabatic conditions and recommended that objective methods should be developed to gain more accurate flow pattern date and that the impact of the physical properties on flow pattern needed to be studied further. Liu (2009) [11] analyzed the flow pattern characteristics of inclined wellbore, considered that the mist flow model is more suitable for wells with high gas–liquid ratio, and established a new mist flow model. Yuichi Murai (2010) [25] designed three types of ultrasound interface detection techniques to capture the interface for two-phase flows, and the echo intensity technique and local Doppler technique were found to be appropriate for turbulent interfaces and bubbles, respectively. Gao (2014) [26] plotted the flow pattern of the vertical wellbore two-phase flow with the mixture Reynolds number and the ratio of the apparent velocity of the gas and liquid phase as the horizontal and vertical coordinates. Swanand (2014) [27] presented new equations for a flow-pattern-independent drift flux model that, in his opinion, is applicable to gas–liquid two-phase flow covering a wide range of pipe diameters. Zhou (2016) [28] studied the characteristic of high gas–liquid two-phase flow and found a new type of flow pattern called oscillatory impulse flow. Montoya (2016) [29] researched the characteristics of churn-turbulent flow and suggested computational fluid dynamics (CFD) models as an appropriate method for predicting highly turbulent flow. Liu (2017) [30] performed comprehensive work to model the flow regime transition criteria for upward two-phase flows in vertical rod bundles and proposed a new flow regime transition criteria model based on the analysis of the underlying physics of the upward two-phase flow behavior. Lu (2018) [31] studied the effects of flow regime, pipe size, and flow orientation on two-phase frictional pressure drop analysis and recommended that Lockhart–Martinelli approach be used to predict pressure drop.

In summary, there are still some defects in the research of multiphase flow. This includes: (1) The previous researches on pressure calculation methods of multiphase pipe flow mainly concentrated on horizontal and vertical pipes, and research on multiphase pipe flow under the conditions of different inclined angles is insufficient. (2) The available pressure calculation methods of multiphase pipe flow were mostly empirical and semiempirical methods and were all obtained within certain experimental scope; the calculation results were not always correct when exceeding the scope, and the application of these methods is therefore largely restricted. In particular, there are few experimental researches on the multiphase pipe flow under the conditions of large output from slanted well bores.

Therefore, in line with multiple well types, complicated casing program, various inclined angles, high production proration of single well, high gas–oil ratio (GOR > 100 m^3/m^3), low viscosity (1.2–5.7 cp), and heavy crude (19–29 API°) in the Middle East, it is necessary to conduct experimental research on multiphase pipe flow at high yield and different inclined angles. This will help find the flow patterns of gas and liquid in oil wells with high yield and different inclined angles accurately and optimize the multiphase flow pressure gradient computation method suitable for high-yield inclined pipes.

2. Experimental System of Multiphase Pipe Flow

2.1. Experimental Apparatus

The experimental apparatus was composed of the following parts: well bore simulation experiment interval, experimental fluid feeding system, experimental fluid measurement and control system, and experimental fluid flow rate and pressure acquisition system. The well bore simulation interval adopted a DN60 straight pipe, which was about 14 m long in total, 9.5 m long effectively, transparent, and visible. With the experimental apparatus, the oil, gas, and water can realize a multiphase flow in a pipe with 0–90° inclined angles, various flow patterns like bubble flow and mist flow can be observed, and parameters like flow rate, pressure differential pressure, liquid holdup, and temperature of multiphase flow fluid in the pipe can be measured. The fluid flow rate and pressure can be controlled manually or automatically. The experimental apparatus and its parameters are listed as in Figure 1 and Table 1 respectively.

Table 1. Parameters of experimental apparatus.

Experimental Pipe	Atmospheric Pressure (0–0.8 × 10⁴ KPa)	DN60 Straight Pipe
Experimental inclined angle	0–90°	
Experimental media	Air, water	
Maximum flow rate of media	water	20 m³/h
	Displacement of air compressor	35 m³/min
Measuring range of flow meter	water	0–20 m³/h
	Air	0–35 m³/min
Measuring accuracy of flow meter	water	±0.5%
	Air	±1%
Working pressure scope	Atmospheric pressure (×10⁴ KPa)	0–0.8
Measuring accuracy of pressure gauge	Ordinary pressure signal: ±0.1%; pressure loss calculation interval: ±0.025–0.04%	
Medium temperature	Atmospheric temperature −90 °C; measuring accuracy of temperature probe: ±0.5%	
High speed camera	500 frame/s, 1920 × 1080 resolution, length of exposure: 1 μs, recording time ≥5 s	

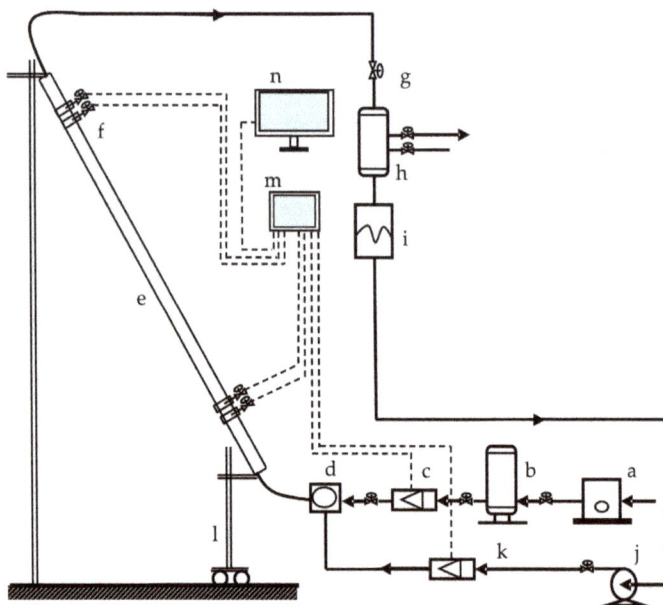

Figure 1. Flow diagram of experiment: (a) air compressor; (b) gas storage tank; (c) gas flow meter; (d) gas–liquid mixer; (e) observation section for flow pattern; (f) temperature/pressure gauge; (g) valve; (h) gas–liquid separator; (i) water tank; (j) centrifugal pump; (k) liquid turbine flow meter; (l) support frame; (m) paperless recorder; (n) computer.

With air and water as media and under conditions of the same pipe diameter, different inclined angles, and different fluid and gas rates, the experiments simulated the flow rules and pressure drawdown variation rules of gas–liquid two-phase flow in the well bore; tested the flow behavior, liquid holdup, and differential pressure at any inclined angle, well fluid, and gas flow rate; and then optimized and obtained the multiphase flow pressure prediction method suitable for middle and high-yield slanted well bores.

2.2. Experimental Contents

2.2.1. Preparation before Experiment

1. Before the commencement of the experiment, the pipeline airtightness and clearance of process pipeline pathway were firstly checked. Then, it was checked whether the compressor, water pump, each instrument, and recording software could normally work.
2. Experimental contents were determined, and the inclined angles of test pipes were selected based on the testing program.
3. The test stand was raised with a hoist; when a required angle was attained, the lifting hoist was stopped.
4. The air inlet and outlet valve switches of the selected test pipe were opened, and the valves of other test pipes were closed.
5. Inspection was repeated to ensure the process was correct, the experimental pipeline ports were open, and there was no pressure buildup phenomenon.

2.2.2. Experimental Procedure

1. The software system of the console was opened, and the testing string inclination was adjusted.
2. The air compressor was started, and the water pump was opened on the console.
3. The fluid and gas volumes entering the string were adjusted. The gas volume was adjusted by slowly adjusting the opening of the air inlet valve, whereas the fluid volume was adjusted by slowly adjusting the pump frequency and the opening of the reflux valve; simultaneously, the instrument readings on the console were observed until the target values were reached.
4. After the target values were reached, the experimental phenomena were observed, and the pressure, differential pressure, temperature, fluid volume, and gas volume displayed on the instrument were recorded.
5. The experimental data recording time range was set, the test data was saved, a high speed camera was used to take photographs, and the fluid flow patterns in the pipe were recorded and saved.
6. After the gas and liquid in test string was stopped with quick closing valve and the liquid was still, the height of liquid in the plexi-glass tubular was read, and the liquid holdup was figured out.
7. The gas volume and the liquid volume in the testing string as well as the inclined angle of the string were readjusted. Then, the above procedures were repeated, and the gas flow rate, liquid flow rate, pressure, differential pressure, temperature, flow behavior, and liquid holdup tested at different inclined angles were recorded.
8. After the experiment was finished, the water pump and air compressor were shut off, the test stand was placed horizontally, and the power switches of the computer and console were turned off.

2.2.3. Experimental Parameters

The influence of inclined angle, liquid and gas volume on two-phase flow was analyzed. And the specific experimental parameters are shown in Table 2.

Table 2. Parameters of experiment.

Variable	Parameter
Pipe diameter (mm)	60
Inclined angle (°)	0, 15, 30, 45, 60, 75, 90
Liquid volume (m³/day)	50, 100, 150, 200, 250, 300, 350, 400
Gas volume (m³/day)	5000, 10,000, 15,000, 20,000, 25,000, 30,000, 48,000

3. Experimental Analysis

3.1. Analysis of Factors Affecting Pressure Drawdown

Figure 2a shows that when the liquid volume ranged 2.10–20.83 m³/h and the gas volume was higher than or equaled 210 m³/h, the pressure drawdown firstly rose and then reduced with the increase of the inclined angle; when the gas volume was 625 m³/h, the pressure drawdown rose with the increase of the inclined angle. The leading cause for this was that when the inclined angle ranged 0–60°, the pressure drawdown of gravity term gradually increased, and the total pressure drawdown increased with the increase of the inclined angle. When the inclined angle was higher than 60°, although the pressure drawdown of gravity term also gradually increased, the liquid carrying capability of gas increased, the liquid holdup reduced (also shown in the liquid holdup map below), and the total pressure drawdown reduced with the increase of the inclined angle.

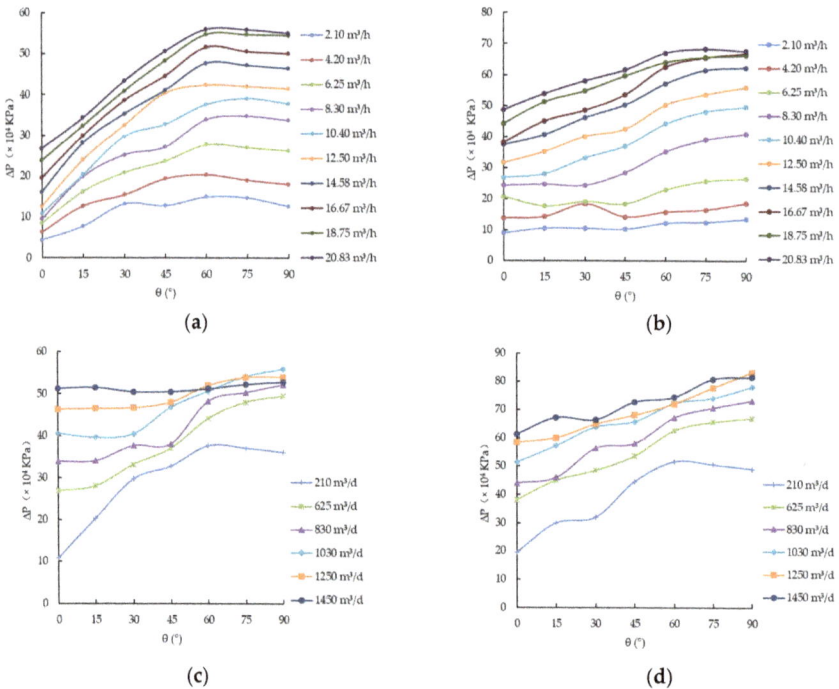

Figure 2. The law of pressure drop with angle: (**a**) when gas rate is 210 m³/h; (**b**) when gas rate is 625 m³/h; (**c**) when liquid rate is 10.40 m³/h; (**d**) when liquid rate is 16.67 m³/h.

3.2. Analysis on Influential Factors of Liquid Holdup

Under the conditions of the same liquid volume and gas volume, the variation rule of liquid holdup with different inclined angles is shown in Figure 3. At the same liquid volume, as the gas volume increased, the liquid holdup fell. In particular, the liquid holdup fell faster when the gas injection rate was 200–700 m^3/h, but the liquid holdup fell more slowly when the gas injection rate was higher than 700 m^3/h. At the same gas injection rate, as the liquid volume increased, the liquid holdup rose. At the same liquid volume and gas volume, the liquid holdup exhibited a trend of firstly rising and then falling with the increase of the inclined angle; however, observed from the total liquid volume (at gas injection rate of more than 210 m^3/h), the liquid holdup changed little at 0–90°.

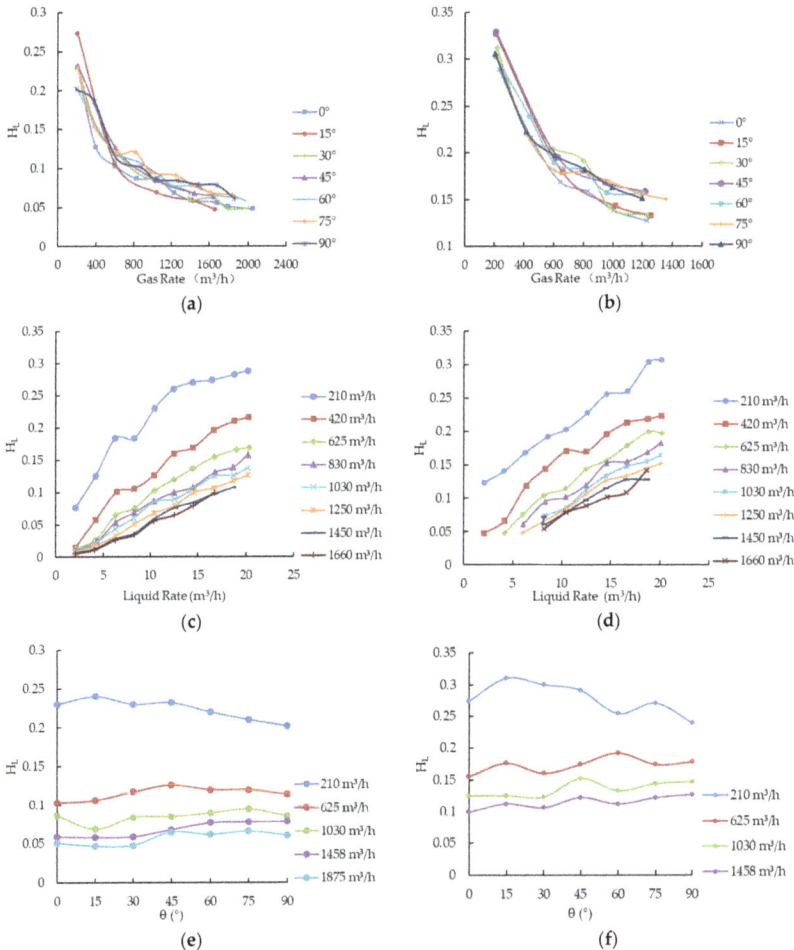

Figure 3. Analysis of influencing factors of liquid holdup: (**a**) when liquid rate is10.40 m^3/h, the law of liquid holding rate with inclination angle and gas flow rate; (**b**) when liquid rate is 20.83 m^3/h, the law of liquid holding rate with inclination angle and gas flow rate; (**c**) when inclined angle is 0°, the law of liquid holding rate with liquid and gas flow rate; (**d**) when inclined angle is 90°, the law of liquid holding rate with liquid and gas flow rate; (**e**) when liquid rate is 10.40 m^3/h, the law of liquid holding rate with inclination angle; (**f**) when liquid rate is 16.67 m^3/h, the law of liquid holding rate with inclination angle.

3.3. Flow Pattern Variation at Different Inclined Angles

Under the experimental conditions, when the gas volume was small, laminar flow occurred, but no laminar flow occurred in the inclined state. In the horizontal and inclined states, the most common flow patterns were slug flow and transition flow; annular flow was rare, and no bubble flow or mist flow occurred.

In the horizontal state, only laminar flow, slug flow and transition flow occurred; furthermore, laminar flow and transition flow occurred only under the conditions of very small or very large gas volume, and slug flow appeared in most cases. At the same liquid volume, as the gas volume increased, in the experiment of horizontal state, the flow pattern converted from laminar flow to slug flow and then to transition flow.

With the increase of inclination angle of the tube, due to the action of gravity, laminar flow no longer occurred at 15°, 30°, 45°, 60°, 75°, and 90°. At the same liquid volume, as the gas volume increased, in the experiment of inclined angles of 15°, 30°, 45°, 60°, 75°, and 90°, the flow pattern tended to convert from slug flow to transition flow and then to annular flow.

The typical flow patterns observed in the experiment are shown as in Figure 4.

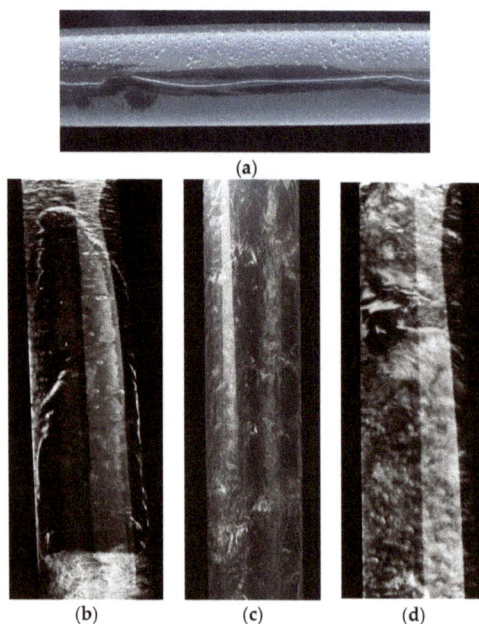

Figure 4. The flow pattern observed in the experiment: (**a**) laminar flow (inclined angle is 0°); (**b**) slug flow (inclined angle is 30°); (**c**) annular flow (inclined angle is 90°); (**d**) transition flow (inclined angle is 90°).

3.4. Verification of Flow Pattern Maps

At present, in the identification of multiphase flow patterns, the commonly used empirical flow pattern maps include Duns and Ros flow pattern map [7], Hewitt and Roberts flow pattern map [32], Aziz flow pattern map [10], Gould flow pattern map [33], Ansari flow pattern map [34], Beggs–Brill flow pattern map [12], and Mukherjee flow pattern map [1]. In this study, based on the experimental data, the identification effects of the above flow pattern maps were verified, and Figure 5 shows the verification results as follows:

- For the annular flow, Beggs–Brill flow pattern map, Mukherjee flow pattern map, and Aziz flow pattern map were the most accurate, i.e., out of the 469 groups of experiments, all the annular flows were consistent with the flow pattern maps; however, in the Ansari flow pattern map and Hewitt and Roberts flow pattern map, most of the experimental data points exceeded their estimation range, indicating that the experiments in this study exceeded the application scope of these two flow pattern maps.
- For the slug flow, the Mukherjee flow pattern map and Duns and Ros flow pattern map were the most accurate, i.e., at inclined angles of 0–90°, the judgment accuracy reached 80–100%; for the transition flow, the Duns and Ros flow pattern map was the most accurate, with an accuracy of 46–66%.

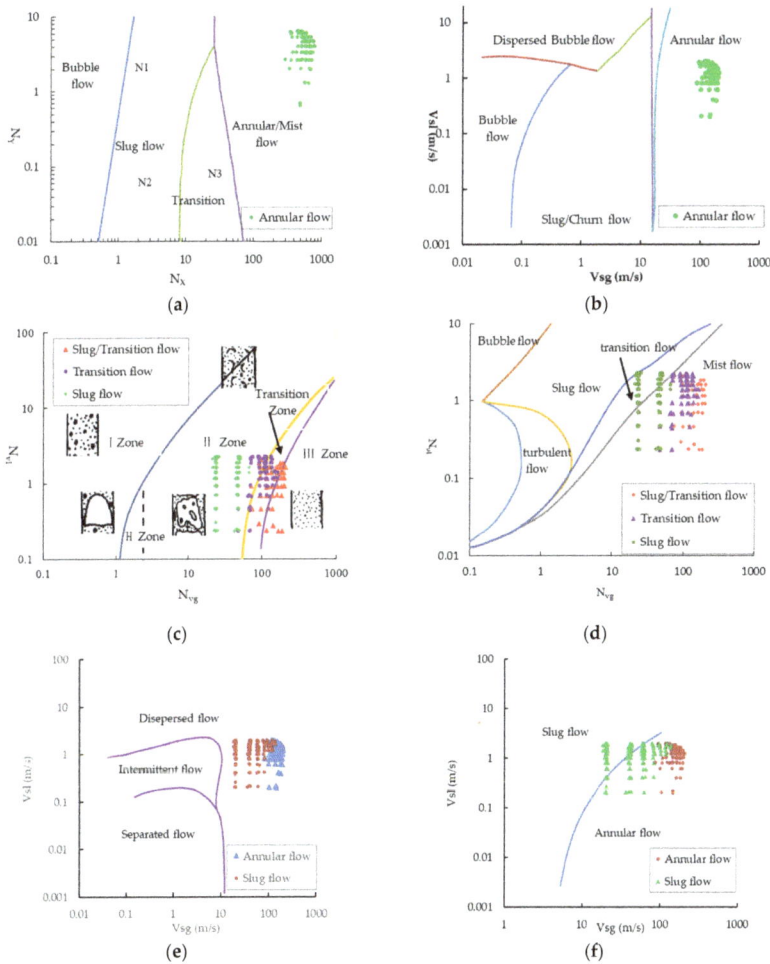

Figure 5. Verification of flow pattern maps: (**a**) Flow pattern map of Aziz; (**b**) Flow pattern map of Ansari; (**c**) Flow pattern map of Duns and Ros; (**d**) Flow pattern map of Gould; (**e**) Flow pattern map of Beggs–Brill; (**f**) Flow pattern map of Mukherrjee. V_{sg}—the superficial gas velocity; V_{sl}—the superficial liquid velocity; N_{vg}—gas phase velocity criterion; N_{vl}—liquid phase velocity criterion; N_x—gas phase correction parameter; N_y—liquid phase correction parameter.

3.5. Verification of Liquid Holdup and Pressure Drawdown

As shown in Figure 6a, for the calculation of liquid holdup, when the inclined angle was between 0° and 15°, the calculation results of Beggs and Brill were more accurate, and when the inclined angle ranged between 30 and 90°, the calculation results of Mukherjee and Beggs were more accurate. As shown in Figure 6b, for the calculation of pressure drawdown, when the inclined angle ranged from 0° to 60°, the calculation results of Aziz were more accurate, when the inclined angle was between 75° and 90°, the calculation results of Hasan [35] were more accurate, and when the inclined angle ranges from 0 to 90°, the JPI model [18] was the most stable.

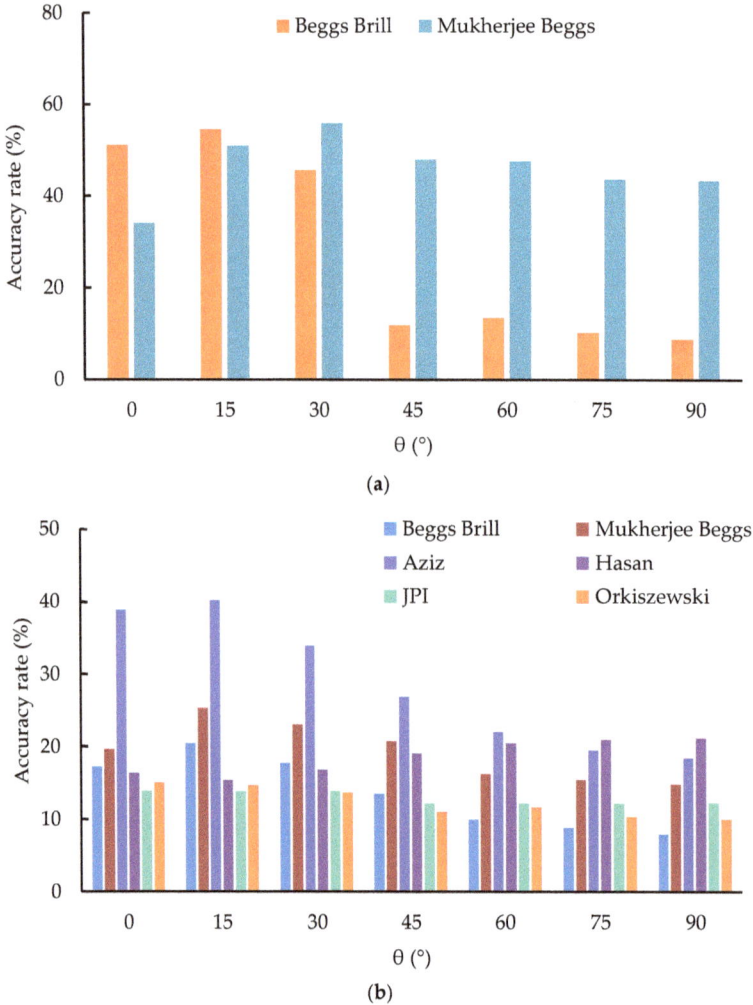

(a)

(b)

Figure 6. Verification of liquid holdup and pressure drawdown: (**a**) verification of liquid holdup; (**b**) verification of pressure drawdown.

3.6. New Model for Calculating Liquid Holdup and Pressure Drawdown

3.6.1. Simulation of the New Model

Beggs–Brill (Mukherjee adopted the same calculation method) [1] presented the calculation equation of total pressure drawdown:

$$-\frac{dp}{dz} = \frac{[\rho_l H_l + \rho_g(1-H_l)]g\sin\theta + \frac{\lambda G v}{2DA}}{1 - \{[\rho_l H_l + \rho_g(1-H_l)]vv_{sg}\}/p} \tag{1}$$

The calculation equation of total pressure gradient shows that at the time of calculating the total pressure gradient of gas–liquid two-phase flow, the calculation of liquid holdup H_l and resistance coefficient λ along the tube must be mastered. In the study, the experimental data was used to match the calculation methods of liquid holdup and resistance coefficient along the tube, respectively.

1. Simulation of liquid holdup

Based on experimental data and regression analysis, Mukherjee presented the law of liquid holdup for gas and liquid inclined pipe flow, i.e., [1],

$$H_l = \exp\left[\left(c_1 + c_2\sin\theta + c_3\sin^2\theta + c_4 N_l^2\right)\frac{N_{vg}{}^{c_5}}{N_{vl}{}^{c_6}}\right] \tag{2}$$

$$N_{vl} = v_{sl}\left(\frac{\rho_l}{g\sigma}\right)^{0.25} \tag{3}$$

$$N_{vg} = v_{sg}\left(\frac{\rho_l}{g\sigma}\right)^{0.25} \tag{4}$$

$$N_l = \mu_l\left(\frac{g}{\rho_l\sigma^3}\right)^{0.25} \tag{5}$$

where v_{sl} is the apparent velocity of liquid phase (m/s); v_{sg} is the apparent velocity of gas phase (m/s); σ is the surface tension of liquid phase (N/m); μ_l is the viscosity of liquid phase (mPa·s); c_1–c_6 are all empirical constants.

The relationship between the experimental liquid holdup H_l and the correlated numbers N_{vl}, and N_l presented by Mukherjee were matched, and the obtained new empirical constants are listed in Table 3.

Table 3. Empirical constants.

c_1	c_2	c_3	c_4	c_5	c_6
−0.592	0.0236	−0.011	0.063	0.470	0.177

2. Simulation of resistance coefficient along the tube

Referring to the Mukherjee's calculation method for resistance coefficient of gas–liquid two-phase flow along the tube [1], the resistance coefficient λ along the tube is the function of no slip resistance coefficient along the tube f_m, i.e., the resistance coefficient along the tube is calculated as:

$$\lambda = f_r f_m \tag{6}$$

In this study, the relation of friction coefficient f_r worked out with experimental data is as follows:

$$f_r = (4.326\sin\theta + 10.284)H_r \tag{7}$$

in which the relative liquid holdup H_r is as follows:

$$H_r = H_l' / H_l \tag{8}$$

$$H_l' = \frac{v_{sl}}{v_{sl} + v_{sg}} \tag{9}$$

where v_{sg} is the apparent velocity of gas (m/s); v_{sl} is the apparent velocity of liquid (m/s); and H_l is the liquid holdup.

The no slip resistance coefficient along the tube f_m is calculated directly using the equation presented by Mukherjee:

$$Re_m \leq 2300, \; f_m = \frac{64}{Re_m} \tag{10}$$

$$Re_m 2300, \; f_m = [1.14 - 21g\left(\frac{k}{D} + \frac{21.25}{Re_m^{0.9}}\right)]^{-2} \tag{11}$$

in which the no slip Reynolds number is as follows:

$$Re_m = \frac{v_m \rho_m D}{\mu_{ns}} \tag{12}$$

The no slip mixture density is as follows:

$$\rho_m = (1 - H_l')\rho_g + H_l'\rho_l \tag{13}$$

where ρ_l is the density of liquid phase (kg/m^3); ρ_g is the density of gas phase (kg/m^3).

Viscosity of no slip mixture is as follows:

$$\mu_{ns} = (1 - H_l')\mu_g + H_l'\mu_l \tag{14}$$

3.6.2. Comparison of Calculation Errors

The new model and six commonly used pressure calculation models were adopted to predict the pressure gradient under experimental conditions, and the experimental results were then compared. Table 4 shows that the error of the new model was 11.3%, suggesting it had a better prediction effect than the six commonly used pressure calculation methods.

Table 4. Comparison of prediction errors of models.

Models	Beggs	Mukherjee	Aziz	Hasan	JPI	Orkiszewski	New Models
Errors (%)	30.3	68.3	71.7	66.8	34.1	38.1	11.3

4. Conclusions

By analyzing and summing up the variation law of flow pattern, pressure drawdown, and liquid holdup with the output and gas injection rate at the tube inclined angles of 0°, 15°, 30°, 45°, 60°, 75°, and 90° as well as conducting verification analysis on several multiphase flow prediction methods, the following conclusions have been drawn:

- At the same liquid volume, as the gas volume increases, the flow pattern in the horizontal state tends to convert from laminar flow to slug flow and then to transition flow, whereas in the inclined state, the flow pattern tends to convert from slug flow to transition flow and then to annular flow.
- Under the experimental conditions, the Beggs–Brill flow pattern map, Mukerherjee flow pattern map, and Aziz flow pattern map are most accurate for the judgment of annular flow, with an accuracy of 100%; the Mukherjee flow pattern map and Duns and Ros flow pattern map have

a 80–100% accuracy in judging slug flow; and the Duns and Ros flow pattern map has a 46–66% accuracy in identifying transition flow.

- The liquid holdup and pressure drawdown are both affected by the gas injection rate, liquid volume, and inclined angle. When the inclined angle ranges 0–60°, the pressure drawdown increases with the increase of inclined angle; when the inclined angle exceeds 60°, the pressure drawdown reduces with the increase of inclined angle.
- Under the experimental conditions, the errors of six pressure drawdown prediction models are all bigger; therefore, a pressure drawdown model with new coefficients has been matched, with an error of only 11.3%.

Author Contributions: Each author has made contribution to the present paper. Conceptualization, D.Q., H.Z. and Y.D.; Data curation, D.Q.; Investigation, D.Q., H.Z., W.L. and J.Y.; Methodology, W.L.; Validation, D.Q. and H.Z.; Writing—original draft, D.Q.; Writing—review & editing, D.Q.

Funding: This research was funded by [Major national science and technology special project] grant number [2017ZX05030-005].

Acknowledgments: We would like to express appreciation to the following financial support: Major national science and technology special project: Research and Application of Key Technologies for Oil Production and Gas Recovery in Complex Carbonate Reservoirs in Central Asia and Middle East (2017ZX05030-005).

Conflicts of Interest: The authors declare no conflict of interest.

References

1. Li, Y. *The Technology of Petroleum Production*, 1st ed.; Petroleum Industry Press: Beijing, China, 2002; pp. 41–43, ISBN 9787502135904.
2. Chen, J. *Oil-Gas Flow in Pipes*, 2nd ed.; Petroleum Industry Press: Beijing, China, 2009; pp. 187–189, ISBN 750210317.
3. Liu, T. Mechanistic Model for Two-Phase Flow in Liquid-Cut Gas Wells. Ph.D. Thesis, Southwest Petroleum University, Chengdu, Sichuan Province, China, June 2014.
4. Han, H.; Zhang, X.; Xu, X. Experimental Study on the Flow Pattern of Two-phase Flow in Inclined Pipe. *Contem. Chem. Ind.* **2015**, *44*, 709–714.
5. Govier, G.W. Developments in Understanding of Vertical Flow of 2 Fluid Phases. *Can. J. Chem. Eng.* **1965**, *43*. [CrossRef]
6. Fancher, J.G.H.; Brown, K.E. Prediction of Pressure Gradients for Multiphase Flow in Tubing. *Soc. Petr. Eng. J.* **1963**, *3*, 59–69. [CrossRef]
7. Duns, J.H.; Ros, N.C.J. Vertical Flow of Gas and Liquid Mixtures in Wells. In Proceedings of the Sixth World Petroleum Congress, Frankfurt am Main, Germany, 19–26 June 1963.
8. Hagedorn, A.R; Brown, K.E. Experimental Study of Pressure Gradients Occurring During Continuous Two-Phase Flow in Small-Diameter Vertical Conduits. *J. Pet. Technol.* **1965**, *17*, 475–484. [CrossRef]
9. Orkiszewski, J. Predicting Two-phase Pressure Drops in Vertical Pipes. *J. Pet. Technol.* **1967**, *19*, 829–838. [CrossRef]
10. Aziz, K.; Govier, G.W.; Fogarasi, M. Pressure Drop in Wells Producing Oil and Gas. *J. Can. Pet. Technol.* **1972**, *11*, 38–47. [CrossRef]
11. Liu, X.; Xu, Y.; Peng, Y. Mechanistic Modeling of Two-phase Flow in Deviated Wells. *Oil Drill. Prod. Technol.* **2009**, *31*, 52–57. [CrossRef]
12. Beggs, H.D.; Brill, J.P. Study of Two Phase Flow in Inclined Pipes. *J. Pet. Technol.* **1973**, *25*, 607–617. [CrossRef]
13. Barnea, D. Transition From Annular-Flow and From Dispersed Bbbble Flow-Unified Models For The Whole Range of Pipe Inclinations. *Int. J. Multiph. Flow* **1986**, *12*, 733–744. [CrossRef]
14. Xiao, J.J.; Shoham, O.; Brill, J.P. A Comprehensive Mechanistic Model for Two-phase Flow in Pipelines. In Proceedings of the SPE Annual Technical Conference and Exhibition, New Orleans, Louisiana, 23–26 September 1990.
15. Chen, J. Comprehensive Mechanical Model of Upward Gas-liquid Two-phase Flow in a Wellbore. *Foreign Oil Field Eng.* **1992**, *6*, 57–64.

16. Han, H.; Wang, Z.; Yang, S. Mathematical Model of Gas-liquid Two-phase Flow Void Fraction in Circular Pipe. *Acta Pet. Sin.* **2002**, *26*, 19–21.

17. Barnea, D.; Taitel, Y. A Model for Slug Length Distribution in Gas-Liquid Slug Flow. *Int. J. Multiph. Flow* **1993**, *19*, 829–838. [CrossRef]

18. Liao, R.; Wang, Q.; Zhang, B. A Method for Predicting Vertical Multiphase-flow Pressure Gradient in Vertical Wellbore. *J. Jianghan Pet. Inst.* **1998**, *20*, 59–63.

19. Hibiki, T.; Ishii, M. Experimental Study on Interfacial Area Transport in Bubbly Two-phase Flows. *Int. J. Heat Mass Transf.* **1999**, *42*, 3019–3035. [CrossRef]

20. Zhang, M.; Webb, R.L. Correlation of Two-phase Friction for Refrigerants in Small-diameter Tubes. *Exp. Therm. Fluid Sci.* **2001**, *25*, 131–139. [CrossRef]

21. Hou, X. The Study on the Wellbore Pressure under Multiphase Flow Condition. *Pet. Drill. Tech.* **2004**, *32*, 32–34.

22. Chen, L.; Tian, Y.S.; Karayiannis, T.G. The Effect of Tube Diameter on Vertical Two-phase Flow Regimes in Small Tubes. *Int. J. Heat Mass Transf.* **2006**, *49*, 4220–4230. [CrossRef]

23. Yu, X.; Li, M.; Li, J. Analogue Caculation of Tempetrature Distribution for Two-Phase Flow in Vertical Pipes. *J. Qiqihar Univ.* **2008**, *24*, 40–44.

24. Cheng, L.X.; Ribatski, G.; Thome, J.R. Two-phase Flow Patterns and Flow-pattern Maps: Fundamentals and Applications. *Appl. Mech. Rev.* **2008**, *61*. [CrossRef]

25. Murai, Y.; Tasaka, Y.; Nambu, Y.; Takeda, Y.; Gonzalez, S.R. Ultrasonic Detection of Moving Interfaces in Gas-liquid Two-phase Flow. *Flow Meas. Instrum.* **2010**, *21*, 356–366. [CrossRef]

26. Gao, Q.; Li, T.; Zhao, Y.; Li, M.; Sun, J. Simulation Experiment of Flow Characteristic with Gas-liquid Two-phase Flow in Wellbore. *J. Yangtze Univ.* **2014**, *11*, 84–87. [CrossRef]

27. Bhagwat, S.M.; Ghajar, A.J. A Flow Pattern Independent Drift Flux Model Based Void Fraction Correlation for a Wide Range of Gas-liquid Two Phase Flow. *Int. J. Multiph. Flow* **2014**, *59*, 186–205. [CrossRef]

28. Zhou, S.; Zhu, L.; Zhang, Y.; Zheng, W. Experimental Study of Two-phase Flow in Inclined Pipe under High Gas-Liquid Flow. *Contemp. Chem. Ind.* **2016**, *45*, 504–510. [CrossRef]

29. Montoya, G.; Lucas, D.; Baglietto, E.; Liao, Y.X. A Review on Mechanisms and Models for the Churn-turbulent Flow Regime. *Chem. Eng. Sci.* **2016**, *141*, 86–103. [CrossRef]

30. Liu, H.; Hibiki, T. Flow Regime Transition Criteria for Upward Two-phase Flow in Vertical Rod Bundles. *Int. J. Heat Mass Transf.* **2017**, *108*, 423–433. [CrossRef]

31. Lu, C.H.; Kong, R.; Qiao, S.X.; Larimer, J.; Kim, S.; Bajorek, S.; Tien, K.; Hoxie, C. Frictional Pressure Drop Analysis for Horizontal and Vertical Air-water Two-phase Flows in Different Pipe Sizes. *Nucl. Eng. Des.* **2018**, *332*, 147–161. [CrossRef]

32. Liu, X.; Li, Z.; Wu, Y.; Lv, J. Effect of Tube Size on Flow Pattern of Air-Water Two-Phase Flow in Vertical Tubes. *Chin. J. Hydrodyn.* **2012**, *27*, 531–536. [CrossRef]

33. Gould, T.L.; Tek, M.R.; Katz, D.L. 2-Phase Flow through Vertical, Inclined, Or Curved Pipe. *J. Pet. Technol.* **1974**, *26*, 915–926. [CrossRef]

34. Ansari, A.M.; Sylvester, N.D.; Sarica, C.; Shoham, O.; Brill, J.P. A Comprehensive Mechanistic Model for Upward 2-Phase Flow in Wellbores. *SPE Prod. Facil.* **1994**, *9*, 143–152. [CrossRef]

35. Hasan, A.R. Void Fraction In Bubbly, Slug and Churn Flow In Vertical 2-Phase Up-Flow. *Chem. Eng. Commun.* **1988**, *66*, 101–111. [CrossRef]

Article

Flow Simulation of Artificially Induced Microfractures Using Digital Rock and Lattice Boltzmann Methods

Yongfei Yang [1,*], Zhihui Liu [1], Jun Yao [1], Lei Zhang [1], Jingsheng Ma [2], S. Hossein Hejazi [3], Linda Luquot [4] and Toussaint Dono Ngarta [1]

[1] Research Centre of Multiphase Flow in Porous Media, China University of Petroleum (East China), Qingdao 266580, Shandong, China; s16020334@s.upc.edu.cn (Z.L.); rcogfr_upc@126.com (J.Y.); zhlei84@163.com (L.Z.); toussaintdonongarta@yahoo.fr (T.D.N.)
[2] Institute of Petroleum Engineering, Heriot-Watt University, Riccarton, Edinburgh EH14 4AS, UK; Jingsheng.ma@pet.hw.ac.uk
[3] Department of Chemical and Petroleum Engineering, University of Calgary, Calgary, Alberta T3A 6C9, Canada; shhejazi@ucalgary.ca
[4] Hydrosciences Montpellier, Université Montpellier, CNRS, IRD, 300 Avenue du Pr. Emile Jeanbrau CC57, 34090 Montpellier, France; linda.luquot@umontpellier.fr
* Correspondence: yangyongfei@upc.edu.cn

Received: 5 July 2018; Accepted: 9 August 2018; Published: 17 August 2018

Abstract: Microfractures have great significance in the study of reservoir development because they are an effective reserving space and main contributor to permeability in a large amount of reservoirs. Usually, microfractures are divided into natural microfractures and induced microfractures. Artificially induced rough microfractures are our research objects, the existence of which will affect the fluid-flow system (expand the production radius of production wells), and act as a flow path for the leakage of fluids injected to the wells, and even facilitate depletion in tight reservoirs. Therefore, the characteristic of the flow in artificially induced fractures is of great significance. The Lattice Boltzmann Method (LBM) was used to calculate the equivalent permeability of artificially induced three-dimensional (3D) fractures. The 3D box fractal dimensions and porosity of artificially induced fractures in Berea sandstone were calculated based on the fractal theory and image-segmentation method, respectively. The geometrical parameters (surface roughness, minimum fracture aperture, and mean fracture aperture), were also calculated on the base of digital cores of fractures. According to the results, the permeability lies between 0.071–3.759 (dimensionless LB units) in artificially induced fractures. The wide range of permeability indicates that artificially induced fractures have complex structures and connectivity. It was also found that 3D fractal dimensions of artificially induced fractures in Berea sandstone are between 2.247 and 2.367, which shows that the artificially induced fractures have the characteristics of self-similarity. Finally, the following relations were studied: (a) exponentially increasing permeability with increasing 3D box fractal dimension, (b) linearly increasing permeability with increasing square of mean fracture aperture, (c) indistinct relationship between permeability and surface roughness, and (d) linearly increasing 3D box fractal dimension with increasing porosity.

Keywords: CT; digital rock; microfractures; Lattice Boltzmann method; pore-scale simulations

---

1. Introduction

The study of fractures has great importance in performing the following tasks: (a) properly characterizing and developing fractured oil and gas reservoirs, (b) nuclear-waste repository performance-assessment studies, (c) geothermal energy development, (d) studies of groundwater contamination, etc. In oil and gas development, microfractures are divided into natural microfractures and induced microfractures. Inducing fractures by hydraulic fracturing is one of the main methods to increase production. The existence of induced fractures will affect the fluid-flow system (expand the drainage radius of production wells), and act as flow paths for the leakage of fluids injected to the wells, and even facilitate depletion in tight reservoirs. Therefore, the characteristic of the flow in artificially induced fractures is of great significance.

Significant attention has been paid to the analysis of the microflow mechanism in fractures. The effect of fracture size and data uncertainties on fractured-rock permeability was analyzed by Sagar et al. [1]; the parameters influencing the interaction among closely spaced hydraulic fractures were reported by Bunger [2]; the hydraulic conductivity of rough fracture surface was studied by References [3–5]. It is well known that the flow in fractures is influenced by many factors [6], such as: (1) flow behavior through single fractures, (2) fracture network pattern, and (3) in situ stress system. It can be found that the induced fractures are composed of many intersected single fractures. The flow and stress-coupling characteristics of a rock depend largely on the behavior of the fluid flowing through a single fracture. Hence, the study of fluid flow through single fractures is essential.

Research on the flow through single fractures has been extensive and has led to a general understanding that the fractures in reservoirs are usually rough and it is impossible to put forward a fixed value of a fracture aperture. Therefore, the validity of the cubic law [7] is inappropriate for this case.

$$Q/\Delta h = \frac{\gamma}{12\mu} \frac{w}{L} b^3 \tag{1}$$

where Q, w, b, L, Δh, γ, and μ represent the linear fluid-flow rate, fracture width, fracture aperture, fracture length, total hydraulic head drop along the length of the fracture, weight of the flowing fluid, and the dynamic viscosity of the flowing fluid, respectively.

For rough fractures, various methods have been proposed. The roughness was incorporated into cubic-law simulations by many studies [8–10]: fractal theory was used to generate fracture models with different roughness [11], and the aperture of the idealized parallel smooth fracture was also replaced by the hydraulic aperture in many reports [12–14]. But so far none of these methods is unanimous.

The Lattice Boltzmann method (LBM) is a discrete approximation of the incompressible Navier–Stokes (N-S) equations based on kinetic theory [15]. This method is becoming a strong tool to simulate fluid flow in complex geometries. Besides, numerous studies extended LBM to perform the simulations from a microscale to a macroscale, which made it possible for LBM to be widely used. LBM has also been used to analyze the influence of wettability on fluid flow based on an ideal model [16] or generated self-affine rough fracture [17]. However, there were few reports on fluid-flow simulations in real three-dimensional (3D)-induced rough fractures based on LBM. A characterization of the flow simulations considering real rough-walled fractures is significant to many studies, such as the mechanical properties and transport properties of fractures [18].

Researchers made use of many techniques to obtain the structure of fractures in rocks, such as stylus profilemeters [19,20], surface laser scanning [21,22], nuclear magnetic-resonance imaging [14,23], and computed tomography (CT) scanning [3,24,25]. CT scanning and nuclear magnetic-resonance imaging are two methods by which we can get the structure of fractures without destroying the core.

The CT scanning method, which is known as X-ray computed tomography, can detect the inner structure of nontransparent objects without damage. Because the different components of the rock have different densities, which result in different X-ray absorption coefficients, the skeleton and pore space of rocks can be distinguished. CT scanning was applied widely on the characterization of fractures [26–30]. OP Wennberg et al. scanned the core samples to investigate the effect of natural open fractures on reservoir flow [31]; precise 3D numerical modeling was coupled with X-ray CT to analyze the heterogeneous fracture flow as well as measure porosity and permeability [32]. The X-ray CT scanning experiment is the most accurate and direct method for establishing a 3D digital core, which provides a basis for quantitative analysis and flow simulation of fractures.

In this study, LBM is used to simulate single-phase fluid flow through nine 3D actual fractures and calculate the equivalent permeability of the digital cores. Based on fractal theory, the fractal dimensions of artificially induced fractures in Berea sandstone are calculated. The geometric parameters (porosity, surface roughness, and mean fracture aperture) of the digital cores were calculated. Besides, the following relations were studied: (a) permeability versus 3D box fractal dimension in induced fractures, (b) permeability versus surface roughness, (c) permeability versus mean fracture aperture, and (d) 3D box fractal dimension versus porosity, etc. Finally, partial least square (PLS) regression was applied to observe the marginal effect of each independent variable (3D box fractal dimension, mean fracture aperture, surface roughness and minimum fracture aperture) in explaining permeability.

2. Methodology

2.1. The 3D Digital Core

The 3D digital core is a 3D digital image of the rock, which reflects the microscopic pore structure of the rock on the pore scale. In 2003, Dvorki et al. put forward the concept of Digital Rock Physics (DRP) technology, which is based on the 3D digital core and uses the numerical simulation algorithm to accurately calculate the acoustic, electrical, nuclear magnetic resonance, and seepage characteristics of the rock [33]. Compared to petrophysics experiments, DRP experiments are fast and low in cost. They can simulate different rock physical properties based on the same 3D digital core, and it is easy to analyze the correlation between different physical attributes.

The common methods for constructing a 3D digital core are divided into two types: an X-ray CT-scanning experiment and reconstruction method based on a couple of two-dimensional (2D) images. The reconstruction method based on 2D images is not accurate when compared with an X-ray CT-scanning experiment because the 2D images contain less information of the pore structure. The X-ray CT-scanning experiment is the most accurate and direct method for establishing a 3D digital core, although the capacity of pore recognition is limited by the resolution of the instrument. The schematic diagram is shown in Figure 1.

An X-ray CT-scanning experiment was conducted to construct the 3D digital cores of the fractured cores, which are Berea sandstones with an induced fracture. The fracture was artificially induced using a modified Brazilian technique, which was created by Karpyn et al. [34].

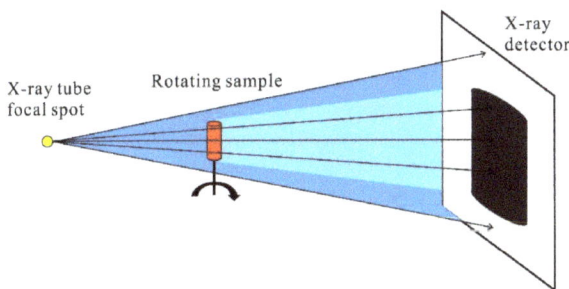

Figure 1. Schematic plot of X-ray micro-CT setup.

2.2. Fractal Theory and Fractal Dimension

2.2.1. The Concept of "Fractals"

In 1967, Mandelbrot published the title of "How long is the British coastline?" in the journal Science [35]. Comparing the photo of 100 kilometers of the coastline in the air to the enlarged 10 kilometers of the coastline, he found that the two pictures were very similar; Mandelbrot called the part, similar to the whole phenomenon, as the "Fractal".

Fractals include regular fractals and irregular fractals. Regular fractals can be generated by simple iterations or by certain rules, such as the Cantor set and Koch curve; the self-similarity and scaling invariance of the regular fractal are theoretically infinite. Irregular fractals refer to randomly generated nonsmooth fractal objects that have statistical self-similarity. Porous media are a kind of irregular fractal object, the fractal research of which has been various [36–40].

2.2.2. The Box-Counting Method and Fractal Dimension

The box-counting method is one of the most familiar methods of calculating the fractal dimension. In the past 20 years, the box-counting method has been applied to various subjects, such as medical studies, physical studies, and chemical studies.

The box-counting method is implemented by covering the function with a grid of identical square boxes and counting the number of boxes intersecting the function as a function of the box size [41,42].

$$N(r) = \frac{1}{r^{D_f}} = r^{-D_f} \tag{2}$$

$$D_f = \frac{\ln N(r)}{\ln(1/r)} \tag{3}$$

In Equation (3), D_f is called the fractal dimension (also called Hausdorff dimension); the fracture is covered with different box sizes r, and the resulting number of boxes $N(r)$, and the corresponding box sizes can be used to compute the fractal dimension.

Fractal dimension is the most important concept and content in fractal theory, and which is proposed by Mandelbrot for the study of complexity and the nondifferentiability (at everywhere) of surface curves. Fractal dimension is the main tool to characterize the complex structure of the fractal object, and the introduction of which is the novelty of fractal theory. It can be easily noticed from existing fractal research that the relationship between the fractal dimensions of the research objects and other physical parameters is the focus of many studies [43,44].

The 2D box fractal dimension was extended to 3D in this work to fully characterize fractures, and then 3D fractal dimensions of artificially induced fractures in Berea sandstone were calculated.

2.3. The Calculating of Permeability by LBM

The LBM is applied to the numerical simulations of different physical phenomena and is the primary domain of application is fluid dynamics. The advantage of the LBM is based on its ability to easily simulate complex practical problems in simulation. To calculate the dimensionless equivalent permeability of fractures by LBM, the binary data of digital cores were the input file of the program.

In the present work, the D3Q19 lattice model was used [45]. Where D represents the space dimension, Q represents the number of discrete velocity vectors. The distribution function, $f_i(x)$ (at each site x, for each lattice vector e_i), stands for the average movement of fluid particles. Figure 2 illustrates the numerical model of LBM. The Lattice Bhatnagar–Gross–Krook (LBGK) approximation was applied for distribution function at time t [45]:

$$f_i(x + e_i\Delta t, t + \Delta t) - f_i(x, t) = -\frac{1}{\tau}[f_i(x, t) - f_i^{eq}(x, t)] \tag{4}$$

The left part of Equation (4) represents the streaming step, while the right represents the collision in the evolution of the distribution function. $f_i^{eq}(x, t)$ is the equilibrium distribution function, and τ is the relaxation parameter.

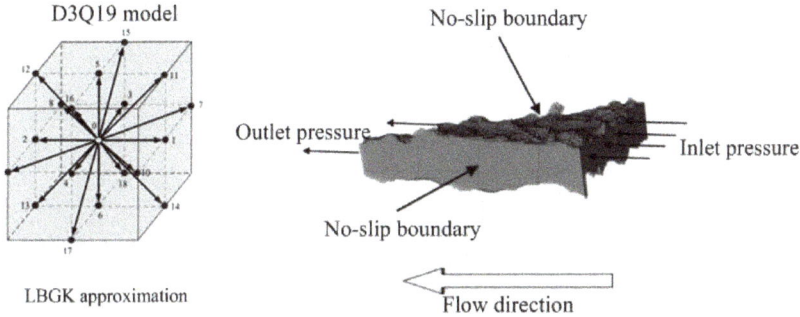

Figure 2. The numerical model of the lattice Boltzmann method (LBM) (The lattice Bhatnagar–Gross–Krook (LBGK) approximation was applied).

$$\tau = \frac{\upsilon}{c_s^2 \Delta t} + 0.5 \tag{5}$$

where υ and c_s represent the kinematic viscosity of fluid and the lattice pseudo-sound-speed, respectively. And the value of c_s is $\frac{1}{\sqrt{3}}$.

According to the incompressible LBM [46], the equilibrium distribution function for model D3Q19 is:

$$f_i^{eq}(x) = \omega_i \left[\rho(x) + \rho_0 \left(3(e_i \cdot u) + \frac{9}{2}(e_i \cdot u) - \frac{3}{2}(u \cdot u) \right) \right] \tag{6}$$

where ρ_0 is the mean density and ω_i is the weight factor in i-th direction. ω_i for D3Q19 is specified as:

$$\omega_i = \begin{cases} 1/3, \ i = 0, \\ 1/18, \ i = 1,2,3,4,5,6, \\ 1/36, \ i = 7,8,9,10,11,12,13,14,15,16,17,18 \end{cases} \tag{7}$$

And the fluid macroscopic density (ρ) and velocity (u) at a node x, are defined as:

$$\rho(x) = \sum_{i=0}^{18} f_i(x) \tag{8}$$

$$u = \frac{1}{\rho(x)} \sum_{i=0}^{18} f_i(x) e_i \tag{9}$$

Besides, the pressure in the calculating of LBM is defined as:

$$p(x) = c_s^2 \rho(x) \tag{10}$$

In this work, the single-phase fluid is driven by pressure difference, which is realized by setting the inlet and outlet pressure as fixed values. As represented in Figure 2, the pressure-boundary condition was set on inlet surface and outlet surface, and the pressure gradient was 0.00005 for all fractures. Using the segmented images as input data, the walls of the porous medium are converted to bounce-back boundary conditions. A bounce-back scheme is a common scheme for dealing with no-slip boundaries. In this scheme, when the particle reaches the wall node, the particle will return to the fluid node along the original path, and the direction is opposite to the incident direction. It is worth noting that particles do not collide on the wall. The bounce-back scheme is very easy to implement, and is suitable for handling systems with complex geometry [47].

The steady state is reached by checking the velocity and density of the fluid in fracture every 1000 steps, until the difference between the 2 adjacent times is less than 10^{-6}. According to Darcy's law, Equation (11), the equivalent permeability of the induced fracture in Berea sandstone was calculated.

$$-\frac{dP}{dx} = \frac{\mu}{k}U \tag{11}$$

3. Results and Discussion

3.1. The Digital Rocks of Fractures

The digital rocks presented in this study are Berea sandstone (with an induced fracture), which were downloaded in the sharing portal [48]. The core samples were prepared in such a way that the images are denoised and smoothed, then fractures were segmented by the multithresholding segmentation method, and the results are shown in Figure 3.

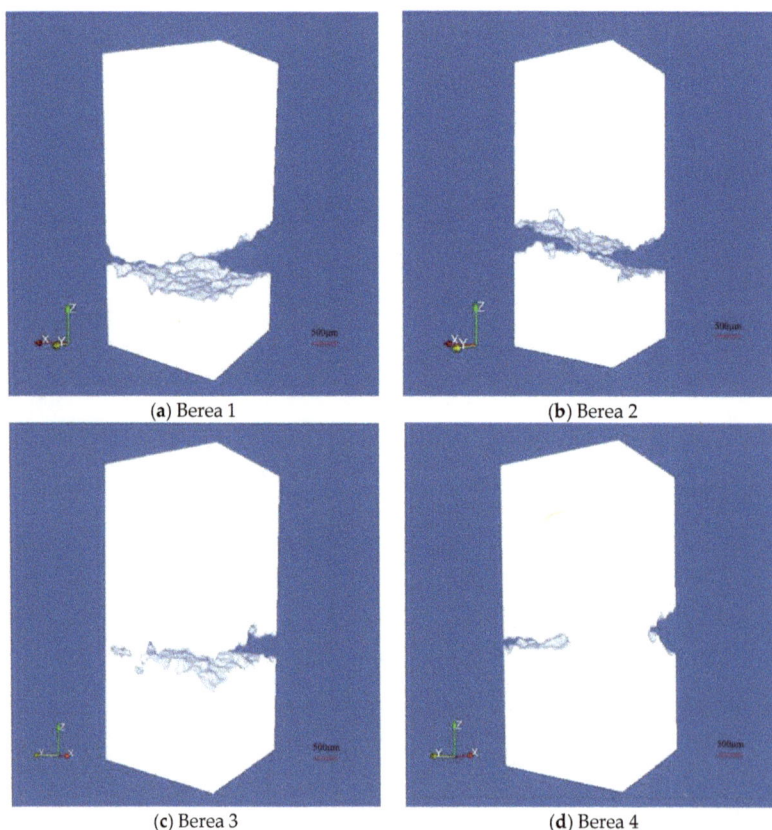

(a) Berea 1

(b) Berea 2

(c) Berea 3

(d) Berea 4

Figure 3. *Cont.*

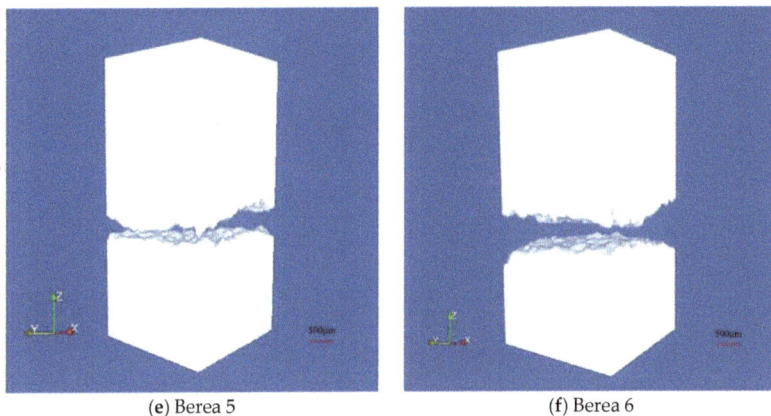

(e) Berea 5   (f) Berea 6

Figure 3. Digital cores of Berea sandstone with induced fracture (the size of every digital core is $2.743 \times 2.743 \times 6.035$ mm^3), (**a**) Berea 1 to (**f**) Berea 6 are samples of all nine digital cores; their aperture and roughness are totally different.

The minimum aperture values, the mean aperture values, surface roughness, and the fracture porosity (which is the ratio of fracture volume to total volume) of all the fractures are shown in Table 1.

Tamura et al. generalized six image texture features related to human visual perception. Roughness is one of them, and its ability to describe texture is very strong. It has more application value in texture synthesis, image analysis and recognition, color migration, and so on. The calculation method of surface roughness came from his article [49].

Table 1. The minimum aperture values, mean aperture values, and fracture porosity of all the fractures.

Core Sample	Porosity	Minimum Fracture Aperture/Pixels	Mean Fracture Aperture/Pixels	Surface Roughness/Pixels
Berea 1	0.111	7	24.4	8.962
Berea 2	0.082	2	18.4	8.568
Berea 3	0.068	0	15.0	7.865
Berea 4	0.067	0	14.7	7.892
Berea 5	0.093	2	20.5	8.424
Berea 6	0.100	8	21.9	8.384
Berea 7	0.117	7	25.7	9.141
Berea 8	0.088	6	19.3	7.972
Berea 9	0.106	8	23.3	8.743

3.2. Fractal Dimension Calculating of Fractures

The digital cores shown in Section 3.1 were used to calculate the fractal dimensions of different fractures. It is noticeable that the 3D box fractal dimension was the fractal dimension of fracture, which was carried out by inverting the pore phase and skeleton phase in Figure 3, and then obtaining the structure of complete fracture, as shown in Figure 4, and Tables 2 and 3:

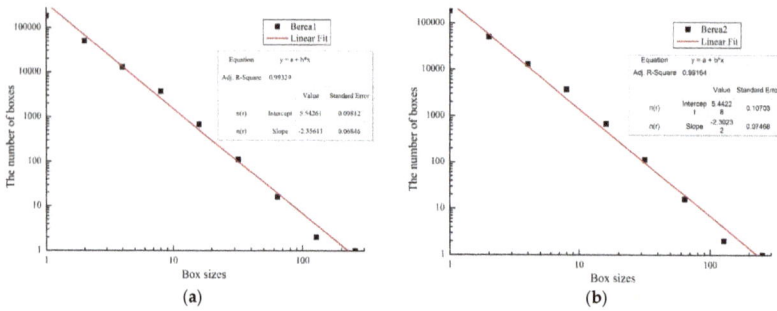

Figure 4. Fractal dimension calculating of (**a**) Berea 1 and (**b**) Berea 2.

Table 2. Statistical results of 3D box fractal dimensions.

	r	1	2	4	8	16	32	64	128	256
	Berea 1	243,739	64,473	14,952	3899	682	112	16	2	1
	Berea 2	184,084	49,754	12,946	3714	679	112	16	2	1
	Berea 3	149,777	40,921	10,970	3302	640	104	16	2	1
	Berea 4	146,858	39,713	9936	2982	637	107	16	2	1
$N(r)$	Berea 5	205,178	54,841	13,687	3751	685	112	16	2	1
	Berea 6	219,441	58,328	14,339	3866	685	112	16	2	1
	Berea 7	256,973	67,771	15,537	4025	686	112	16	2	1
	Berea 8	193,450	51,953	13,339	3734	686	112	16	2	1
	Berea 9	232,925	61,929	14,801	3926	686	112	16	2	1

Table 3. 3D box fractal dimension of Berea with an induced fracture.

Core Sample	Berea 1	Berea 2	Berea 3	Berea 4	Berea 5	Berea 6	Berea 7	Berea 8	Berea 9
3D Box Fractal Dimension	2.356	2.302	2.259	2.247	2.323	2.337	2.367	2.312	2.349

3.3. The Permeability of Fractures

The streamlines of induced fractures are shown in Figure 5, and the permeability of induced fractures in Berea sandstone is obtained in Table 4. It should be noted that all units in this work are nondimensional LB units.

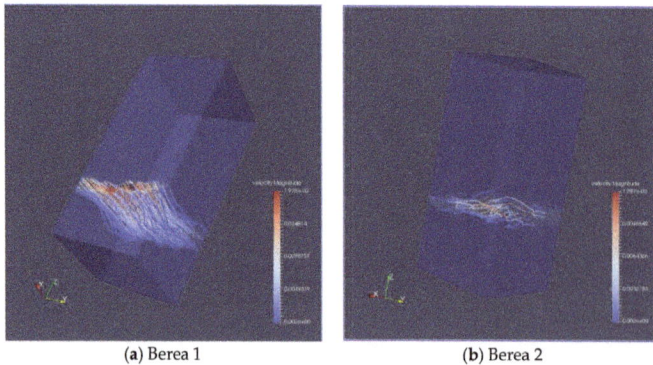

Figure 5. *Cont.*

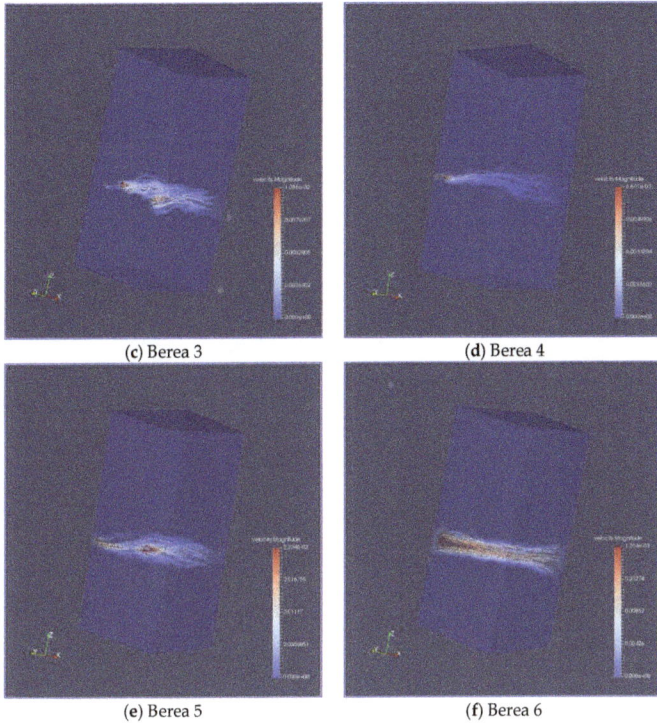

(c) Berea 3   (d) Berea 4

(e) Berea 5   (f) Berea 6

Figure 5. The streamlines of induced fractures in Berea.

Table 4. The permeability of Berea with an induced fracture.

Core Sample	Berea 1	Berea 2	Berea 3	Berea 4	Berea 5	Berea 6	Berea 7	Berea 8	Berea 9
Permeability	1.903	0.820	0.300	0.071	1.183	2.108	3.759	1.492	2.852

3.4. The Relations of Permeability and Other Parameters

Fluid flow through a single fracture was determined by spatial distribution of fracture aperture, which can be quantified by fractal parameter, mean aperture, or surface roughness. So we study the functional relations between (**a**) permeability and 3D box fractal dimension, (**b**) permeability and surface roughness, (**c**) permeability and mean fracture aperture, (**d**) 3D box fractal dimension and porosity, and (**e**) 3D box fractal dimension and mean fracture aperture.

3.4.1. Permeability versus 3D Box Fractal Dimension

The correlation between permeability and 3D box fractal dimension is widely studied; Ju et al. investigated the mechanism of fluid flow through a single rough fracture of rocks, and a nonlinear relationship between the fractal equivalent permeability of a single fracture and the fractal dimension D of its rough structure [11]. They generated a single fracture based on the Weierstrass–Mandelbrot fractal function. The relationship between permeability and 3D box fractal dimension obtained in this work is different from theirs, which is shown in Figure 6:

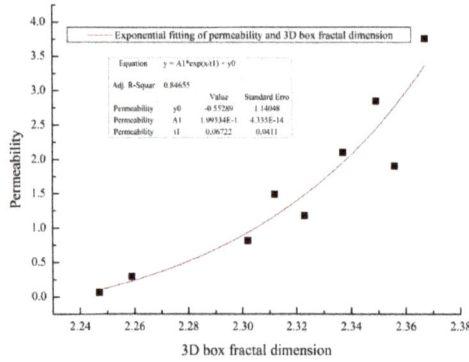

Figure 6. The relationship between permeability and 3D box fractal dimension.

The equivalent permeability (dimensionless LB units) of induced fracture, k_{LB}^{equ}, increase exponentially with the increasing of 3D box fractal dimension, FD_{3d}. The equation between permeability and 3D box fractal dimension obtained is:

$$k_{LB}^{equ} = e^{-10.32+14.92FD_{3d}} - 0.553 \tag{12}$$

3.4.2. Permeability versus Surface Roughness

The relationship between permeability and surface roughness is shown in Figure 7. The roughness of natural rock fractures is a topic of interest to many researchers, including rock hydrogeologists, geochemists, and geophysicists. The complexity of fracture morphology (the roughness and variability in fracture aperture) is the fundamental property that keeps fractures open and makes flow through fractures significant and important to subsurface fluid flow. Nevertheless, this parameter is hard to measure and determine [49].

Tayfun et al. found that surface roughness of fracture walls had a critical effect on the hydraulic conductivity of a fracture [50]. The hydraulic conductivity ratio also shows decreasing trend with increasing fracture surface roughness.

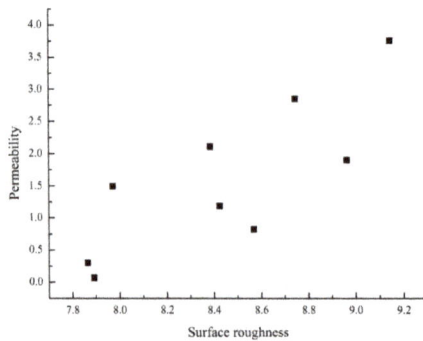

Figure 7. The relationship between permeability and surface roughness.

However, in this study, the relationship between permeability and surface roughness is indistinct in Berea sandstone with an induced fracture and the permeability shows an increasing trend with the increase of surface roughness. This is because of the difference between real fractures and the artificially generated fracture by the Weierstrass–Mandelbrot fractal function. Real fractures have no fixed aperture, and the roughness of fractures with a different aperture is varied, which results in increasing permeability with the increase of surface roughness. In large part, it is caused by the influence of other factors, for instance, the mean fracture aperture on permeability. As shown in Section 3.6, the influence of surface roughness on a dependent variable is much weaker compared with the square of mean fracture aperture and minimum fracture aperture.

3.4.3. Permeability versus Mean Fracture Aperture

Although it is impossible to put forward a fixed value of fracture aperture, the relationship between permeability and mean fracture aperture is still worth analyzing, as illustrated in Figure 8. It is generally known that the relationship between permeability and the square of mean fracture aperture for a horizontal flat-plate model is:

$$k^{equ} = \frac{b^2}{12} \tag{13}$$

where k^{equ} is the equivalent permeability of the horizontal flat-plate model.

So the relationship between permeability and the square of mean fracture aperture is studied in the following figure:

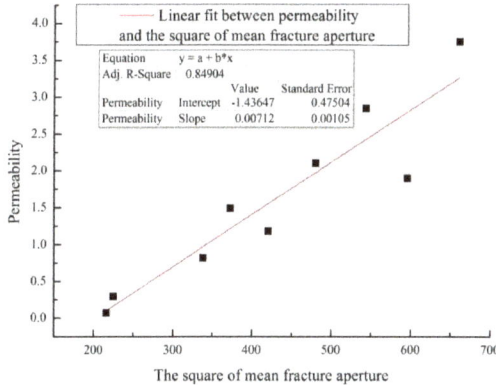

Figure 8. The relationship between permeability and the square of mean fracture aperture of Berea sandstone with an induced fracture.

Figure 8 shows a very strong linear relation between permeability and mean fracture aperture, which illustrates that the equivalent permeability of real fractures is proportional to the square of the mean fracture aperture. High R^2 also indicates that there is a strong correlation between permeability and the square of mean fracture aperture.

3.4.4. Permeability versus Minimum Fracture Aperture

In Figure 9, the relationship between permeability and minimum fracture aperture is shown. Minimum fracture aperture has a significant impact on the flow because, the smaller the minimum fracture aperture is, the greater the pressure drops. From Figure 9 we can also conclude that permeability increases with the increasing minimum fracture aperture.

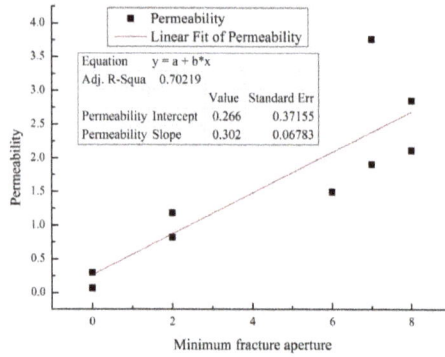

Figure 9. The relationship between permeability and minimum fracture aperture of Berea sandstone with an induced fracture.

3.5. The Relationship of 3D Box Fractal Dimension and Other Parameters

To characterize the spatial distribution of fracture aperture, the 3D box fractal dimension, FD_{3d}, was used to characterize 3D artificially induced fractures in Berea sandstone. The relationship of 3D box fractal dimension and other parameters was analyzed below.

3.5.1. 3D Box Fractal Dimension versus Porosity

Figure 10 displays the curve of the 3D box fractal dimension versus porosity, which shows that the 3D box fractal dimension increases linearly with increasing porosity.

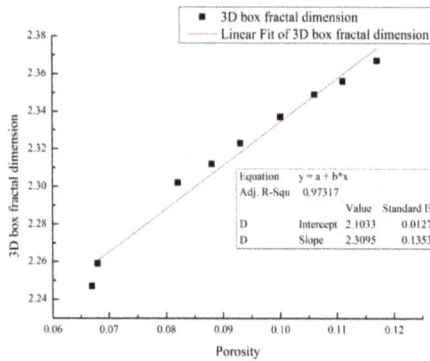

Figure 10. The relationship between the 3D box fractal dimension and porosity of Berea sandstone with an induced fracture.

The figure shows a very strongly linear relation between the 3D box fractal dimension and porosity (R^2 value is greater than 0.97). The standard errors of slope and intercept of the linear relationship lie in a small range.

3.5.2. 3D Box Fractal Dimension versus Mean Fracture Aperture

In the previous sections, it was found that the relationship between 3D box fractal dimension and porosity is strongly linear, and the relationship between permeability and mean fracture aperture is almost the same as the relationship between permeability and porosity, as shown in Figure 11.

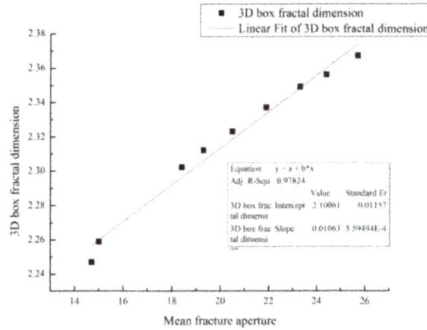

Figure 11. The relationship between 3D box fractal dimension and mean fracture aperture.

It is not difficult to see that the R^2 value of the relationship between 3D box fractal dimension and mean fracture aperture is even larger than that of 3D box fractal dimension and porosity, which indicates that mean fracture aperture has a great impact on 3D box fractal dimension.

3.5.3. 3D Box Fractal Dimension versus Surface Roughness

Surface roughness may have a great impact on 3D box fractal dimension because more boxes are counted by the box-counting method if surface roughness is larger. The relationship between 3D box fractal dimension and surface roughness of Berea sandstone with an induced fracture is shown in Figure 12.

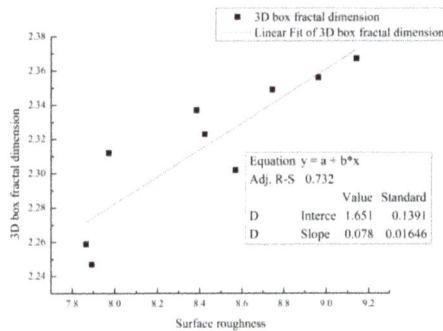

Figure 12. The relationship between 3D box fractal dimension and surface roughness of Berea sandstone with an induced fracture.

The Figure 12 shows that 3D box fractal dimension increases with the increasing surface roughness in Berea sandstone; however, a good linear fit was not achieved ($R^2 = 0.733$).

3.6. PLS Regression

In order to observe the marginal effect of each independent variable in explaining permeability more quickly and intuitively, we can draw a regression coefficient map, as shown in Figure 13, which is based on regression equation for standardized data.

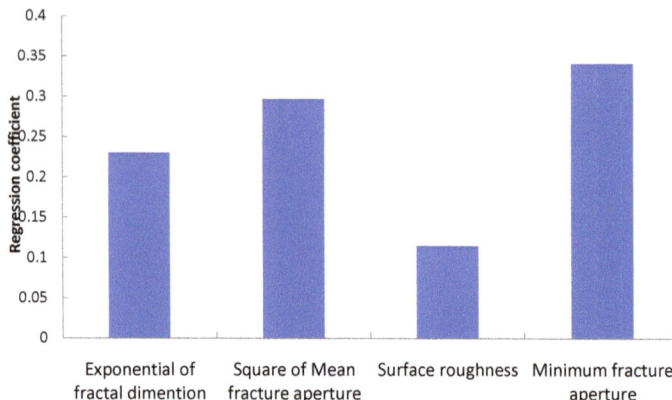

Figure 13. Regression coefficient map based on regression equations for standardized data.

It can be observed from the regression coefficient diagram that the minimum fracture aperture and square of mean fracture aperture play a very important role in explaining the regression equation. At the same time, it can be seen from the Figure 13 that the influence of surface roughness on dependent variable is weaker than other variables, which indicates that the permeability of real fractures is more strongly affected by other parameters.

In order to investigate the accuracy of the regression equation model, we used ($\hat{k}_{LB}^{equ}$, k_{LB}^{equ}) as the coordinate values to draw the prediction map for all the sample points. $\hat{k}_{LB}^{equ}$ was the predicted value of the sample point (k_{LB}^{equ}). As shown in Figure 14, all points are evenly distributed near the diagonal of the graph, the difference between the fitting value of the equation and the original value is very small, and the fitting effect of the equation is satisfactory.

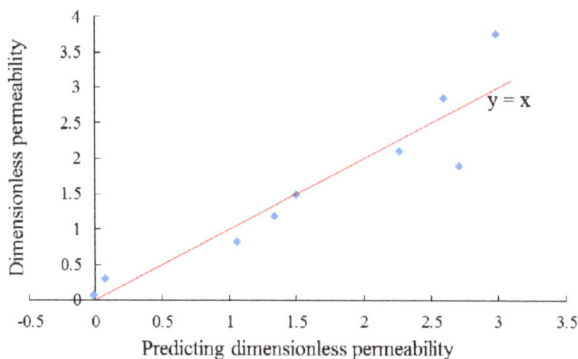

Figure 14. The fitting value of the equation versus the original value.

The regression equation is:

$$k_{LB}^{equ} = -9.0185 + 0.6541 \exp(FD_{3d}) + 0.0023\bar{b}^2 - 0.2945Ra + 0.1198b_{\min} \qquad (14)$$

where $\bar{b}$, Ra and $b_{\min}$ represent mean fracture aperture, surface roughness, and minimum fracture aperture, respectively.

4. Conclusions

(1) Artificially induced fractures were studied in this work, and their geometrical parameters (porosity, surface roughness, minimum fracture aperture, and mean fracture aperture) were calculated. 3D box fractal dimension of fractures was also calculated; it was found that a 3D box fractal dimension is between 2.247–2.367, which shows that the artificially induced fractures have the characteristics of self-similarity.

(2) LBM was used to compute the permeability of artificially induced fractures; the permeability was between 0.071–3.759 (dimensionless LB units). The difference in permeability indicates that artificially induced fractures have complex structures.

(3) The following relations have been identified: (a) exponentially increasing permeability with increasing 3D box fractal dimension in induced fractures, (b) linearly increasing permeability with increasing square of mean fracture aperture, (c) indistinct relationship between permeability and surface roughness, and (d) linearly increasing 3D box fractal dimension with increasing porosity, mean fracture aperture, and surface roughness.

(4) By PLS regression, it can be concluded that minimum fracture aperture and mean fracture aperture play a significant role in explaining the fracture permeability. The influence of surface roughness on dependent variables is much weaker compared with the square of mean fracture aperture and minimum fracture aperture. Besides, the regression equation was obtained, and the fitting result of which was satisfactory.

Author Contributions: Each author has made contribution to the present paper. Y.Y., Z.L. and L.Z. conceived and designed the experiments; Z.L., T.D.N. and S.H.H. performed the experiments; Y.Y. and L.Z. processed and analyzed the experimental data; J.Y. provided the experimental support. General supervision was provided by J.M. and L.L. All authors have read and approved the final manuscript.

Funding: We would like to express appreciation to the following for their financial support: the National Natural Science Foundation of China (No. 51674280, 51711530131,51490654), Key Research and Development Plan of Shandong Province (2018GSF116009), Applied Basic Research projects of Qingdao Innovation Plan (16-5-1-38-jch), the Fundamental Research Funds for the Central Universities (No. 17CX05003, 18CX02031A), National Science and Technology Major Project (2016ZX05010002005), and Program for Changjiang Scholars and Innovative Research Team in University (IRT_16R69). Hossein Hejazi acknowledges partial support from the University of Calgary Beijing Research Site, a research initiative associated with the University of Calgary Global Research Initiative in Sustainable Low Carbon Unconventional Resources, and the Kerui Group.

Conflicts of Interest: The authors declare no conflict of interest.

References

1. Sagar, B.; Runchal, A. Permeability of fractured rock: Effect of fracture size and data uncertainties. *Water Resour. Res.* **1982**, *18*, 266–274. [CrossRef]
2. Bunger, A.P.; Zhang, X.; Jeffrey, R.G. Parameters Affecting the Interaction Among Closely Spaced Hydraulic Fractures. *SPE J.* **2012**, *17*, 292–306. [CrossRef]
3. Akhavan, A.; Shafaatian, S.M.H.; Rajabipour, F. Quantifying the effects of crack width, tortuosity, and roughness on water permeability of cracked mortars. *Cem. Concr. Res.* **2012**, *42*, 313–320. [CrossRef]
4. Neuman, S.P. Trends, prospects and challenges in quantifying flow and transport through fractured rocks. *Hydrogeol. J.* **2005**, *13*, 124–147. [CrossRef]
5. Or, D.; Tuller, M. Flow in unsaturated fractured porous media: Hydraulic conductivity of rough surfaces. *Water Resour. Res.* **2000**, *36*, 1165–1177. [CrossRef]
6. Samuel, F.; Pejman, T.; Mohammed, P. Interaction Between Fluid and Porous Media with Complex Geometries: A Direct Pore-Scale Study. *Water Resour. Res.* **2018**. [CrossRef]
7. Snow, D.T. A Parallel Plate Model Of Fractured Permeable Media. Ph.D. Thesis, University of California, Oakland, CA, USA, 1965.
8. Ge, S. A governing equation for fluid flow in rough fractures. *Water Resour. Res.* **1997**, *33*, 53–61. [CrossRef]
9. Genabeek, O.V.; Rothman, D.H. Critical behavior in flow through a rough-walled channel. *Phys. Lett. A* **1999**, *255*, 31–36. [CrossRef]

10. Konzuk, J.S.; Kueper, B.H. Evaluation of cubic law based models describing single-phase flow through a rough-walled fracture. *Water Resour. Res.* **2004**, *40*, 389–391. [CrossRef]
11. Yang, J.; Zhang, Q.G.; Yang, Y.M.; Xie, H.P.; Gao, F.; Wang, H.J. An experimental investigation on the mechanism of fluid flow through single rough fracture of rock. *Sci. China Technol. Sci.* **2013**, *56*, 2070–2080.
12. Bertels, S.P.; Dicarlo, D.A.; Blunt, M.J. Measurement of aperture distribution, capillary pressure, relative permeability, and in situ saturation in a rock fracture using computed tomography scanning. *Water Resour. Res.* **2001**, *37*, 649–662. [CrossRef]
13. Brown, S.R. Transport of fluid and electric current through a single fracture. *J. Geophys. Res. Solid Earth* **1989**, *94*, 9429–9438. [CrossRef]
14. Renshaw, C.E.; Dadakis, J.S.; Brown, S.R. Measuring fracture apertures: A comparison of methods. *Geophys. Res. Lett.* **2000**, *27*, 289–292. [CrossRef]
15. Wolf-Gladrow, D.A. *Lattice Gas Cellular Automata and Lattice Boltzmann Models*; Springer: Tokyo, Japan, 2000.
16. Zhang, L.; Yao, J.; Zhao, J.L.; Li, A.; Sun, H.; Wan, Y.; Su, Y. The influence of wettability and shut-in time on oil recovery through microscale simulation based on an ideal model. *J. Nat. Gas Sci. Eng.* **2017**, *48*, 178–185. [CrossRef]
17. Dou, Z.; Zhou, Z.; Sleep, B.E. Influence of wettability on interfacial area during immiscible liquid invasion into a 3D self-affine rough fracture: Lattice Boltzmann simulations. *Adv. Water Resour.* **2013**, *61*, 1–11. [CrossRef]
18. Chen, Y.F.; Zhou, J.Q.; Hu, S.H.; Hu, R.; Zhou, C.B. Evaluation of Forchheimer equation coefficients for non-Darcy flow in deformable rough-walled fractures. *J. Hydrol.* **2015**, *529*, 993–1006. [CrossRef]
19. Tsang, Y.W. The Effect of Tortuosity on Fluid Flow Through a Single Fracture. *Water Resour. Res.* **1984**, *20*, 1209–1215. [CrossRef]
20. Alyaarubi, A.H.; Pian, C.C.; Grattoni, C.A.; Zimmerman, R.W. Navier-stokes simulations of fluid flow through a rock fracture. *Geophys. Monogr.* **2013**, *162*, 55–64.
21. Hans, J.; Boulon, M. A new device for investigating the hydro-mechanical properties of rock joints. *Int. J. Numer. Anal. Meth. Geomech.* **2003**, *27*, 513–548. [CrossRef]
22. Koyama, T.; Neretnieks, I.; Jing, L. A numerical study on differences in using Navier–Stokes and Reynolds equations for modeling the fluid flow and particle transport in single rock fractures with shear. *Int. J. Rock Mech. Min. Sci.* **2008**, *45*, 1082–1101. [CrossRef]
23. Brown, S.; Caprihan, A.; Hardy, R. Experimental observation of fluid flow channels in a single fracture. *J. Geophys. Res. Solid Earth* **1998**, *103*, 5125–5132. [CrossRef]
24. Berkowitz, B. Characterizing flow and transport in fractured geological media: A review. *Adv. Water Resour.* **2002**, *25*, 861–884. [CrossRef]
25. Liu, Z.; Yang, Y.; Yao, J.; Zhang, Q.; Ma, J.; Qian, Q. Pore-scale remaining oil distribution under different pore volume water injection based on CT technology. *Adv. Geo-energ. Res.* **2017**, *1*, 171–181. [CrossRef]
26. Cai, Y.; Liu, D.; Mathews, J.P.; Pan, Z.; Elsworth, D.; Yao, Y.; Li, J.; Guo, X. Permeability evolution in fractured coal—Combining triaxial confinement with X-ray computed tomography, acoustic emission and ultrasonic techniques. *Int. J. Coal. Geol.* **2014**, *122*, 91–104. [CrossRef]
27. Heriawan, M.N.; Koike, K. Coal quality related to microfractures identified by CT image analysis. *Int. J. Coal. Geol.* **2015**, *140*, 97–110. [CrossRef]
28. Mazumder, S.; Wolf, K.-H.A.A.; Elewaut, K.; Ephraim, R. Application of X-ray computed tomography for analyzing cleat spacing and cleat aperture in coal samples. *Int. J. Coal. Geol.* **2006**, *68*, 205–222. [CrossRef]
29. Montemagno, C.D.; Pyrak-Nolte, L.J. Fracture network versus single fractures: Measurement of fracture geometry with X-ray tomography. *Phys. Chem. Earth Part A* **1999**, *24*, 575–579. [CrossRef]
30. Yang, Y.; Liu, Z.; Sun, Z.; An, S.; Zhang, W.; Liu, P.; Yao, P.; Ma, J. Research on Stress Sensitivity of Fractured Carbonate Reservoirs Based on CT Technology. *Energies* **2017**, *10*, 1833. [CrossRef]
31. Wennberg, O.P.; Rennan, L.; Basquet, R. Computed tomography scan imaging of natural open fractures in a porous rock; geometry and fluid flow. *Geophys. Prospect.* **2009**, *57*, 239–249. [CrossRef]
32. Watanabe, N.; Ishibashi, T.; Hirano, N.; Ohsaki, Y.; Tsuchiya, Y.; Tamagawa, T.; Okabe, H.; Tsuchiya, N. Precise 3D Numerical Modeling of Fracture Flow Coupled With X-Ray Computed Tomography for Reservoir Core Samples. *SPE J.* **2011**, *16*, 683–691. [CrossRef]
33. Jack, D.; Qian, F.; Naum, D. Etudes in computational rock physics: Alterations and benchmarkingEtudes in computational rock physics. *Geophysics* **2012**, *77*, D45–D52.

34. Karpyn, Z.T.; Grader, A.S.; Halleck, P.M. Visualization of fluid occupancy in a rough fracture using micro-tomography. *J. Colloid. Interface Sci.* **2007**, *307*, 181–187. [CrossRef] [PubMed]
35. Mandelbrot, B.B. How long is the coastline of Britain. *Science* **1967**, *156*, 3775. [CrossRef] [PubMed]
36. Cai, J.; Wei, W.; Hu, X.Y.; Liu, R.H.; Wang, J.J. Fractal characterization of dynamic fracture network extension in porous media. *Fractals* **2017**, *25*, 1750023. [CrossRef]
37. Cai, J.; Yu, B.; Zou, M.; Luo, L. Fractal characterization of spontaneous co-current imbibition in porous media. *Energy Fuels* **2010**, *24*, 1860–1867. [CrossRef]
38. King, P.R. The fractal nature of viscous fingering in porous media. *J. Phys. Math. Gen.* **1987**, *20*, L529. [CrossRef]
39. Xie, S.; Cheng, Q.; Ling, Q.; LI, B.; Bao, Z.; Fan, P. Fractal and multifractal analysis of carbonate pore-scale digital images of petroleum reservoirs. *Mar. Pet. Geol.* **2010**, *27*, 476–485. [CrossRef]
40. Zheng, Q.; Yu, B. A fractal permeability model for gas flow through dual-porosity media. *J. Appl. Phys.* **2012**, *111*, 024316. [CrossRef]
41. Mandelbrot, B.B.; Wheeler, J.A. The Fractal Geometry of Nature. *Am. J. Phys.* **1983**, *51*, 468. [CrossRef]
42. Li, J.; Liu, Z.; Li, J.; Lu, S.F. Fractal characteristics of continental shale pores and its significance to the occurrence of shale oil in china: A case study of biyang depression. *Fractals* **2018**, *26*, 1840008. [CrossRef]
43. Yuliang, S.; Sheng, G.; Wang, W.; Zhang, Q.; Lu, M.; Ren, L. A mixed-fractal flow model for stimulated fractured vertical wells in tight oil reservoirs. *Fractals* **2016**, *24*, 1650006.
44. Wang, F.; Liu, Z.; Jiao, L.; Wang, C.; Guo, H. A fractal permeability model coupling boundary-layer effect for tight oil reservoirs. *Fractals* **2017**, *25*, 1750042. [CrossRef]
45. Qian, Y.H.; D'Humières, D.; Lallemand, P. Lattice BGK Models for Navier-Stokes Equation. *Europhys. Lett.* **1992**, *17*, 479.
46. He, X.; Luo, L.S. Lattice Boltzmann Model for the Incompressible Navier–Stokes Equation. *J. Stat. Phys.* **1997**, *88*, 927–944. [CrossRef]
47. Qisu, Z.; He, X. On pressure and velocity boundary conditions for the lattice Boltzmann BGK model. *Phys. Fluids* **1997**, *9*, 1591–1598.
48. Induced Rough Fracture in Berea Sandstone Core Digital Rocks Portal. 2016. Available online: http://www.digitalrocksportal.org/projects/31 (accessed on 10 August 2018).
49. Tamura, H.; Mori, S.; Yamawaki, T. Textural Features Corresponding to Visual Perception. *IEEE Trans. Syst. Man Cybern.* **1978**, *8*, 460–473. [CrossRef]
50. Babadagli, T.; Ren, X.; Develi, K. Effects of fractal surface roughness and lithology on single and multiphase flow in a single fracture: An experimental investigation. *Int. J. Multiph. Flow* **2015**, *68*, 40–58. [CrossRef]

![energies](energies logo)

MDPI

Article

Evaluation of the Vertical Producing Degree of Commingled Production via Waterflooding for Multilayer Offshore Heavy Oil Reservoirs

Fei Shen [1,2,*], Linsong Cheng [1], Qiang Sun [3] and Shijun Huang [1]

[1] College of Petroleum Engineering, China University of Petroleum, Beijing 102249, China; lscheng@cup.edu.cn (L.C.); fengyun7407@163.com (S.H.)

[2] Research Institute of Petroleum Exploration and Development, Zhongyuan Oil Field, Sinopec, Zhengzhou 450046, China

[3] Tianjin Branch of CNOOC Ltd., Tianjin 300452, China; sunqiang19@cnooc.com.cn

[*] Correspondence: daniel3274@163.com; Tel.: +86-188-1045-9506

Received: 15 July 2018; Accepted: 10 September 2018; Published: 13 September 2018

Abstract: Recently, commingling production has been widely used for the development of offshore heavy oil reservoirs with multilayers. However, the differences between layers in terms of reservoir physical properties, oil properties and pressure have always resulted in interlayer interference, which makes it more difficult to evaluate the producing degree of commingled production. Based on the Buckley–Leverett theory, this paper presents two theoretical models, a one-dimensional linear flow model and a planar radial flow model, for water-flooded multilayer reservoirs. Through the models, this paper establishes a dynamic method to evaluate seepage resistance, sweep efficiency and recovery percent and then conducts an analysis with field data. The result indicates the following: (1) the dynamic difference in seepage resistance is an important form of interlayer interference during the commingled production of an offshore multilayer reservoir; (2) the difference between commingled production and separated production is small within a certain range of permeability ratio or viscosity ratio, but separated production should be adopted when the ratio exceeds a certain value.

Keywords: multilayer reservoir; interlayer interference; producing degree; seepage resistance

1. Introduction

As we know, waterflooding is the most widely and effectively used method in secondary recovery. For offshore conventional heavy oil reservoirs, with a viscosity of 50–200 mPa·s, waterflooding is still the best process available to produce crude oil. However, because of the variation in the depositional environment, the formational rocks may exhibit huge variations in their petrophysical properties, especially permeability. In order to increase the optimum oil production rates in offshore reservoirs, commingling techniques have been used in many oilfields. As the rock petrophysical properties and fluid parameters change in multilayers, earlier water breakthrough occurred in layers of higher permeability during commingling production, which may be called thief zones. Thus, the interlayer contradictions become prominent and the oilfield development is seriously affected. Many cases on the developmental laws of waterflooding performance via commingling production in multilayer reservoirs have been recorded in different oilfields in the U.S. [1–3], China [4–11], Canada [12,13] and Australia [14].

At present, there are few evaluative studies of the vertical producing degree of multilayer reservoirs from the perspective of reservoir engineering. In this case, the traditional frontal advance theory of Buckley–Leverett cannot be easily used for multilayers [15]. Stiles assumed the displacement velocity in a layer to be proportional to its absolute permeability, neglecting the effect of the mobility ratio [16]. Dykstra and Parsons developed a model (the D-P method) for non-communicating layers without crossflow between layers [17], whereas Hiatt presented a model for communicating layers with complete crossflow [18]. Based on the piston waterflooding theory, Osman and Dyes et al. studied the influence of the mobility ratio of vertical heterogeneous reservoirs on the development performance, but the deviation between the piston waterflooding theory and actual situations was large [19,20]. Johnson simplified the D-P method using graphic treatment to predict waterflooding recovery [21]. For multilayer disconnected reservoirs, Lefkovits [22] studied the dynamic changes in oil wells in multilayer disconnected reservoirs, mainly studying the influence of the different layer properties on the bottom hole pressure, and the influence of different pressure drops on the pressure buildup curve. Lefkovits and Kucuk [23] used analytical methods to study the pressure changes in a commingling system under the condition of no crossflow.

Bourdet [24] calculated the pressure curves and the pressure derivative of multilayer oil reservoirs with the skin effect and wellbore storage. Tompang et al. [25] analyzed the change in crossflow in multilayer reservoirs through mathematical derivation, and the crossflow index was mainly related to the effective length-to-height ratio and vertical-to-horizontal pressure gradient ratio. Tian et al. [26] studied the influence of strata pinch out and lens on vertical interlayer interference tests by establishing a three-layer model. Using the theory of B-L displacement mathematical model, Noaman et al. [27–29] analyzed the influence of gravity number, mobility ratio, permeability variation coefficient and liquidity on waterflooding performance in inclined multilayer reservoirs, and compared the finding with the D-P method. Guo [30] established the low permeability gas reservoir model for multilayer commingling production, considering the Darcy flow and the starting pressure gradient in both cases. The average pressure, bottom hole pressure and speed of the oil production rate of each layer were derived.

According to the theory of oil and water two-phase seepage, Li et al. concluded that the essence of the interlayer contradiction was the difference in the velocity of water displacing oil in each layer [31]. Through the establishment of a one-dimensional multilayer reservoir non-piston waterflood model, Zhou and An studied the influence of the multilayer reservoir heterogeneity on production indexes, such as the recovery percent, liquid producing rate and water cut [32,33]; however, the research was limited because the model could only calculate the development index of the water breakthrough time in each layer. With the establishment of a one-dimensional multilayer reservoir waterflood model, Zhang et al. studied the variation in the vertical sweep efficiency of water flooding in multilayer commingling production. This implicit method is not widely applicable because this model used a step length with the increase in the average water saturation of the layer after the water breakthrough in each layer [34].

Although many studies have been conducted on the production performance of multilayer heavy oil reservoirs with commingling production, some research was limited to the calculation of development parameters at the water breakthrough time, or some models were difficult to solve for widespread use because of the choice of an average water saturation increase rate as the step length. In order to simplify the calculation, this paper establishes a non-piston multilayer reservoir water flooding model with two-phase oil and water based on the Buckley–Leverett theory, which sets the time microelement, Δt, as the step length to calculate the seepage resistance, liquid production rate and position of the water flooding front of each layer. Then, through a circular iterative solution, the production performance and vertical producing degree were obtained. In addition, this model is successfully validated using the previous model, showing that this calculation model is feasible and reliable. Later, through case analysis from the perspective of permeability and viscosity, the root cause of interlayer contradiction is highlighted, which also reflects the results of the synergistic effects of the reservoir properties and fluid parameters. Finally, comparing the ratio of liquid production in different water cut stages, the best time for adjustment measures to be undertaken is obtained. Overall, the calculated results of the model can be used to determine the separated production boundary, which also provides theoretical and technological support for the subdivision of multilayer offshore heavy oil reservoirs.

2. Multilayer Reservoir Commingling Production Waterflood Model

2.1. Multilayer Water Flooding Model under One-Dimensional Linear Seepage Flow

2.1.1. Assumptions

With the linear rowed well pattern, the seepage flow follows the linear flow method [35]; a schematic diagram of the water flooding model is shown in Figure 1. The following assumptions are made:

1. The left boundary is the supply boundary with constant injection, and the right boundary is the production ends, which creates a balance between injection and production.
2. The media is rigid and porous, and the fluid is incompressible.
3. There are stable interlayers between layers, regardless of inter-layer cross flow.
4. Non-piston water displacement oil is present, and there are two phases of oil and water.

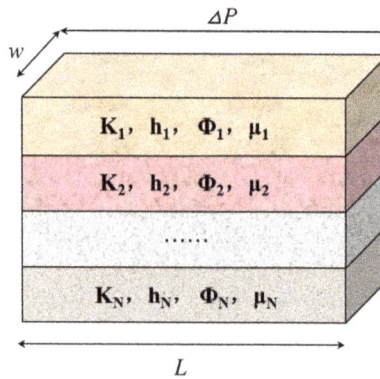

Figure 1. Schematic of the one-dimensional multilayer water flood model.

2.1.2. Modeling

Suppose that the water breaks through vertically from the first layer to the N layer in turn, where M is the number of water-breakthrough layers ($0 \leq M \leq N$), and the physical properties of the reservoir and fluid parameters of each layer are different. Based on the Buckley–Leverett theory [35], for planar radial flow, the isosaturation level movement equation of the waterflood front of layer i ($M < i \leq N$) can be expressed as

$$x_{fi} = \frac{f'_w \left(S_{wfi} \right)}{\phi_i w h_i} \int_0^t Q_i dt \tag{1}$$

Correspondingly, the liquid production rate in layer i can be expressed as

$$Q_i = \frac{k_i w h_i \Delta P}{\int_0^{x_{fi}} \frac{1}{\frac{k_{ro}}{\mu_o} + \frac{k_{rw}}{\mu_w}} dx + \mu_{oi} \left(L - x_{fi} \right)} \tag{2}$$

The seepage resistance of layer i is

$$R_i = \left[\int_0^{x_{fi}} \frac{1}{\frac{k_{ro}}{\mu_{oi}} + \frac{k_{rw}}{\mu_w}} dx + \mu_{oi} \left(L - x_{fi} \right) \right] / (k_i w h_i) \tag{3}$$

In particular, when the water breakthrough happens in the layer i, the liquid production rate becomes

$$Q_i = \frac{k_i w h_i \Delta P}{\int_0^L \frac{1}{\frac{k_{ro}}{\mu_o} + \frac{k_{rw}}{\mu_w}} dx} \tag{4}$$

The total fluid production rate of the multilayer commingling production is

$$Q_i = \sum_{i=1}^{M} \frac{k_i w h_i \Delta P}{\int_0^L \frac{1}{\frac{k_{ro}}{\mu_{oi}} + \frac{k_{rw}}{\mu_w}} dr} + \sum_{j=M+1}^{N} \frac{k_i w h_i \Delta P}{\int_0^{x_{fi}} \frac{1}{\frac{k_{ro}}{\mu_{oi}} + \frac{k_{rw}}{\mu_w}} dx + \mu_{o_i} \left(L - x_{fi} \right)} \tag{5}$$

The sweep efficiency of the multilayer commingling production is as follows:

$$E_v = \frac{\sum\limits_{i=1}^{M} w h_i \phi_i L + \sum\limits_{j=M+1}^{N} w h_i \phi_i x_{fi}}{\sum\limits_{i=1}^{N} w h_i \phi_i L} \tag{6}$$

The reservoir recovery percent of the multilayer commingling production is as follows:

$$\eta = \frac{\sum\limits_{i=1}^{M} w h_i L \phi_i (\bar{s}_{wi} - s_{wci}) + \sum\limits_{j=M+1}^{N} \int_0^t Q_i dt}{\sum\limits_{i=1}^{N} w h_i L \phi_i (1 - s_{wci})} \tag{7}$$

where the following formula can be used to solve $\bar{s}_w$:

$$\bar{s}_w = \frac{\int_0^L s_w dx}{L} = \frac{\sum\limits_{i=1}^{N} s_w \Delta x}{L} \tag{8}$$

In order to facilitate derivation calculation, the relative permeability needs to be processed as the Corey type.

2.2. Multilayer Water Flooding Model under Planar Radial Flow

2.2.1. Assumptions

For the common inverted nine-spot area well pattern, the seepage flow follows the planar radial flow method [35], and the schematic diagram of the water flooding model is shown in Figure 2. The following assumptions are made:

1. The boundary is the supply boundary with constant injection and creates a balance between injection and production.
2. The media is rigid and porous, and the fluid is incompressible.
3. There are stable interlayers between layers, regardless of inter-layer cross flow.
4. Non-piston water displacement oil is present, and there are two phases of oil and water.

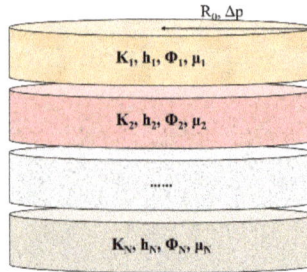

Figure 2. Schematic of planar radial multilayer water flood model.

2.2.2. Modeling

Suppose that the water breaks through vertically from the first layer to the N layer in turn, where M is the number of water-breakthrough layers ($0 \leq M \leq N$), and the physical properties of the reservoir and fluid parameters of each layer are different. Based on the Buckley–Leverett theory [34], for planar radial flow, the isosaturation level movement equation of the waterflood front of layer i ($M < i \leq N$) can be expressed as

$$\frac{dr}{dt} = \frac{-Q_i}{\phi_i \cdot 2\pi r h_i} f'_w\left(s_{wfi}\right) \tag{9}$$

Integral to get

$$\int_{R_o}^{r} 2\pi \phi_i h_i r dr = f'_w(s_{wfi}) \int_0^t Q_i dt \tag{10}$$

$$R_o^2 - r_{fi}^2 = \frac{f'_w(s_{wfi})}{\pi \phi_i h_i} \int_0^t Q_i dt \tag{11}$$

Correspondingly, the liquid production rate in layer i can be expressed as

$$Q_i = \frac{2\pi k_i h_i \Delta P}{\int_{r_f}^{R_o} \frac{1}{(\frac{k_{ro}}{\mu_o} + \frac{k_{rw}}{\mu_w})r} dr + \mu_{o_i} \ln(\frac{r_{fi}}{r_w})} \tag{12}$$

The seepage resistance of layer i is

$$R_i = [\int_{r_{fi}}^{R_o} \frac{1}{(\frac{k_{ro}}{\mu_{oi}} + \frac{k_{rw}}{\mu_w})r} dr + \mu_{oi} \ln \frac{r_{fi}}{r_w}] / (2\pi k_i h_i) \tag{13}$$

The total fluid production rate of the multilayer commingling production is

$$Q_i = \sum_{i=1}^{M} \frac{2\pi k_i h_i \Delta P}{\int_{r_{fi}}^{R_o} \frac{1}{(\frac{k_{ro}}{\mu_{oi}} + \frac{k_{rw}}{\mu_w})r} dr} + \sum_{j=M+1}^{N} \frac{2\pi k_i h_i \Delta P}{\int_{r_f}^{R_o} \frac{1}{(\frac{k_{ro}}{\mu_{oi}} + \frac{k_{rw}}{\mu_w})r} dr + \mu_{o_i} \ln(\frac{r_{fi}}{r_w})} \tag{14}$$

The sweep efficiency of the multilayer commingling production is as follows:

$$E_v = \frac{\sum\limits_{i=1}^{M} \pi R_o^2 h_i \phi_i + \sum\limits_{j=M+1}^{N} \pi (R_o^2 - r_{fi}^2) h_i \phi_i}{\sum\limits_{i=1}^{N} \pi R_o^2 h_i \phi_i} \tag{15}$$

The reservoir recovery percent of the multilayer commingling production is as follows:

$$\eta = \frac{\sum_{i=1}^{M} \pi R_o^2 h_i \phi_i (\bar{s}_{wi} - s_{wci}) + \sum_{j=M+1}^{N} \int_o^t Q_i dt}{\sum_{i=1}^{N} \pi R_o^2 h_i \phi_i (1 - s_{wci})} \tag{16}$$

Of these, the pressure drop of near borehole zones under planar radial flow is very large [35], so the seepage resistance changes greatly. When calculating the seepage resistance it is necessary to convert the radial coordinate r of unequal distance to the x-coordinate of equal distance [36]; thus, $\Delta x = \ln\left(\frac{R_o}{r_f}\right)/n$, and $r = r_f e^{i\Delta x} = r_f e^x$.

Equation (3) can be converted to the following summation formula:

$$R = (\sum_{i=1}^{n} \frac{\Delta x}{\frac{k_{ro}}{\mu_o} + \frac{k_{rw}}{\mu_w}} + \mu_o \ln \frac{r_f}{r_w}) / (2\pi k h) \tag{17}$$

The method for solving the average water saturation after water breakthrough in the layers is as follows:

$$\bar{s}_w = \frac{\int_0^{V_p} s_w dv}{\pi \phi (R_o^2 - r_w^2)} = \frac{\pi h \phi \int_{r_w}^{R_o} 2 r s_w dr}{\pi \phi (R_o^2 - r_w^2)} = \frac{-\sum_{i=0}^{n} f_w''(s_w) s_w \Delta s_w}{\pi h \phi (R_o^2 - r_w^2)} \int_0^t Q dt \tag{18}$$

where V_p is the total pore volume of the reservoir, m^3, and

$$\Delta s_w = (1 - s_{or} - s_{we})/n \tag{19}$$

For the convenience of the derivation calculation, the relative permeability curve should be copied with a Corey type relative permeability curve [37].

3. Model Solving

The mathematical model established on the basis of the reservoir engineering method, using the time microelement Δt as a step length, could be solved to obtain the parameters such as seepage resistance, recovery degree and sweep efficiency at different times through iterative calculation. In the previous mathematical model, the increase of average water saturation was used as the step length, and the solution method was much too complex. However, in this derivation the time step length is used, which was relatively easier to solve than before, and it was convenient to calculate the injection pore volume and water cut. The iteration steps are described as follows: First, the seepage resistance of each layer at the initial time t0 was calculated, and the liquid production rate and position of the water flooding front of each layer were obtained. Second, the seepage resistance was calculated by integrating the position of the water flooding front, and the liquid production rate at t1 was obtained by the seepage resistance of each layer. Third, this liquid production rate was added to that in the last time step. Thus, the cumulative amount of liquid production was obtained. Then, the position of the later water flooding front was obtained again. This process was iteratively repeated until the first layer water broke through.

The solution flow diagram is as in Figure 3:

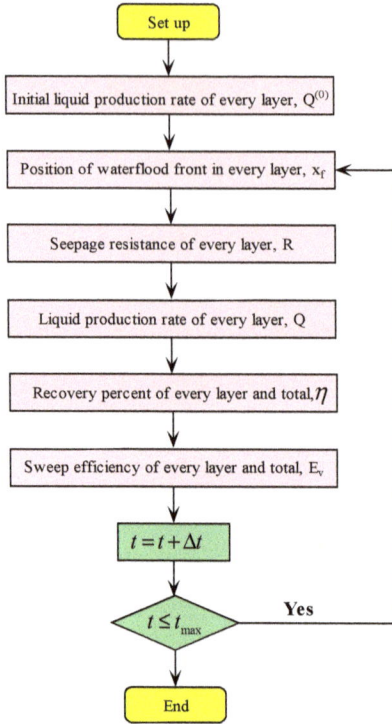

Figure 3. Flow chart for solving the multilayer water flooding model.

4. Model Validation

To demonstrate the accuracy of the model for multilayer water flooding, combined with detailed reservoir data from the model in Zhang [34], the model was calculated and the calculation results were compared with the calculation results from the reference. The results are shown in Figure 4. It can be seen that the daily oil production (Model X) of different permeability layers coincided with the data provided by Zhang's model, and the results of the two calculations were basically consistent. The figure also showed that the model results presented in this paper are reliable.

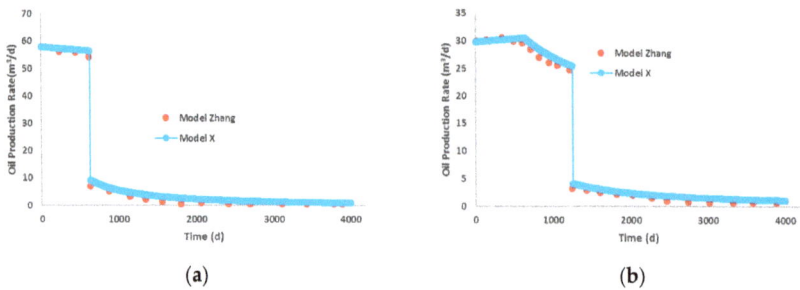

(a)

(b)

Figure 4. *Cont.*

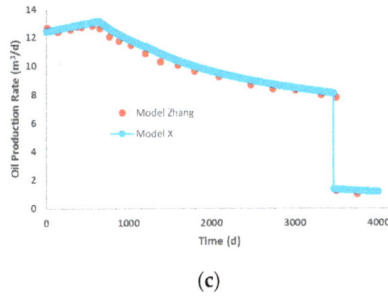

(c)

Figure 4. Comparison of our calculated model and Zhang's model [34] for the daily oil production rate in different permeability layers. Red dots denote Zhang's model; blue dotted line denotes the calculated model. (**a**) High permeability layer; (**b**) Middle permeability layer; (**c**) Low permeability layer.

Due to differences in the physical properties, it can be seen from the calculation results that the production rate curve at the output end presented characteristics with more obvious periodic changes, which reflected the interlayer interference between the layers before and after the water breakthrough.

5. Model Application and Discussion

According to the specific data of reservoir Q in the Bohai oilfield, an example is provided to establish a multilayer water-flooding model with planar radial flow, and the calculation and analysis of the waterflooding performance in the multilayer reservoir are conducted. This model only considers the difference of the horizontal permeability in layers, and the other physical parameters are the same. The parameters are shown in Table 1.

Table 1. Main parameters of the model.

Model Parameters	Value	Model Parameters	Value
Reservoir radius (m)	350	Layer 1 permeability (10^{-3} μm^2)	3000
Reservoir thickness (m)	5	Layer 2 permeability (10^{-3} μm^2)	1800
Reservoir porosity (%)	25	Layer 3 permeability (10^{-3} μm^2)	600
Water injection rate (m^3/d)	500	Oil viscosity (mPa·s)	50

This model is suitable for the calculation of constant liquid production and a balance state of injection and production. It is used for the black oil model calculation, and the layers are composed of rigid and porous rock, which is not suitable for condensate oil or gas reservoirs. To simplify the solving, there is no fluid flowing from one layer to another except the wells. The water flooding feature in the media is non-piston-like displacement, and there are two-phase regions for oil and water together.

When the multilayer heavy oil reservoir produced with general water injection and commingling production, the difference in the seepage resistance of each layer constantly changes, as shown in Figure 5. The difference in the seepage resistance between layers is an important cause of interlayer interference. For the multilayer water flooding model with planar radial flow, the seepage resistance in the near borehole zones was relatively large, thus the seepage resistance of each layer rapidly decreases before the water breakthrough but slowly decreases after the water breakthrough.

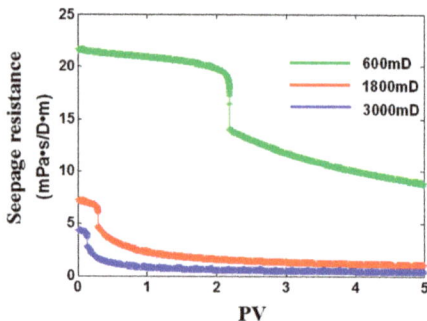

Figure 5. Comparison of the seepage resistance with different permeability in each layer of the planar radial flow model ranging from 0 to 5 pore volume (PV).

As shown in Figure 6, the total recovery degree of the reservoir rises like stairs with the increase of the injection pore volume, and the rising speed slows down when the water breaks through in each oil layer. The recovery degree of each layer rises quite fast before the water breakthrough, while the rising speed slows down after the water breakthrough. The difference between layers is mainly reflected in the performance of the low permeability layer, which is shown in Figure 6b. The solid line represents the commingling production, while the dotted line represents the separated production. Commingling production has a great influence on the recovery degree of the low permeability layer. Especially in the stage of the first pore volume (PV), in which there is no water produced in the low permeability layer, the gap between the separated and the commingling production is the largest. At this time, the water had broken through in the high permeability layer, which is reflected in the curves and belongs to the stage of rapid water rise. Therefore, at this time, if separated production measures are taken, the low permeability layer will be better used.

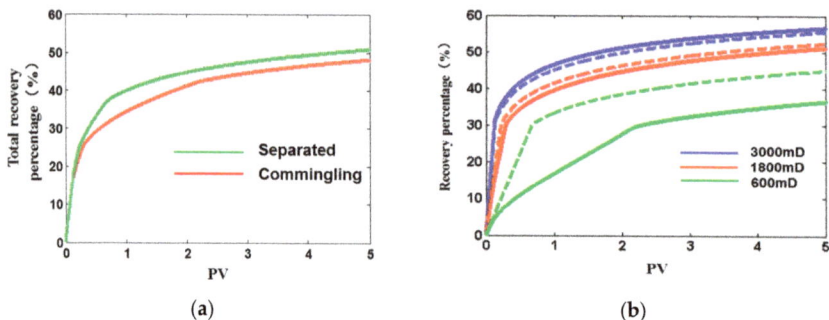

(a)

(b)

Figure 6. Comparison of recovery percentage in the reservoir ranging from 0 to 5 PV (**a**) separated and commingling production; (**b**) each different permeability layer (solid line, commingling production; dotted line, separated production).

As shown in Figure 7, from the perspective of fluid production, the differences between layers is reflected in the following aspects: (a) when the reservoir exhibits commingling production, the difference of interlayer use caused by the permeability ratio first rises and then decreases with the increase of the water injection pore volume and water cut; (b) the liquid production difference between layers is always greater than the initial one; and (c) the largest difference between layers caused by permeability occurs in the production middle and later stages, thus the middle and later stages of development are the best time for measure adjustment.

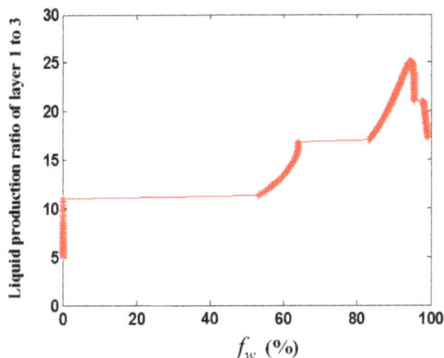

Figure 7. Ratio of liquid production between high and low permeability layers (best to the worst) as the water cut ranging from 0 to nearly 100%.

With the changing model parameters in Table 1, commingling production and separated production can be compared in terms of the recovery degree calculated for the different permeability ratio or different viscosity ratio models. The parameters are showed in Table 2.

Table 2. Main parameters of the model (for viscosity ratio).

Model Parameters	Value	Model Parameters	Value
Reservoir radius (m)	350	Oil viscosity in Layer 1 (mPa·s)	17
Reservoir thickness (m)	5	Oil viscosity in Layer 2 (mPa·s)	50
Reservoir porosity (%)	25	Oil viscosity in Layer 3 (mPa·s)	85
Water injection rate (m^3/d)	500	Layers' permeability (10^{-3} μm^2)	1800

Another example is provided to compare the previous model that was affected by the permeability ratios. Thus this model only considers the difference in the viscosity, and the other physical parameters remain unchanged. In order to compare the effect of the permeability ratio and viscosity ratio, the permeability and viscosity in the middle layer were set as the standard model, and the viscosity was changed in the first and third layers.

As shown in Figure 8, taking the daily liquid production rate as an example, the comingling production dynamic differences between layers were considerably different from the permeability-to-viscosity ratios. Compared to the viscosity ratios, the interlayer interference caused by the permeability ratios was more serious, and the interference lasted much longer. The fundamental reason for this is that the interference caused by permeability comes from the physical properties of the layers, whereas the interference caused by fluid viscosity changes with changes in the water cut. Thus, when measure adjustment is implemented, the interference caused by these two reasons requires different coping mechanisms.

Taking the permeability ratio as an example, the variation graph of the difference in the value of the recovery degree between the separated production and the commingling production with the permeability ratio rising was obtained using the model calculation. As shown in Figure 9, we found that there was little difference between the separated production and the commingling production when the permeability ratio was less than three. However, when the permeability ratio was greater than three, the total recovery of the commingling production significantly worsened, thus separated production should be adopted. The larger the permeability ratio, the better the effect obtained by separated production. In terms of an increase in the recovery degree rate by 5% after the implementation of separated production measures, the limiting line should be approximately three.

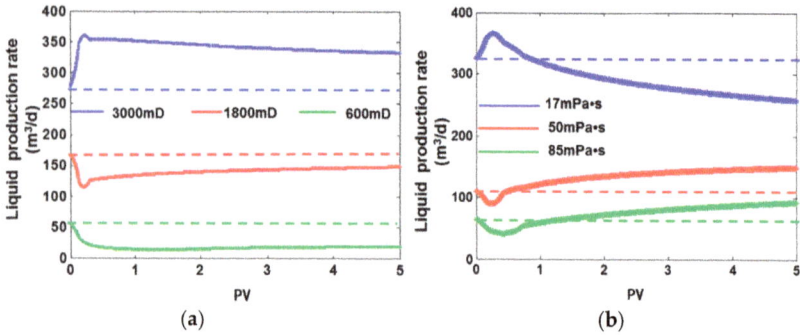

Figure 8. Comparison of the liquid production rate with a permeability-to-viscosity ratio ranging from 0 to 5 PV. (**a**) Permeability ratio = 5; (**b**) Viscosity ratio = 5.

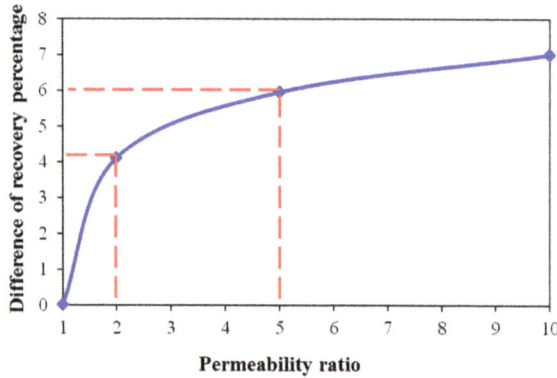

Figure 9. Recovery degree increase of separate production with a permeability ratio ranging from 1 to 10.

For the multilayer oil reservoir with waterflooding development, the root cause of the interlayer interference is the difference between the reservoir properties and fluid parameters. The internal causes of the performance change in the interlayer interference are the viscosity differences between oil and water, displacement performance, the distance to the waterflood front, and the seepage resistance change affected by the oil–water transition zone. For heavy oil reservoirs, the most rapid increase in the water cut was approximately 20–80%, and the total water cut with commingling production was most affected by the high permeability layer. Therefore, suppressing the water production rate in the high permeability layer can further improve the interference between layers and improve the development effect.

In the water injection well group in the Dongying Formation multilayer reservoir of oilfield A in Bohai Bay, which has an inverted nine-spot area well pattern, the average permeability ratio of the multilayer was approximately 4.39. A field test of the subdivision of the layer series was carried out in August 2013, which means that commingling production turned into separated production. After nine months, as shown in Figure 10, the average daily liquid production rate of the single well in the well group was reduced by 17%, the average daily oil production rate of the single well was rated at 22%, and the average water cut was reduced by 9%. The production situation obviously improved.

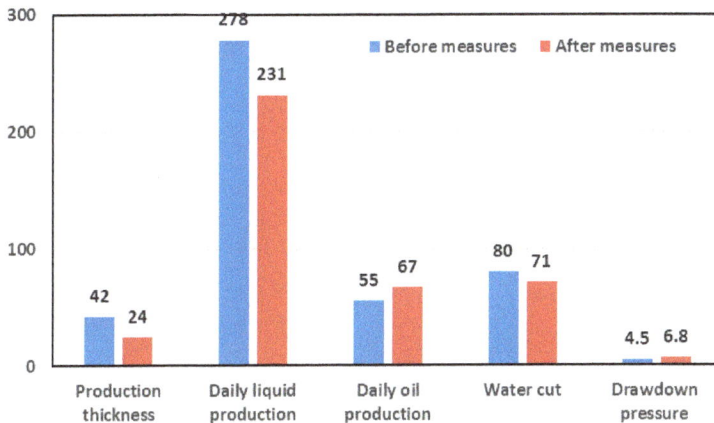

Figure 10. Comparison of the single well production index after separated production for nine months.

6. Conclusions

Four conclusions can be drawn from this study. Firstly, based on the Buckley–Leverett theory, two theoretical models from the one-dimensional linear flow and planar radial flow aspects were established in this paper for waterflood multilayer reservoirs. With model validation, the models were showed to be reliable and accurate.

Secondly, the models can be used to solve and analyze the liquid or oil production rate, the liquid or oil production index and the seepage resistance, sweep efficiency and recovery percentages in each layer under different development stages. The conditions of separated and commingling production were obtained.

Thirdly, with the help of the model, the performance of water flooding with different fluid parameters and reservoir properties was also studied. During commingling production, the interlayer interference caused by permeability and viscosity were quite different. They interacted with each other, leading to dynamic differences in seepage resistance and then resulting in interlayer contradictions.

Lastly, for conventional heavy oil reservoirs, when the water cut was approximately between 40% and 80%, separated production measures were carried out. At this point, the water cut increased the most. Suppressing the water production rate in the high permeability layer can further improve the interference between layers and improve the development effect.

Author Contributions: The research study was carried out successfully with contribution from all authors. F.S. contributed the main research idea and established the numerical model; Q.S. and F.S. wrote the original manuscript, performed the correlative simulation and studied the applicability of the simulation method; Q.S. implemented the model validation and data correction; L.C. and S.H. gave several suggestions from the on-site production perspectives. All authors revised and approved the publication of the paper.

Funding: This research was funded by [National Science and Technology Major Projects of China] grant number [2011ZX05024-002-006].

Acknowledgments: The technical support from the CNOOC Research Institute, CNOOC are gratefully acknowledged.

Conflicts of Interest: The authors declare no conflict of interest.

Nomenclature

r_{fi}	position of waterflood front in layer i, m;
R_o	initial oil edge radius, m;
h_i	thickness of layer i, m;
ϕ_i	porosity of layer i, dimensionless;
s_{wfi}	water saturation of the waterflood front of layer i, dimensionless;
$f'_w\left(s_{wfi}\right)$	the derivative of the fractional flow corresponding to the water saturation of the waterflood front in layer i, dimensionless;
Q_i	liquid production rate in layer i, m^3/d;
Q_l	total liquid production rate, m^3/d;
K_i	the permeability of layer i, $10^{-3}\,\mu$m^2;
K_{ro}	relative permeability of oil, fraction;
K_{rw}	relative permeability of water, fraction;
r_w	wellbore radius, m;
ΔP	displacement pressure in layer i, MPa;
μ_{oi}	oil viscosity in layer i, mPa·s;
V_p	total pore volume of the reservoir, m^3;
R_i	resistance of layer i, mPa·s/(d·m);
η	reservoir recovery percent of the multilayer commingling production, dimensionless;
E_v	sweep efficiency of the multilayer commingling production, dimensionless;
$\bar{s}_w$	average water saturation in two-phase region, fraction;
s_{wc}	irreducible water saturation, fraction;
M	number of water-breakthrough layers, dimensionless;
N	total number of model layers, dimensionless;
i, j	serial number of layers.

References

1. Warner, G.E. Waterflooding a Highly Stratified Reservoir. *J. Pet. Technol.* **1968**, *10*, 1179–1186. [CrossRef]
2. Hatzignatiou, D.G.; Ogbe, D.O.; Dehghani, K.; Economides, M.J. Interference Pressure Behavior in Multilayered Composite Reservoirs. In Proceedings of the SPE Annual Technical Conference and Exhibition, Dallas, TX, USA, 27–30 September 1987; pp. 245–260.
3. Pan, Y.; Kamal, M.; Kikani, J. Field Applications of a Semianalytical Model of Multilateral Wells in Multilayer Reservoirs. *SPE Reserv. Eval. Eng.* **2010**, *13*, 861–872.
4. Tang, Z.X.; Liu, H.; Jiang, C.Z. Technical Measures for Improving Oilfield Development by Waterflooding in a Large Multi-Layered Heterogeneous Sandstone Reservoir. In Proceedings of the International Meeting on Petroleum Engineering, Beijing, China, 17–25 March 1986; pp. 287–298.
5. Hong, L.; Jin, P.; Xiaolu, W.; Xinan, Y.; Qing, L. Analysis of Interlayer Interference and Research of Development Strategy of Multilayer Commingled Production Gas Reservoir. *Energy Procedia* **2012**, *16*, 1341–1347. [CrossRef]
6. Du, Q.L.; Zhao, Y.; Lu, Y.; Zhu, L. Study on Genetic Type and Potential Tapping Measures of the Remaining Oil in Multi-layered and Heterogeneous Sandstone Reservoir. *J. Dermatol. Surg. Oncol.* **1999**, *17*, 865–868.
7. Huang, S.J.; Kang, B.T.; Cheng, L.S.; Zhou, W.S.; Chang, S.P. Quantitative Characterization of Interlayer Interference and Productivity Prediction of Directional Wells in the Multilayer Commingled Production of Ordinary Offshore Heavy Oil Reservoirs. *Pet. Explor. Dev.* **2015**, *42*, 533–540. [CrossRef]
8. Su, Y.C. A Set of Techniques to Describe Quantitatively Remaining Oil in Offshore Heavy Oilfields under Commingling Production in Large Well Spacing. *China Offshore Oil Gas* **2012**, *24*, 82–85.
9. Wang, F.Y.; Liu, Z.C.; Jiao, L.; Wang, C.L.; Guo, H. A Fractal Permeability Model Coupling Boundary-layer Effect for Tight Oil Reservoirs. *Fract. Complex Geom. Patterns Scaling Nat. Soc.* **2017**, *25*, 1750042. [CrossRef]
10. Li, C.X.; Shen, Y.H.; Ge, H.K.; Su, S.; Yang, Z. Analysis of Spontaneous Imbibition in Fractal Tree-like Network System. *Fract. Complex Geom. Patterns Scaling Nat. Soc.* **2016**, *24*, 1650035. [CrossRef]

11. Qiu, L.; Yang, S.; Qu, C.; Xu, N.; Gao, Q.; Zhang, X.; Liu, X.; Wang, D. A Comprehensive Porosity Prediction Model for the Upper Paleozoic Tight Sandstone Reservoir in the Daniudi Gas Field, Ordos Basin. *J. Earth Sci.* **2017**, *28*, 1086–1096. [CrossRef]

12. Brice, B.W.; Renouf, G. Increasing Oil Recovery from Heavy Oil Waterfloods. In Proceedings of the SPE International Thermal Operations and Heavy Oil Symposium, Calgary, AB, Canada, 20–23 October 2008; pp. 1–13.

13. Miller, K.A. Improving the State of the Art of Western Canadian Heavy Oil Waterflood Technology. *J. Can. Pet. Technol.* **2006**, *45*, 7–11. [CrossRef]

14. Pari, M.N.; Kabir, A. Viability Study of Implementing Smart/Intelligent Completion in Commingled Wells in an Australian Offshore Oil Field. *Soc. Pet. Eng.* **2009**, *49*, 441.

15. Buckley, S.E.; Leverett, M.C. Mechanism of Fluid Displacement in Sands. *Trans. AIME* **1942**, *146*, 107–116. [CrossRef]

16. Stiles, W.E. Use of Permeability Distribution in Water-flood Calculations. *Trans. AIME* **1949**, *186*, 9–13. [CrossRef]

17. Dykstra, H.; Parsons, R.L. *The Prediction of Oil Recovery by Waterflooding in Secondary Recovery of Oil in the United States*, 2nd ed.; API: Washington, DC, USA, 1950; pp. 160–174.

18. Hiatt, N.W. Injected–fluid Coverage of Multi–well Reservoirs with Permeability Stratification. *Drill Prod. Prac.* **1958**, *165*, 165–194.

19. Dyes, A.B.; Caudle, B.H. Oil Production after Breakthrough as Influenced by Mobility Ratio. *AIME* **1954**, *201*, 81–86. [CrossRef]

20. Osman, M.E.; Tiab, D. Waterflooding Performance and Pressure Analysis of Heterogeneous Reservoirs. In Proceedings of the Middle East Technical Conference and Exhibition, Manama, Bahrain, 9–12 March 1981; pp. 773–779.

21. Johnson, C.E. Prediction of Oil Recovery by Waterflood—A Simplified Graphical Treatment of the Dykstra-Parsons Method. *J. Pet. Technol.* **2013**, *8*, 55–56. [CrossRef]

22. Lefkovits, H.C.; Hazebroek, P.; Allen, E.; Matthews, C.S. A Study of the Behavior of Bounded Reservoirs Composed of Stratified Layers. *Soc. Pet. Eng. J.* **1961**, *1*, 43–58. [CrossRef]

23. Kucuk, F.; Ayestaran, L. Well Testing and Analysis Techniques for Layered Reservoirs. *SPE Form. Eval.* **1984**, *8*, 342–354. [CrossRef]

24. Bourdet, D. Pressure Behavior of Layered Reservoirs with Crossflow. In Proceedings of the SPE California Regional Meeting, Bakersfield, CA, USA, 27–29 March 1985; pp. 405–412.

25. Tompang, R.; Kelkar, B.G. Prediction of Waterflood Performance in Stratified Reservoirs. *Chest* **1988**, *92*, 657–662.

26. Tang, X.; Liu, S. A New Vertical Interference Test Method in a Three Layer Reservoir with an Unstable Impermeable Interlayer. In Proceedings of the SPE Annual Technical Conference and Exhibition, Denver, CO, USA, 6–9 October 1996; pp. 725–736.

27. El-Khatib, N.A.F. The Application of Buckley-Leverett Displacement to Waterflooding in Non-Communicating Stratified Reservoirs. In Proceedings of the SPE Middle East Oil Show, Manama, Bahrain, 17–20 March 2001; pp. 1–12.

28. El Khatib, N.A.F. Waterflooding Performance in Inclined Communicating Stratified Reservoirs. *SPE J.* **2012**, *17*, 31–42. [CrossRef]

29. El-Khatib, N.A.F. The Modification of the Dykstra-Parsons Method for Inclined Stratified Reservoirs. *SPE J.* **2012**, *17*, 1029–1040. [CrossRef]

30. Guo, P.; Wang, J.; Liu, Q.; Zhang, M. Study on Development Mechanism of Commingling Production in Low-Permeability Gas Reservoirs. In Proceedings of the International Oil and Gas Conference and Exhibition in China, Beijing, China, 8–10 June 2010; pp. 1–20.

31. Li, L.R.; Yuan, S.Y.; Hu, Y.L. Quantitative Rules for the Subdivision and Recombination of Layer Series in Water-displacement Reservoirs, The Innovation and Practice of Seepage Mechanics and Engineering. In Proceedings of the 11th National Conference on Seepage Mechanics, Chongqing, China, 28–30 April 2011; pp. 216–222.

32. Zhou, Y.F.; Fang, Y.J.; Wang, X.D. Study on non-piston water displacement efficiency for layered reservoir. *J. Pet. Geol. Recov. Effic.* **2009**, *16*, 86–89.

33. An, W.Y. Interference Mechanism of Multilayer Heterogeneous Reservoir in High Water cut Stage. *J. North East Pet. Univ.* **2012**, *36*, 72–75.

34. Zhang, S.K.; Liu, B.G.; Zhong, S.Y. Theoretical Model for Waterflooding Development of Multi-Zone Reservoir. *J. Xinjiang Pet. Geol.* **2009**, *30*, 734–737.

35. Zhang, J.G.; Lei, G.L.; Zhang, Y.Y. *Seepage Mechanics*; China University of Petroleum Press: Beijing, China, 1998; pp. 148–160.

36. Li, S.X.; Gu, J.W. *Foundation of Reservoir Numerical Simulation*; China University of Petroleum Press: Beijing, China, 2009; pp. 147–148.

37. Corey, A.T. The Interrelation between Gas and Oil Relative Permeabilities. *Prod. Mon.* **1954**, *19*, 38–41.

energies

MDPI

Article

Numerical Study on the Characteristic of Temperature Drop of Crude Oil in a Model Oil Tanker Subjected to Oscillating Motion

Guojun Yu [1,2,*], Qiuli Yang [1], Bing Dai [3], Zaiguo Fu [4] and Duanlin Lin [5]

[1] Merchant Marine College, Shanghai Maritime University, Shanghai 201306, China; yangql_smu@sina.com
[2] School of Naval Architecture, Ocean & Civil Engineering, Shanghai Jiao Tong University, Shanghai 200240, China
[3] China Petroleum Technology & Development Corporation, Beijing 100009, China; daib_offshore@126.com
[4] College of Energy and Mechanical Engineering, Shanghai University of Electric Power, Shanghai 200090, China; fuzaiguo@shiep.edu.cn
[5] Hubei Subsurface Multi-Scale Imaging Key Laboratory, Institute of Geophysics and Geomatics, China University of Geosciences, Wuhan 430074, China; lindl@cug.edu.cn
*   Correspondence: gjyu@shmtu.edu.cn; Tel.: +86-21-3828-2960

Received: 14 April 2018; Accepted: 9 May 2018; Published: 11 May 2018

Abstract: During tanker transportation, crude oil is heated occasionally to ensure its good flowability. Whether the heating scheme is scientific or not directly influences the safety and economy of the tanker transportation. The determination of a scientific heating scheme requires fully understanding of the characteristic of oil temperature drop during tanker transportation. However, the oscillation caused by the marine environment leads to totally different thermal and hydraulic characteristic from that of the static cases. Therefore, a systematic investigation of thermal and hydraulic process of the motion system is more than necessary. Since the marine is subjected to rotational and/or translational motion, the essence of the temperature drop process is an unsteady mixed convection process accompanied with free liquid surface movement. In this study, the movement of the free liquid surface and the characteristic of the temperature drop of the crude oil in the cargo when the tanker is subjected to rotational motion were investigated using ANSYS FLUENT (15.0, Ansys, Inc., Canonsburg, PA, USA) with user defined functions. The research result shows that the oscillating motion leads to the motion of the free surface, converting the natural convection for the static case to forced convection, and thus significantly enhancing the temperature drop rate. It is found that the temperature drop rate is positively related to the rotational angular velocity.

Keywords: oil tanker; temperature drop; oscillating motion; numerical simulation

---

1. Introduction

Crude oil, the lifeblood of the national economy, is the most important energy in the world at present and will be in the long future. China's demand for crude oil is large, and is still growing [1]. In 2015, China became the world's largest oil importer, when the net annual import of crude oil amounted to 328 million tons. The foreign dependence of crude oil exceeded 60% [1] for the first time, and is expected to reach 75% by 2035 [2,3]. In China, about 90% of crude oil import depends on tanker transport [4]. During tanker transportation, crude oil should be heated occasionally, which consumes a large amount of energy and discharges large quantities of pollutants. Making a scientific heating plan to heat the crude oil reasonably is the precondition to guarantee the normal operation of the tanker and to realize the energy saving and emission reduction. However, formulating the heating scheme requires fully understanding of the temperature drop characteristic of crude oil in the cargo during the transportation. Therefore, the study of the thermal and hydraulic characteristic of the crude oil in oil tankers is necessary and bears great significance.

During the tanker transportation, the tanker will be subjected to oscillating motion due to the special nature of the marine environment, as shown in Figure 1. Accordingly, the crude oil in the cargo will be forced to flow occasionally. Oscillating involves six degrees of freedom motion, including three degrees of freedom rotational motion and three degrees of freedom translational motion. Therefore, under oscillating conditions, the process of oil temperature drop is an unsteady mixed convection process accompanied with free liquid surface movement. The thermal and hydraulic characteristic of this mixed convection is more complicated than that of static condition. To formulate the scientific scheme for oil heating, the thermal and hydraulic characteristic of oil temperature drop must be thoroughly investigated.

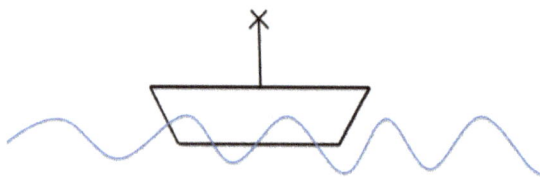

Figure 1. Schematic of tanker navigation at sea.

Due to the late development of oil tanker, there are limited studies in this field in China. Yue [5], Zhang [6,7] and Zhang [8] adopted the lumped parameter method to study the average temperature drop and heat source required to heat the crude oil. The lumped parameter method treats the crude oil as a whole without considering the internal temperature gradient. Although it is convenient for calculating, it does not accurately reflect the thermal and hydraulic characteristic of crude oil. Shi et al. [9] set up a two-dimensional analog "resistance capacitance" network to calculate the characteristic of temperature field change. However, the heat transfer process of crude oil was treated as heat conduction during calculation. Jin [10] employed ANSYS FLUENT to analyze the velocity and temperature distribution of crude oil during heating and naturally cooling process in the tanker cargo. The heat transfer process was considered as the natural convection. Most of the research in China did not consider the effect of oscillation motion on the temperature drop characteristic of crude oil.

Since the tanker business is mainly monopolized by western countries, more research regarding the thermal and hydraulic characteristic of tanker oil have been conducted. However, the calculation is still not accurate enough. Akagis [11] experimentally studied the oil heating by a steam coil in the tanker, focusing on the heating efficiency and the heat loss from the bulkhead. Suhara [12] experimentally measured the heat loss during the heating process of a 33,000 DWT (Dead Weight Tonnage) tanker. According to the principle of heat conservation, Chen [13] calculated the heat loss during the heating and storage cooling process of crude oil in the tank cargo. However, the research mentioned above did not consider the influence of oscillation motion on the thermal and hydraulic process of crude oil in the tank cargo. To fully investigate the effect of oscillating on the thermal and hydraulic process of crude oil, Kato [14] experimentally studied the heat transfer process of crude oil in the tank cargo subjected to oscillation, obtaining the heat transfer formula for the bulkhead and the top of the tanker, finding that the heat transfer coefficient increases linearly with respect to the oscillating angle and frequency. Doerffer et al. [15] studied the correlation between heat transfer in the flow boundary layer and the external disturbance with the analytic method, taking the forced convection caused by the oscillating motion of the ship in actual navigation into consideration. However, the analytic method is only suitable for small perturbations, and is only flexible for heat flow in the boundary layer. It is impossible to obtain the complete thermal and hydraulic characteristic of crude oil in the tank cargo. Akagi et al. [16] considered the forced convection caused by the oscillation, established a mathematical model based on body fitted coordinate system by introducing an inertia force in the momentum equation. With the proposed model, the influence of related parameters on the thermal and hydraulic characteristic was studied. However, the gas–liquid interface movement, which seriously influences the thermal process, was not considered in the study.

To sum up, there has not been a comprehensive study which accurately considered the influence of oscillation of the tanker. The related previous research results cannot meet the requirements of precise design. Therefore, in this study, the thermal and hydraulic process of crude oil under the condition of oscillating motion will be studied in detail, and the thermal and hydraulic characteristic of crude oil under oscillating condition will be clarified. This study will provide theoretical basis for the related design and heating program formulation.

2. Physical and Mathematical Model

2.1. Physical Model

Modern large oil tankers are usually double hull vessels, with special ballast tanks at the sides of cargo tanks. The cross section of the tanker holds is shown as Figure 2. When the ship is sailing in ballast, the heat of crude oil in the tank is released to the sea through the inside shell, ballast water and outer shell. When the ship is not sailing in ballast, the heat of crude oil in the tank is released to the sea through the shell, the air and the shell. In addition, the crude oil in the tank exchanges heat with the inert gas at the top of the tank by convection. The inert gas occupies a small part of the oil tank, and the heat of the inert gas is released to the atmosphere through the upper deck. A heating coil is installed at the bottom of the cabin to avoid the oil temperature naturally cooling below freezing point due to the influence of the above factors. Considering the economy and practicability, the crude oil in the tank is heated to a temperature which is 10 °C above the gel point.

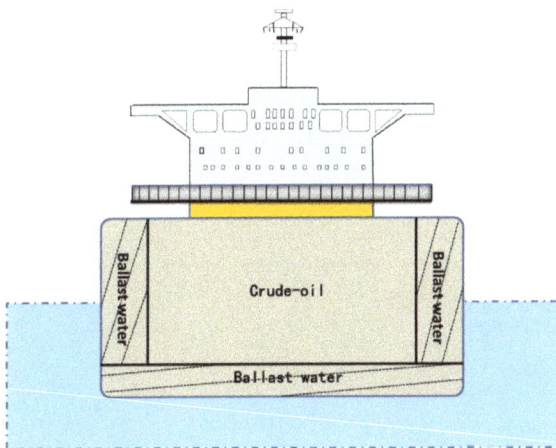

Figure 2. Schematic of the transverse section of a tanker.

Since the temperature gradient of the crude oil in the longitudinal direction is very small, and the temperature gradient in the transverse direction is very large, the three-dimensional physical model is simplified to a two-dimensional one. This paper focuses on studying the influence of oscillating motion on the thermal and hydraulic characteristic of crude-oil, so the ballast water tank and inert gas tank are not considered. In addition, to save computation time, we only study a 40 cm × 30 cm model tank cargo. As mentioned above, the physical model of the thermal and hydraulic process of the crude oil in the tank cargo is shown in Figure 3.

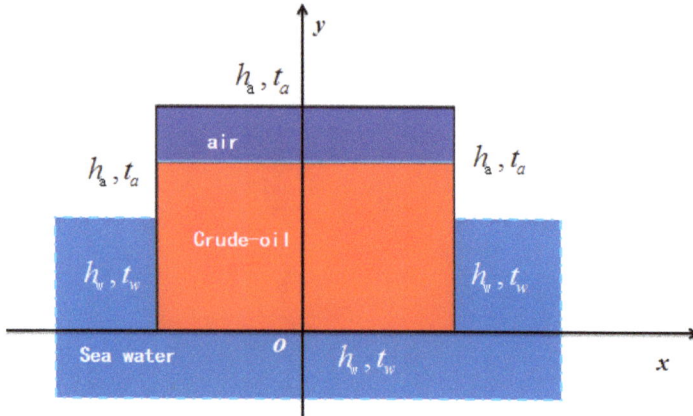

Figure 3. The physical model of the tank cargo system.

The red section in Figure 3 represents the crude oil, the blue part indicates the air above the crude oil, and there is a phase interface between the crude oil and the air. The upper boundary is the deck of the tanker, exposed to the air and is subjected to the third boundary condition. The left and right sides are two side walls of the ship respectively, the lower part is immersed in the sea water, and the upper part is exposed to the air. Thus, the left and right boundaries are also subjected to third boundary condition. The lower boundary is the bottom of the ship and immersed in sea water, which is subjected to the third boundary condition. When the oil tanker is subjected to an oscillating motion, it is assumed that the tanker rotates with an imagined z axis which is perpendicular to x and y axis at the original point *o* of the coordinate system in Figure 3. To simplify calculations, the following assumptions are introduced:

(1) The liquid oil does not evaporate, and the total volume does not change with temperature and time.
(2) The oil tanker is only subjected to rotational motion.
(3) The change of density with temperature is described by the Boussinesq approximation in the momentum equation, while the change of other physical properties with temperature is not considered, i.e., only the average value in the range of temperature change is used.
(4) There is no phase change in the crude oil during temperature drop.

2.2. Mathematical Model

2.2.1. Governing Equations

The nature of the physical problem studied in this paper is the mixed convection heat transfer under external disturbances, so the governing equation is a convection diffusion equation. Since the computation domain is changing its position all the time, the governing equations is provided in the framework of dynamic grid system.

(1) Volume Fraction Equation

The tracking of the interface between the phases is accomplished by solution of a continuity equation for the volume fraction of one of the phases. For the *q*th phase, this equation has the following form:

$$\frac{\partial}{\partial t}\left(\alpha_q \rho_q\right) + \nabla \cdot \left(\alpha_q \rho_q \mathbf{U}_q\right) = S_{aq} + \sum_{p=1}^{n}\left(\dot{m}_{pq} - \dot{m}_{qp}\right) \tag{1}$$

In Equation (1), α_q denotes the volume fraction of qth phase in the cell; U_q is the velocity of qth phase, m/s; $\dot{m}_{pq}$ indicates the mass transfer from phase p to phase q, kg/(m^3·s); and S_{aq} is the source term and is zero in this research, kg/(m^3·s). The volume fraction equation is not solved for the primary phase; the primary-phase volume fraction is computed by $\sum\limits_{q=1}^{n} \alpha_q = 1$.

The properties appearing in the transport equations are determined by the presence of the component phases in each control volume. In a gas–liquid system, if the phases are represented by the subscripts 1 and 2, and if the volume fraction of the second of these is being tracked, the density in each cell is given by

$$\rho = \alpha_2\rho_2 + (1 - \alpha_2)\rho_1 \tag{2}$$

All other properties (for example, viscosity) are computed in this manner.

(2) Momentum conservation equation

$$\frac{\partial}{\partial t}(\rho U) + \nabla\cdot[\rho(U - U_g)V] = -\nabla p + \nabla\cdot\tau - \rho g \tag{3}$$

where U_g is the velocity of moving mesh, m/s; p is static pressure, Pa; and τ is stress vector, Pa. U is the volume-averaged velocity, m/s, which is calculated by

$$U = \sum_{q=1}^{n} \alpha_q U_q \tag{4}$$

The density ρ in the last term in Equation (3) is described by Boussinesq approximation, i.e., $\rho = \rho_c[1 - \beta(T - T_c)]$.

(3) Energy conservation equation

$$\frac{\partial}{\partial t}(\rho h) + \nabla\cdot[\rho(U - U_g)h] = \nabla\cdot\left(k_{eff}\nabla T\right) + S_h \tag{5}$$

In Equation (5), volume-averaged value of h is calculated by

$$h = \frac{\sum\limits_{q=1}^{n} \alpha_q\rho_q h_q}{\sum\limits_{q=1}^{n} \alpha_q\rho_q} \tag{6}$$

where h is the enthalpy, J/kg; k_{eff} is the effective conductivity, m^2/s; and S_h is the defined volume source term, W/m^2.

Because of the oscillation, forced convection heat transfer will occur. Since the Reynolds number for a real-size oil tanker is large, the turbulence model should be employed for establishing a general mathematical model. In this paper, k-epsilon model is used to describe the turbulence effect. The equations of turbulent energy k and turbulent dissipation rate ε are as follows:

$$\frac{\partial}{\partial t}(\rho k) + \nabla\cdot[\rho k(U - U_g)] = \nabla\cdot\left[\left(\mu + \frac{\mu_t}{\sigma_k}\right)\nabla k\right] + G_k + G_b + \rho\varepsilon \tag{7}$$

$$\frac{\partial}{\partial t}(\rho\varepsilon) + \nabla\cdot[\rho\varepsilon(U - U_g)] = \nabla\cdot\left[\left(\mu + \frac{\mu_t}{\sigma_\varepsilon}\right)\nabla\varepsilon\right] + \rho C_1 S\varepsilon \\ -\rho C_2\frac{\varepsilon^2}{k + \sqrt{v\varepsilon}} + C_{1\varepsilon}\frac{\varepsilon}{k}C_{3\varepsilon}G_b \tag{8}$$

In these equations, μ_t is turbulent viscosity ($\mu_t = c_\mu k^2/\varepsilon$), Pa·s. G_k represents the generation of turbulence kinetic energy caused by mean velocity gradients, kg/(m·s^3). G_b is the generation of turbulence kinetic energy brought by buoyancy, kg/(m·s³). C_1, C_2, $C_{1\varepsilon}$ and $C_{3\varepsilon}$ are constants. σ_k and σ_ε are the turbulent Prandtl numbers for k and ε, respectively.

With respect to dynamic meshes, the integral form of the conservation equation for a general scalar, ϕ, on an arbitrary control volume, V, whose boundary is moving can be written as

$$\frac{d}{dt} \int_V \rho\phi dV + \int_{\partial V} \rho\phi(\boldsymbol{U} - \boldsymbol{U_g}) \cdot d\boldsymbol{A} = \int_{\partial V} \Gamma \nabla \phi \cdot d\boldsymbol{A} + \int_V S_\phi dV \tag{9}$$

where Γ is the diffusive coefficient, m²/s; S_ϕ is the source term of ϕ; and ∂V indicates the boundary of control volume V.

2.2.2. Boundary Conditions

As indicated in Figure 3, all the boundaries are subjected to the third-type boundary condition. The temperature of sea water is set to be 290.4 K, which is the annual average temperature of sea water; the temperature of air is chosen to be 293 K, which is also the annual average temperature. The initial temperature of the crude oil in the cabin is set at 323 K. It is provided in [17] that the forced heat transfer coefficient of water is 1000–1500 W/(m²·K), and that of air is 20–100 W/(m²·K). In this paper, the forced heat transfer coefficient of water is 1250 W/(m²·K), and that of air is 50 W/(m²·K). In conclusion, the detailed information about the boundary conditions is provided in Table 1.

Table 1. Boundary conditions.

Boundaries	Convective Heat Transfer Coefficient	Fluid Temperature
$X = -20$ cm, $Y \geq 15$ cm	$h_f = 50$ W/(m²·K)	$T_f = 293.15$ K
$X = 20$ cm, $Y \geq 15$ cm	$h_f = 50$ W/(m²·K)	$T_f = 293.15$ K
$X = -20$ cm, $Y \leq 15$ cm	$h_f = 1250$ W/(m²·K)	$T_f = 290.4$ K
$X = 20$ cm, $Y \leq 15$ cm	$h_f = 1250$ W/(m²·K)	$T_f = 290.4$ K
-20 cm $\leq X \leq 20$ cm, $Y = 0$ cm	$h_f = 1250$ W/(m²·K)	$T_f = 290.4$ K

3. Numerical Method

The computational domain is mapped by structured quadrilateral mesh generated by ICEM (The Integrated Computer Engineering and Manufacturing code) combined in ANSYSY FLUENT 15.0 [18], and the grid system is sketched in Figure 4. The rational grid density, which is 200 × 150, is determined after grid independent testing shown in Figure 5.

The governing equations are discretized in the framework of the finite volume method. The convection terms are discretized with the QUICK scheme. The volume fraction is discretized by QUICK scheme. The unsteady term is discretized by the first order forward difference. The coupling between velocity and pressure is calculated by semi-implicit pressure correction algorithm (SIMPLE algorithm). Since the problem studied involves the motion of the region and the free surface, the time step is set as 0.01 s. The discretized equations were solved by ANSYS FLUENT 15.0 solver.

Since the computational domain is changing its position with time, so a dynamic mesh technique is used in this research. The movement of the grid system is defined by a user defined function (DEFINE_CG_MOTION).

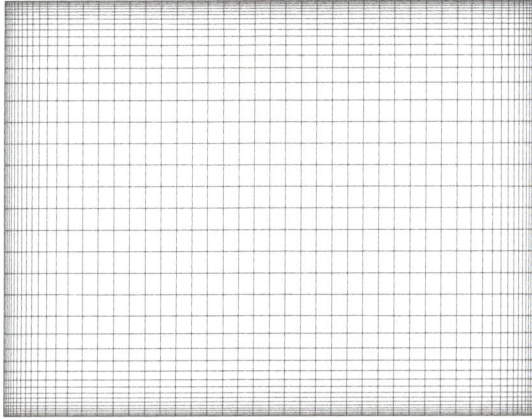

Figure 4. Schematic of the grid system.

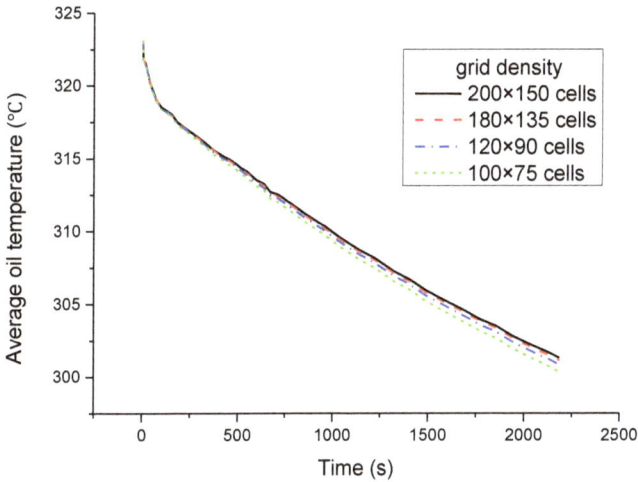

Figure 5. Grid-independence test with Case 3.

4. Results and Discussion

In this section, the temperature drop characteristic of the crude oil in the cargo tank at different oscillating frequencies is studied. To save computation time, only a small-size model tank cargo, which is 40 cm × 30 cm, is studied and shown in Figure 3. The depth of the crude oil is 22.5 cm (i.e., the thickness of air layer is 7.5 cm). The initial temperature in the tank cargo is 323.15 K uniformly. The physical properties of crude oil and air in the calculation are shown in Table 2. Non-Newtonian behavior of crude oil is not considered in this research, if anyone who wants to include this behavior, a power law [19] can be applied to characterize the non-Newtonian behavior of crude oil. With the parameters provided, the maximum Raleigh number for the static case can be calculated, which is 5×10^8.

Table 2. Physical properties of crude oil and air.

Materials	Density (kg/m³)	Thermal Conductivity (W/m·°C)	Specific Heat Capacity (J/kg·°C)	Dynamic Viscosity (Pa·s)	Volume Expansion Coefficient (1/°C)
crude oil	850	0.14	2000	0.004	1.0×10^{-5}
air	1.225	0.0242	1006.43	1.7894×10^{-5}	0.00272

In this study, three different cases named Case 1, Case 2 and Case 3 will be tested for clarifying the influence of the oscillation on the temperature drop. The variables of the three cases are the rotational angular velocity. In Cases 1, the rotational angular velocity is 0, i.e., the tanker does not oscillate. In Cases 2 and 3, the rotational angular velocity is described by Equation (10).

$$\omega = A \cdot \cos(Bt) \tag{10}$$

In Equation (9), ω is the angular velocity of the oscillation motion, and varies by cosine. The period of angular velocity is the same as that of the tanker oscillation ($t_c = \frac{2\pi}{B}$); A and B are both constants. The amplitude of the tanker wobble can be calculated by $\Theta = \int_0^{\frac{t_c}{4}} \omega dt$, and $\Theta = \int_0^{\frac{t_c}{4}} \omega dt = \frac{A}{B}$ can be obtained with ω and t_c substituted. In the two oscillating cases, the time cycles of Case 2 and Case 3 are $t_{c1} = 10$ s and $t_{c2} = 20$ s, respectively. Therefore, $B = \frac{2\pi}{t_c}$ in Equation (9) gives $B_1 = 0.628$ and $B_2 = 0.314$, respectively. In the two oscillating cases, the amplitudes are equal, $\Theta = 18.2\,°$. Thus, in Equation (9), the values of $A = B\Theta$ are $A_1 = 0.2$ and $A_2 = 0.1$ respectively.

For easy reference, the information for Case 1 to Case 3 is listed in Table 3.

Table 3. Information for the three test cases.

Case	Time Cycle of Oscillation	Amplitude of Oscillation
Case 1	∞	0
Case 2	10 s	18.2 °
Case 3	20 s	18.2 °

Under the above-mentioned calculation conditions, the thermal and hydraulic processes of the crude oil in the tanker cargo in three cases are calculated numerically. The hydraulic characteristics (including liquid surface movement and stream function) and the thermal characteristics (including boundary Nusselt number [20] and temperature field evolution) are investigated in detail. Afterwards, the sensitivity of temperature drop to rolling frequency is analyzed.

Figures 6 and 7 show the gas liquid phase distribution at different time instants for Case 1 (static) and Case 2 (oscillation), respectively. In Case 1, the interface of gas–liquid phase does not change with time, and the phase distribution at different time is shown in Figure 6. Different from Case 1, the interface of the gas–liquid phase in oscillating case changes with time. Figure 7 shows the four typical time instants in Case 2 ($t_c = 10$ s). As shown in the figure, there is a small amount of wave on the surface at the beginning of the oscillation, and the liquid level tends to the horizontal surface as time goes by. This is because the oscillation frequency is small and the liquid surface has enough time to return to the horizontal plane.

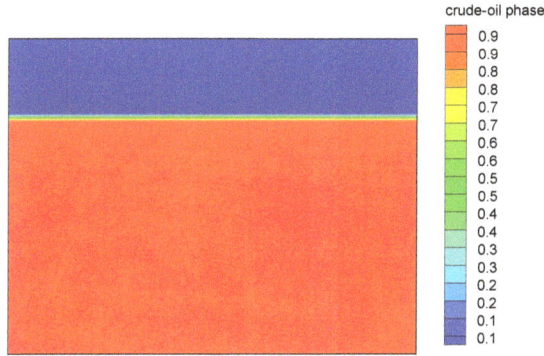

Figure 6. Instantaneous phase distribution in Case 1 (non-oscillation case).

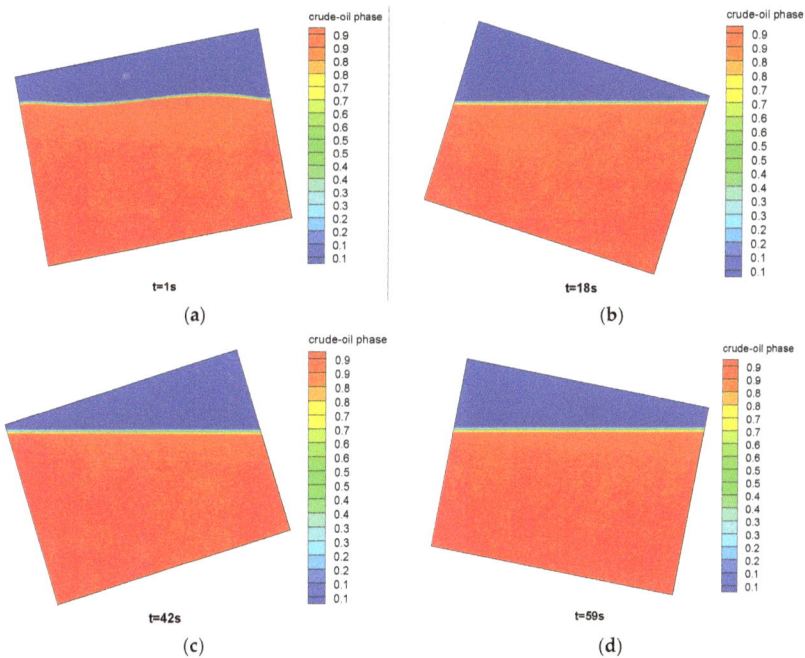

Figure 7. Instantaneous phase distribution in Case 2: (**a**) t = 1 s; (**b**) t = 18 s; (**c**) t = 42 s; and (**d**) t = 59 s.

Figure 8 shows four typical instantaneous stream functions in Case 1. As can be seen from the figure, the stream function is basically symmetrical. There are local vortices generated by natural convection at different position of the fluid region. There are two opposite vortices in the crude oil region, which are generated by natural convection on the left lower boundary and the right lower boundary. There are also a few pairs of vortices in the air region above the liquid level. In this static case, the flow field is dominated by local natural convection at different position.

Figure 9 presents four typical instantaneous stream functions in Case 3, in which Figure 9a,b shows two states at which the domain is rotating to the left, and Figure 9c,d, shows two states at which the domain is rotating to the right. Different from Case 1, the flow field in Case 3 is dominated by integral flow generated by rotation. Because of the effect of rotation, the distribution of stream function is asymmetric, running towards the direction of rotation, and the direction of the flow is also alternately changing. In Figure 9, under the oscillating effect, the streamlines near the solid boundary are running consistent with the solid boundary trend, while there is a central vortex in the core area of the fluid.

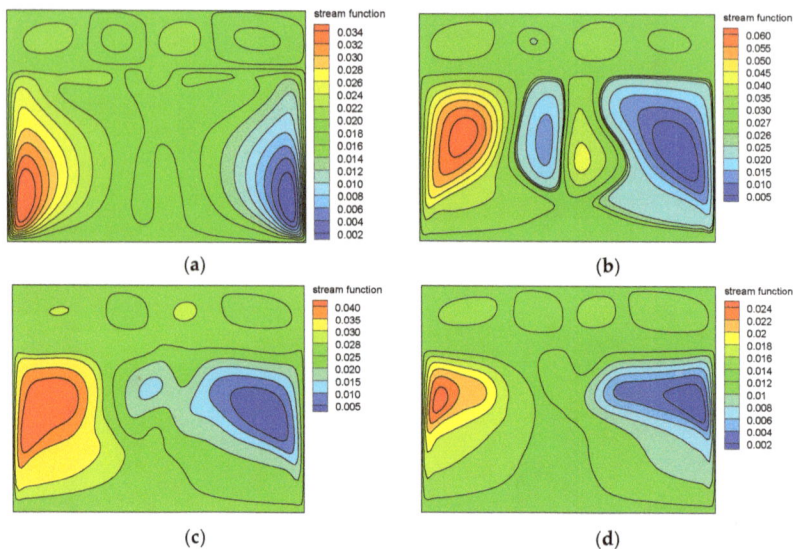

Figure 8. Instantaneous steam function in Case 1: (**a**) $t = 60$ s; (**b**) $t = 660$ s; (**c**) $t = 1020$ s; and (**d**) $t = 2106$ s.

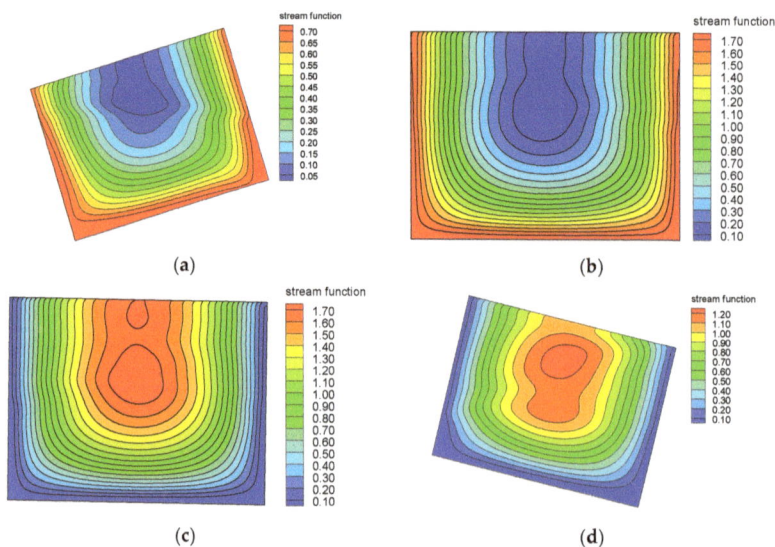

Figure 9. Instantaneous steam function in Case 3: (**a**) $t = 264$ s (moving to the left); (**b**) $t = 660$ s (moving to the left); (**c**) $t = 990$ s (moving to the right); and (**d**) $t = 2112$ s (moving to the right).

To clarify the thermodynamic characteristics, the *Nu* distributions on the four boundaries are studied in this part. Figure 10 shows the *Nu* distributions on the four boundaries of the four typical time instants in Case 1. It can be seen from the figure that the left and right boundary have the identical *Nu*; and the *Nu* on the lower part of left and right boundaries is larger than that on the upper part, since the lower boundary immersed in the water and the upper part exposed in the air, and thus the convective coefficient between sea water and solid wall is larger than that between air and solid wall. On the lower boundary, *Nu* is larger in the middle and lower on both sides. As time proceeds, the *Nu* decreases significantly, and eventually tends to be stable. For the upper boundary, because there are pairs of vortices in the air region, the *Nu* of the upper boundary shows a wavy distribution. In general, the *Nu* on different boundaries gradually decreases with time. In the static case, the heat in the crude oil is mainly lost from the lower part of the left and right boundaries.

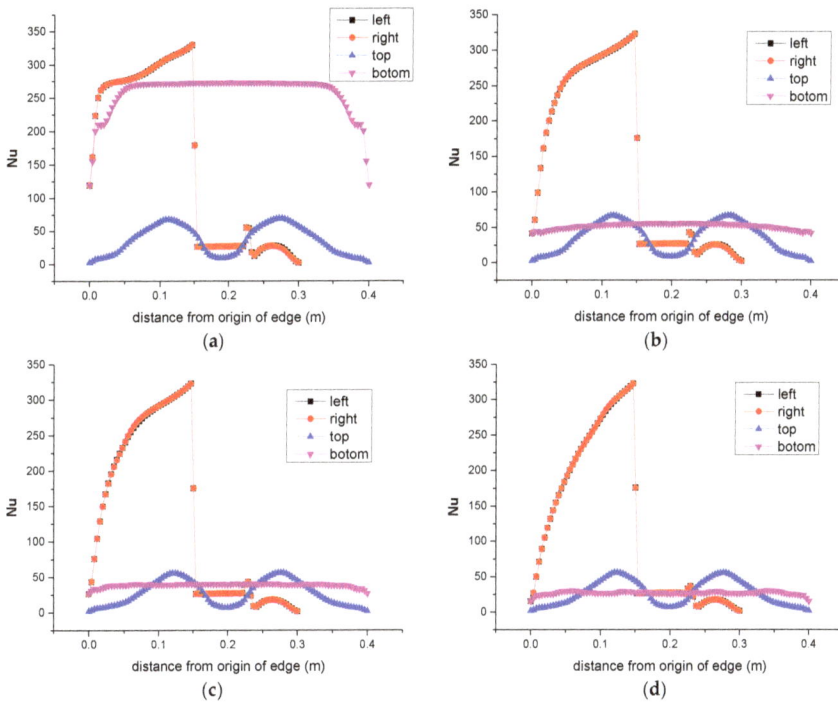

Figure 10. Instantaneous boundary Nusselt number in Case 1: (**a**) $t = 60$ s; (**b**) $t = 660$ s; (**c**) $t = 1020$ s; and (**d**) $t = 2106$ s.

Figure 11 shows the *Nu* distribution on four boundaries at different instances in Case 3. Different from Figure 10, the length of the part immersed in the sea water varies during oscillation in Case 3, so the *Nu* distributions around the left and right boundaries are different and the distribution characteristic change with oscillation. With the oscillation, the flow field is in a complex change, so there is no clear characteristic on the *Nu* difference between the left and right boundaries. The *Nu* of the upper boundary is the smallest, and the *Nu* on the left, right and the lower boundaries are larger. As time proceeds, the *Nu* on each boundary decreases slightly. Unlike the static case, the *Nu* on the lower boundary does not decrease very much. The reason is that oscillation causes the fluid to move along the solid wall all the time, while, in Case 1, the flow near the lower boundary is very weak after a period of temperature drop. Thus, it can be concluded that the loss of heat from the left, right and lower boundary is greater, and that from the upper boundary is relatively small, for the oscillation situation.

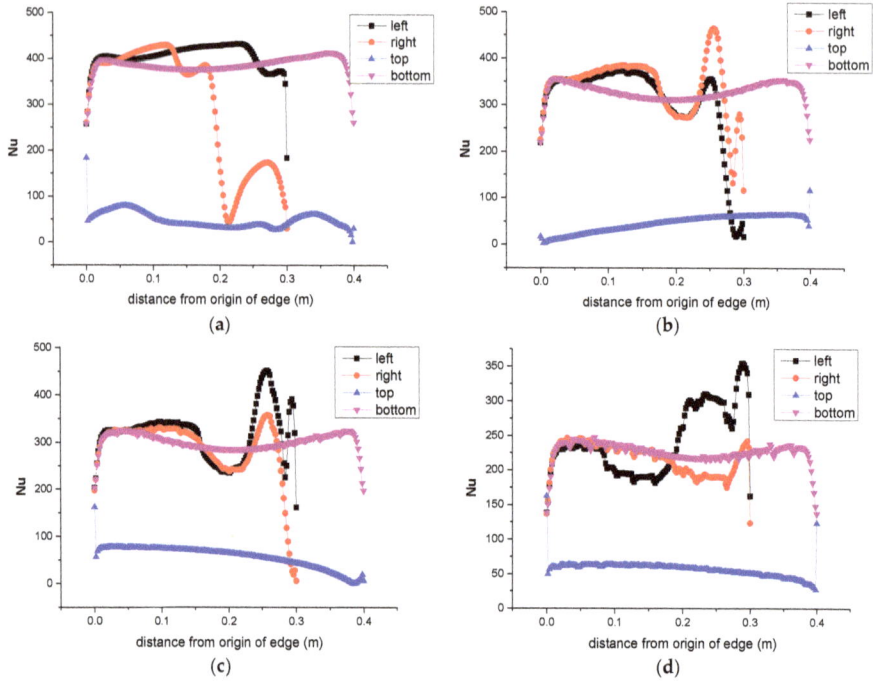

Figure 11. Instantaneous boundary Nusselt number in Case 3: (**a**) *t* = 264 s (moving to the left); (**b**) *t* = 660 s (moving to the left); (**c**) *t* = 990 s (moving to the right); and (**d**) *t* = 2112 s (moving to the right).

Based on the analysis of hydraulic characteristics, the thermal process of the system is investigated. Figure 12 shows the temperature distribution at four typical time instants in the Case 1 (stationary condition). It is found that the temperature field is symmetrical. With the decrease of temperature, the temperature field of the crude oil shows stratified distribution that the lower part has a lower temperature and the upper part has a high temperature. This is the typical stratified temperature distribution formed by the natural convection of the crude oil in the tank. The temperature distribution in the air region presents a vortex distribution. This is because the natural convection of air causes the upper cold air to sink, while the hot air near the oil surface rises and transfers heat upwards. The air zone has multiple adjacent vortices in opposite directions.

The oscillation converts the natural convection in the stationary condition to a mixed convection, and thus the temperature drop characteristic changes dramatically. Figure 13 shows the temperature field distribution at four representative time instants in case 3 (t_c = 20 s). As can be seen in Figure 13, due to the role of oscillation, the temperature field presents an asymmetric distribution, and the temperature field has a tendency to shift toward the rotation. In the early phase of temperature drop, the temperature of the crude oil shows an approximately uniform distribution, and the temperature drop is very slight. However, the temperature drop in the air zone is obvious, and the oscillation makes the temperature distribution in the air zone asymmetric. With the decrease of temperature, the temperature near the four boundaries decreases rapidly, and the low-temperature zone expands gradually to the center. The oil temperature shows the phenomenon that the middle center is higher and the circumference is lower.

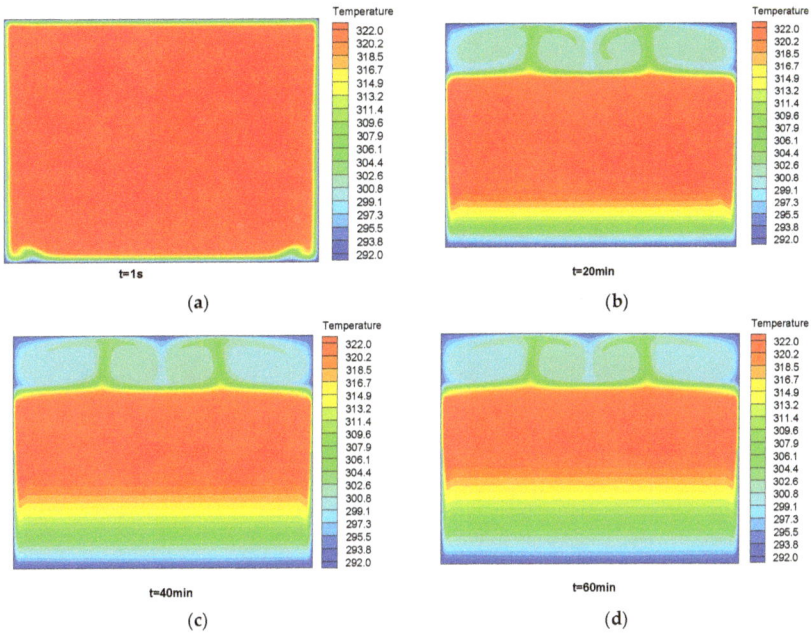

Figure 12. Instantaneous temperature fields in Case 1 (non-osculating condition): (**a**) $t = 1$ s; (**b**) $t = 20$ min; (**c**) $t = 40$ min; and (**d**) $t = 60$ min.

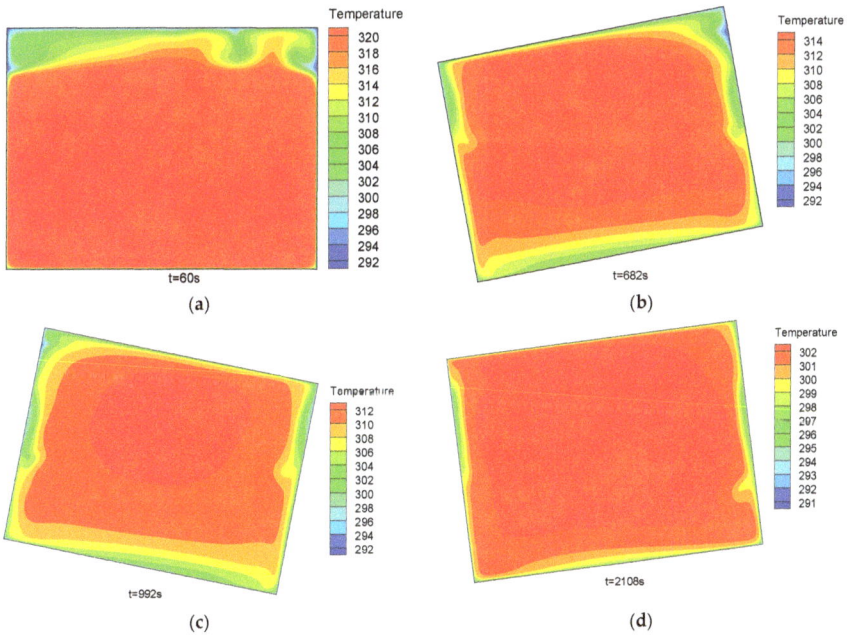

Figure 13. Instantaneous temperature fields in Case 3: (**a**) $t = 60$ s (The tanker is rotating to the left); (**b**) $t = 682$ s (The tanker is rotating to the left); (**c**) $t = 992$ s (The tanker is rotating to the right); and (**d**) $t = 2108$ s (The tanker is rotating to the right).

Comparing Figures 12 and 13, it is found that the oscillation changes the temperature field, and the temperature field shifts to the direction of rotation. From the temperature data, the temperature drop rate under oscillating condition is generally higher than that for the static condition, and the greater the oscillating frequency, the greater the temperature drop rate. The quantitative contrast is shown in Figure 14.

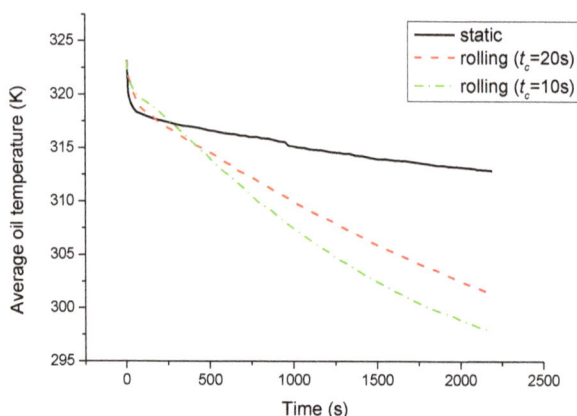

Figure 14. Comparison of average temperature drop under different rolling frequency.

Figure 14 compares the average temperature changes versus time in the cargo in Case 1 (static), Case 2 ($t_c = 10$ s) and Case 3 ($t_c = 20$ s). The temperature drop is obviously enhanced by the oscillating, and the larger the oscillating rate (i.e., the smaller the time cycle), the greater the temperature drop rate. This is because temperature drop of the crude oil under the stationary condition mainly depends on the natural convection, but, in oscillating condition, it depends on mixed convection (natural convection + forced convection), which enhances the heat transfer effect.

Since a small-size model oil tanker is discussed in this research, the thermal process will change if the size of the tanker is changed. If the size is increased to the real size, the Reynolds number, which is defined as $Re = \frac{vd}{v}$, and the Grashof number, which is defined by $Gr = \frac{g\beta\Delta Tl^3}{v}$, will increase dramatically. The Reynolds number and the Grashof number determine the strength of the forced convection and natural convection, respectively. Thus, the larger the tanker size, the more the oscillation will affect the thermal process.

5. Conclusions

In this paper, the thermal and hydraulic characteristic of the crude oil in the cargo tanker under the oscillating condition is numerically studied, with consideration of the free liquid surface motion under oscillating condition. The following conclusions are obtained:

(1) In the case of oscillation, the temperature drop process of crude oil is dominated by mixed convection. The temperature drop process is greatly enhanced and the temperature drop rate is remarkably increased compared with the natural convection for the static case.

(2) According to the actual oscillating period of the tanker, the wave of free surface is not obvious, and the horizontal liquid surface distribution is dominant.

(3) The larger is the oscillation frequency (i.e., the smaller the cycle), the greater is the temperature drop rate.

(4) For the static case, the heat in the crude oil is mainly lost from the lower part of the left and right boundaries, while, in the oscillating case, the loss of heat from the left, right and lower boundary is greater, and that from the upper boundary is relatively small.

Energies **2018**, *11*, 1229

Since only a 2D simplified model is studied in this research, future work will be focused on the real 3D model. In addition, the translational motion should be included in the future work as well.

Author Contributions: G.Y. conceived and designed the research case; Z.F. established the numerical model and performed the numerical simulation; B.D. and D.L. analyzed the solution data; Q.Y. wrote the paper.

Acknowledgments: The study is financially supported by National Science Foundation of China (Nos. 51606117 and 51606114), and China Postdoctoral Science Foundation funded project (2017M621473).

Conflicts of Interest: The authors declare no conflicts of interest

References

1. Sun, X.; Qian, X.; Jiang, X. *Report on the Development in Domestic and Foreign Oil and Gas Industry in 2015*; Petroleum Industry Press: Dongying, China, 2016.

2. BP. *BP Energy Outlook, 2017 Ed.*; BP p.l.c.: Orlando, FL, USA, 2017; Available online: http://www.bp.com/content/dam/bp/pdf/energy-economics/energy-outlook-2017/bp-energy-outlook-2017.pdf (accessed on 25 January 2017).

3. Yin, J.; Tang, B. Analysis of Influence Factors in China Oil Imports Security. *China Energy* **2016**, *11*, 29–33.

4. Li, H. Discussion on current situation and Countermeasures of crude oil transportation market in China. *Chem. Eng. Equip.* **2015**, *6*, 211–212.

5. Yue, D. Calculation of Optimum Heating Time of Tanker Cargo Oil. In *Excellent Collection of Academic Exchange Conference in 2006*; China Institute of Navigation: Beijing, China, 2007; pp. 25–30.

6. Zhang, C. Study on Heat Transfer Mechanism of Crude Oil Heating and Heat Preservation Process. Master's Thesis, Dalian Maritime University, Dalian, China, 2007.

7. Zhang, C.; Liang, Y.; Li, K.; Yue, D. The influence of steam parameters on the heating economy of cargo oil. *J. Dalian Marit. Univ.* **2005**, *4*, 38–40.

8. Zhang, K. Development of Operation System for Oil Heating Process of Tanker. Master's Thesis, Dalian Maritime University, Dalian, China, 2008.

9. Shi, J.; Yin, P.; Zhang, L. Numerical calculation of temperature field in tanker cargo. *J. Dalian Marit. Coll.* **1989**, *3*, 69–74.

10. Jin, Z. Research on Heating and Insulation Process of Tanker Cargo Based on FLUENT Platform. Master's Thesis, Dalian Maritime University, Dalian, China, 2006.

11. Akagi, S. Studies on the Heat Transfer of Oil Tank Heating of a Ship. *J. Kansai Soc. Nav. Arch. Jpn.* **1967**, *124*, 26–36.

12. Suhara, J. Studies of heat transfer on tank heating of tankers. *Jpn. Shipbuild. Mar. Eng.* **1970**, *5*, 5–16.

13. Chen, B.C.M. Cargo oil heating requirements for and FSO vessel conversion. *Mar. Technol.* **1996**, *33*, 58–68.

14. Kato, H. Effects of Rolling on the Heat Transfer from Cargo Oils of Tankers. *J. Soc. Nav. Arch. Jpn.* **2009**, *126*, 421–430.

15. Doerffer, S.; Mikielewicz, J. The influence of oscillations on natural convection in ship tanks. *Int. J. Heat Fluid Flow* **1986**, *7*, 49–60. [CrossRef]

16. Akagi, S.; Kato, H. Numerical analysis of mixed convection heat transfer of a high viscosity fluid in a rectangular tank with rolling motion. *Int. J. Heat Mass Transf.* **1987**, *30*, 2423–2432. [CrossRef]

17. Yang, S. *Heat Transfer*, 4th ed.; Higher Education Press: Beijing, China, 2006.

18. Fluent, A. *ANSYS Fluent Theory Guide 15.0*; ANSYS, Inc.: Canonsburg, PA, USA, 2013.

19. Huang, Z.; Zhang, X.; Yao, J.; Wu, Y.; Yu, T. Non-Darcy displacement by a non-Newtonian fluid in porous media according to the Barree-Conway model. *Adv. Geo-Energy Res.* **2017**, *1*, 74–85. [CrossRef]

20. Luo, L.; Tian, F.; Cai, J.; Hu, X. The convective heat transfer of branched structure. *Int. J. Heat Mass Transf.* **2018**, *116*, 813–816. [CrossRef]

energies

MDPI

Article

An Investigation of Parallel Post-Laminar Flow through Coarse Granular Porous Media with the Wilkins Equation

Ashes Banerjee [1,*], Srinivas Pasupuleti [1], Mritunjay Kumar Singh [2] and G.N. Pradeep Kumar [3]

[1] Departmentof Civil Engineering, Indian Institute of Technology (Indian School of Mines), Dhanbad 826004, Jharkhand, India; vasu77.p@gmail.com
[2] Department of Applied Mathematics, Indian Institute of Technology (Indian School of Mines), Dhanbad 826004, Jharkhand, India; drmks29@rediffmail.com
[3] Department of Civil Engineering, SVU College of Engineering, Sri Venkateswara University, Tirupati 517502, Andhra Pradesh, India; saignp@gmail.com
* Correspondence: ashes@cve.ism.ac.in; Tel.: +91-887-7801831

Received: 17 December 2017; Accepted: 30 January 2018; Published: 2 February 2018

Abstract: Behaviour of flow resistance with velocity is still undefined for post-laminar flow through coarse granular media. This can cause considerable errors during flow measurements in situations like rock fill dams, water filters, pumping wells, oil and gas exploration, and so on. Keeping the non-deviating nature of Wilkins coefficients with the hydraulic radius of media in mind, the present study further explores their behaviour to independently varying media size and porosity, subjected to parallel post-laminar flow through granular media. Furthermore, an attempt is made to simulate the post-laminar flow conditions with the help of a Computational Fluid Dynamic (CFD) Model in ANSYS FLUENT, since conducting large-scale experiments are often costly and time-consuming. The model output and the experimental results are found to be in good agreement. Percentage deviations between the experimental and numerical results are found to be in the considerable range. Furthermore, the simulation results are statistically validated with the experimental results using the standard 'Z-test'. The output from the model advocates the importance and applicability of CFD modelling in understanding post-laminar flow through granular media.

Keywords: Wilkins equation; non-laminar flow; turbulence modelling; porous media

1. Introduction

The porous media flow is commonly characterized by the linear relation between the superficial velocity and hydraulic gradient, proposed by Henry Darcy as:

$$V = ki \tag{1}$$

where V is the superficial velocity (m/s); i is the hydraulic gradient (head loss per unit length in the direction of flow), and k is the coefficient of permeability (m/s).

However, it is observed that Equation (1) predicts the flow satisfactorily only when the Reynolds number is less than 10 [1,2]. At higher Reynolds numbers, the velocity and hydraulic gradient do not exhibit a linear relationship. Therefore, use of Darcy's law can cause considerable error in the post-laminar conditions such as the calculation of well productivity [3] and the discharge measurement and design of pumping wells, hydraulic structures such as rock fill dams [4,5], water treatment filters and fissured rocks. Therefore, the characterization and modeling of flow through porous media is pertinent since the issue aptly addresses the abovementioned challenges. Furthermore, the studies and knowledge about the porous media flow can also largely influence efficiency and understanding of the oil and gas exploration which are directly related to the energy sector.

The flow, in the post-laminar regime is mostly described using the Forchheimer equation, as:

$$i = aV + bV^2 \tag{2}$$

where a (s/m) and b (s^2/m^2) are the Darcy and Non-Darcy coefficients, respectively.

From a thorough review of the literature, one can find enough evidence confirming the complex variation pattern of the Darcy and non-Darcy coefficients with field and media conditions such as porosity, media size, convergent angle, permeability, and so on [6–11]. Efforts are made to relate the Darcy and non-Darcy coefficients with the parameters influencing the flow, but they are observed to be empirical therefore of limited use [12–17]. The complex variation pattern of the coefficients makes it difficult to predict their values for a given set of media and field conditions. Therefore, modelling of the post-laminar flow, using the Forchheimer type equations often prove to be very difficult.

Since there is no proper equation for representing flow in the non-laminar regime, the search for a single constitutive relation continues even now [18,19].

The Wilkins equation was developed to represent flow in both laminar and non-laminar flow regimes. The equation can be presented as [20]:

$$V = C_1 \mu^\alpha r^\beta i^\gamma \tag{3}$$

where r is characteristic length; α and β are constants; C_1 is a coefficient; γ is an exponent having a value between 0.5 and 1.0, depending on the nature of the regime; and μ is the dynamic viscosity of the fluid. For a constant viscosity, Equation (3) can be written as:

$$V = Wr^\beta i^\gamma \text{ with, } W = C_1 \mu^\alpha \tag{4}$$

where W is the modified Wilkins coefficient for a constant viscosity. The coefficients of the Wilkins equation are reported to be relatively non-deviating for media with different hydraulic radius (discussed in Section 2) [21–23], unlike the Forchheimer equation [6,20]. The non-deviating nature of Wilkins coefficients encouraged the authors to further investigate the behaviour of these coefficients when subjected to independently varying media sizes and porosities. Further, an effort is made in the present study to simulate the post-laminar flow through porous media with a comprehensive CFD model using ANSYS FLUENT. The model outputs are compared with the experimentally obtained results and further validated statistically using the standard 'Z-test'. The model outputs and its correlation with the experimental results may advocate the use of such CFD models to analyse and understand the nature and characteristics of post-laminar flow through granular porous media.

2. Experiments and Methodology

A specially conceived parallel flow permeameter with a diameter of 250 mm and a length of 1100 mm is used after packing it with irregularly shaped media having a volume diameter (diameter of a sphere having the same volume as the irregular media) of 29.8 mm, 34.78 mm, and 41.59 mm (Figure 1). To investigate the effect of porosity variation on the flow resistance, velocity behaviour; the same media is repacked three times to achieve separate porosities. The media is packed between two perforated plates and water is allowed to flow at the maximum possible rate for 4 to 5 h to make sure no reorientation of media occurs during the experimentation.

Water is supplied through a header tank (1300 mm × 300 mm × 300 mm) to the permeameter at a constant volumetric flow rate (m^3/s). The flow rate is measured as the average of three volumetric flow measurements (m^3/s) with an accuracy of ±2.35%. The measured volumetric flow rate m^3/s) is divided by the cross sectional area (m^2) of the permeameter to obtain the superficial velocity (m/s). For every flow rate, eight separate piezometric head differences are noted from tapings placed vertically over the permeameter at a regular interval of 50 mm. The average hydraulic gradient calculated from the recorded head differences is used for further analysis with an accuracy of ±1.89%. Such an arrangement eliminates any error due to the non-uniformity of packing.

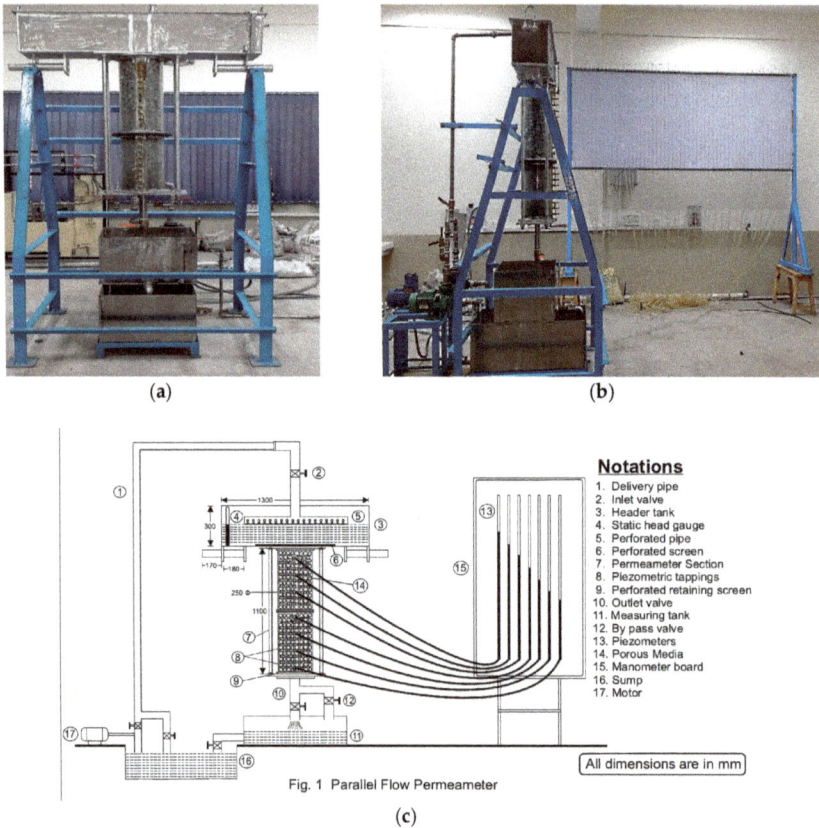

(a)   (b)

Fig. 1 Parallel Flow Permeameter

(c)

Figure 1. The experimental set-up with the parallel flow permeameter. (**a**) the front view; (**b**) the side view; and (**c**) the schematic diagram.

To ensure a post laminar flow condition, experimentations are performed at a higher range of Reynolds number (1736–7194) defined by Kovacs [1,2] as:

$$\mathrm{Re} = \frac{V d_k}{\nu} \tag{5}$$

where ν = kinematic viscosity (m^2/s); d_k = characteristic length, defined as $d(1 - f)\alpha_s/4f$; d and α_s are the volume diameter (m) and shape factor of the media, respectively.

The 'hydraulic radius' (r) [24] (ratio of the void ratio (e) to specific surface (S_0)) is used as the characteristic length while calculating the Wilkins coefficients since it includes both parameters (porosity and media size) that influence the flow. The average surface areas and volumes which are prerequisites to calculate the specific surface (surface area per unit volume) are measured as described by Banerjee et al. [23] and presented in Table 1. After measuring the surface area and volume, the volume diameter (diameter of the sphere having the same volume of the media) and the specific surface is calculated.

Table 1. Properties of the porous media used.

Passing and Retaining Sieve Sizes (cm)	Volume (cm³)	Volume Diameter (cm)	Avg. Surface Area (cm²)	Specific Surface (/cm)
2.50–3.15	14.46	2.98	44.68	2.88
3.15–3.75	23.10	3.48	58.44	2.53
3.75–5.00	38.55	4.16	82.11	2.13

3. Correction Factors

The experimental setups used in the laboratory are confined in nature, which creates uneven packing in the vicinity of the wall and in the interior of the bed. However, in reality, most of the flows are unconfined in nature. Similarly, during the experimentation, the flow path is assumed to be linear whereas, in reality, it is very tortuous. As a consequence of such constraints, an exact simulation of the field condition in the laboratory is very difficult. In view of the fact that the results obtained in the laboratory cannot be directly applicable to the field, the wall, tortuosity and porosity corrections are applied to the obtained results.

3.1. Wall Correction

The confined boundary wall of the permeameters used in the laboratory causes uneven packing between the media near the walls, which is loosely packed, compared to the media inside. Therefore, the velocity of the flow inside the media and near the wall becomes different. Given that, the media used in the present study are larger in size and the ratio of the permeameter to media diameter is low. Therefore, a wall correction factor developed by Thiruvengadam and Kumar [6] is used to correct the data from any wall effect as follows:

$$V_w = \frac{V}{C_w} \text{ with } C_w = \left[\frac{(D + 4.83 \times \frac{d}{2})(D - 0.83 \times \frac{d}{2})}{D^2} \right]^{-1} \tag{6}$$

where V_w is the corrected velocity after wall correction, D is the diameter of the permeameter section (mm), and d is the diameter of the porous media used.

3.2. Tortuosity Correction

Generally, the hydraulic gradient is calculated as the ratio of the total head loss to the distance between the tapings, assuming that the flow path is linear in the direction of the pressure loss. However in reality, the fluid flows through the interlinked void spaces within the media. Owing to that, the flow path is mostly non-linear which can cause inaccuracies in the calculation of hydraulic gradient. Therefore, tortuosity corrections are helpful to reduce that error. By definition, tortuosity (τ) is presented as the ratio between the actual lengths of flow between two tapings to the linear distance between them.

The complexity of the porous structure poses a very challenging task while deriving an expression of tortuosity correction factor. Therefore, most tortuosity corrections are empirical in nature obtained from experiments. The Yu and Li [25] model is used here because it theoretically represents a well-defined relationship between tortuosity and porosity. The model presents the tortuosity correction as:

$$\tau = 1 + \sqrt{1-f}\left(\frac{1}{2} + \frac{\sqrt{(\frac{1}{\sqrt{1-f}} - 1)^2 + \frac{1}{4}}}{1 - \sqrt{1-f}}\right) \tag{7}$$

where f is the porosity of the media.

3.3. Porosity Correction

The velocities obtained from the test are calculated by dividing discharge with the cross-section area of the permeameter (superficial velocity), which is different from the velocity inside the media (pore velocity). In order to estimate the actual pore velocity, porosity corrections [26] are introduced as follows:

$$V_v = \frac{V}{f} \tag{8}$$

where V_v is the pore velocity and f is the porosity.

The results are analysed after incorporating all the corrections to their respective parameters as:

$$V_c = \frac{V.\tau}{fC_w} \text{ and } i_c = \frac{i}{\tau} \tag{9}$$

where V_c and i_c are the corrected velocity and hydraulic gradients, respectively.

4. Data Analysis

Before the analysis of the data, the following assumptions were made: the flow is single phase and unidirectional, the medium is homogeneous, and the porosity of the medium is uniform. After applying all the corrections in to the experimentally obtained superficial velocities (*V*) and hydraulic gradients (*i*), the corrected velocities (*V_c*) and corrected hydraulic gradients (*i_c*) are plotted in Figure 2 and a relationship in the form of a power law is obtained. This type of equations are also referred as Izbash equation [27] or Missbach equation [20] in the literature:

$$i_c = PV_c^j \tag{10}$$

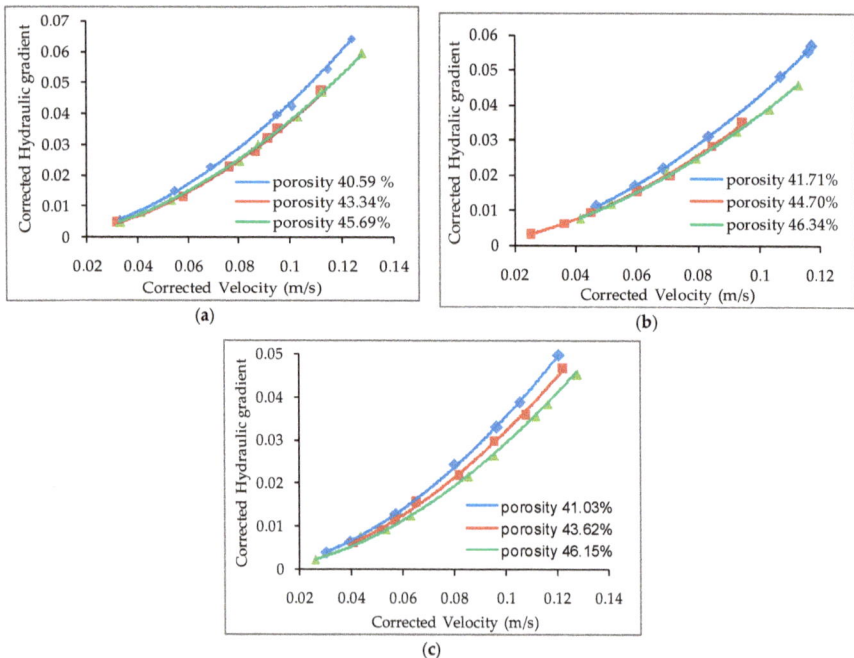

Figure 2. The variation of corrected hydraulic gradient with a corrected velocity for the media size being (**a**) 29.8 mm; (**b**) 34.78 mm; and (**c**) 41.59 mm.

Values of the coefficient P and the exponent j from Equation (10) are calculated from Figure 2. The equation can be modified as follows:

$$V_c = \left(\frac{i_c^{\left(\frac{1}{j}\right)}}{\sqrt[j]{P}} \right) \tag{11}$$

The Wilkins equation for a constant viscosity can be written as:

$$V_c = W r^\beta i_c^\gamma \tag{12}$$

Thus, judging by the similarity of Equations (11) and (12):

$$\sqrt[j]{P} = \frac{1}{W r^\beta} \text{ with } \gamma = \frac{1}{j} \tag{13}$$

$$\frac{1}{j} \log P = -(\log W + \beta \log r) \tag{14}$$

When Equation (14) is plotted in linear form (Figure 3) with $-(1/j)\log P$ on the y axis and $\log r$ on the x axis, the slope represents the value of the coefficient β and the intercept represents the value of $\log W$. The obtained values of W and β are compared for different media sizes and porosities to understand their effect on the Wilkins coefficients.

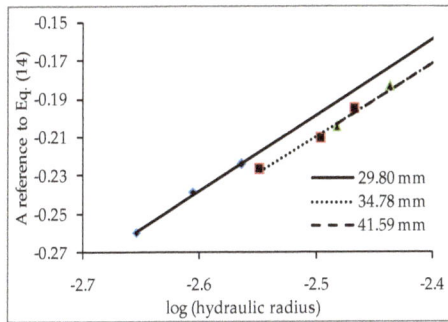

Figure 3. Equation (14) plotted in a linear form to find out the values of W and β.

5. The Behavior of the Wilkins Coefficients with Independently Varying Media Size and Porosity

Experimentally obtained values for the Wilkins coefficients (W and β) are presented in Table 2. The values of the coefficients are found to be rather constant for all the media experimented with. However, values of γ depend on the flow regime, as mentioned earlier. When compared with the results obtained from earlier reported studies (Table 3), almost similar values of the coefficients are observed. However, it is worth mentioning that the coefficients in the present study vary a little from the ones presented by Banerjee et al. [23]. This may be due to the application of the correction factors. To compare the present experimental results with some of the earlier reported data [28], correction factors are directly applied to the measured velocity and hydraulic gradient instead of modifying the Wilkins equation itself. Since some of the studies have used correction factors where others did not, and the minor differences in the presented results can be attributed to application of correction factors and their different values used to correct the velocity and hydraulic gradient along with the uncertainty and complexity of the porous structure and the human errors during experimentation and calculation. However, one can agree that, for all the reported data, the coefficients of the Wilkins equation present similar values. Furthermore, an attempt is made in the present study to understand the nature of the Wilkins coefficient (W) when the size of the media (volume diameter) and porosity vary independently, after having the coefficients calculated for different hydraulic radius. The values of W are calculated from Equation (12) and plotted for different media sizes whilst keeping the porosity of the packing constant (Figure 4a) and similarly, for different porosities whilst keeping the media size constant (Figure 4b). Figure 4a,b indicate a constant nature of W for media size variation, whereas it shows a slightly increasing trend with porosity when media size is constant.

Table 2. The values of the Wilkins coefficient with variation in media size.

Media Size (mm)	Porosity (%)	W (m-s)	β	γ
29.80	40.59 43.34 45.69	6.15	0.39	0.54
34.78	41.72 44.70 46.34	5.52	0.38	0.56
41.59	41.03 43.62 46.15	5.55	0.38	0.55

Table 3. The values of the Wilkins coefficients as reported by earlier researchers.

Proposed by	Media	Volume Diameter (mm)	Porosity (%)	W (m/s)	β	γ
Wilkins (1956) [21]	Crushed stone	51.00	40.00	5.24	0.50	0.54
Garga et al. (1990) [22]	Crushed stone	24.60	47.00	5.39	0.50	0.53
Pradeep Kumar (1994) [27]	Crushed stone	13.10 20.10 28.90 39.50	47.00 45.88 48.73 48.26	4.94	0.51	0.52

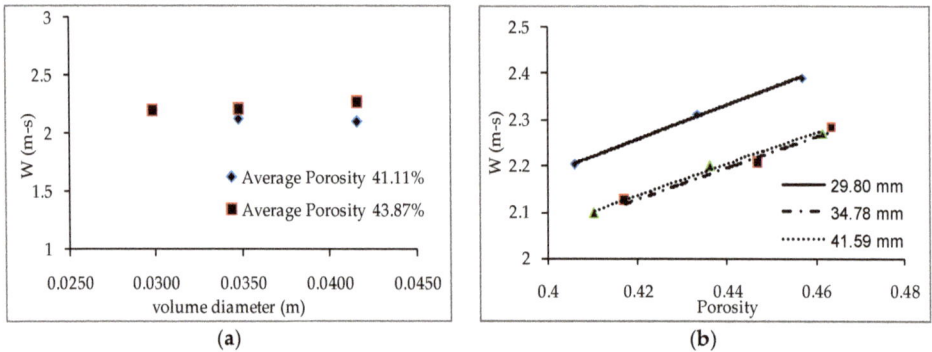

Figure 4. The variation of W with (a) varying volume diameters of the media with constant porosities and (b) with varying porosity with constant media size.

6. Numerical Modelling

The CFD approach is a very powerful tool for modelling and visualising flows under different conditions. The CFD approach numerically solves three basic equations: the continuity (mass conservation) equation, the motion equation and the energy equation at each point of the computational domain. The ANSYS FLUENT CFD solver, which is based on the Finite Volume method, is used to create and simulate laboratory conditions and to solve the CFD problem.

6.1. Turbulence Modelling

The continuity equation and the momentum conservation (Navier–Stokes) equation for incompressible flow are presented as Equations (15) and (16) respectively:

$$\frac{\partial u_i}{\partial x_j} = 0 \tag{15}$$

where u_i is the velocity component in the x_i direction:

$$\rho u_j \frac{\partial u_i}{\partial x_j} = \frac{\partial p}{\partial x_j} + \frac{\partial}{\partial x_j}(\mu \frac{\partial u_i}{\partial x_j}) + \rho g_i \tag{16}$$

where ρ is the density of the fluid, p the pressure, μ is the molecular viscosity and g is the gravitational acceleration.

The turbulent flow is usually simulated after modification of the continuity equation and the momentum conservation (Navier–Stokes) equation for laminar flow with Reynolds decomposition.

The generated quantities known as "Reynolds Stresses" ($-\rho \overline{u_i' u_j'}$) can be defined by the Boussinesq hypothesis using turbulent viscosity (μ_t) and turbulent kinetic energy (k) [29].

Finally, two additional transport equations need to be incorporated in order to close the equation system representing the turbulent kinetic energy and the turbulence dissipation rate (the k-ε model). The "realizable" k-ε model [30] with a standard wall function is used after a bibliographical study concerning the validation of the CFD model including the realizable k-ε model for a wide range of flows, as well as flows that include the boundary layers and flows over obstacles [30,31].

The two equations for turbulent modelling kinetic energy (k) and turbulence dissipation rate (ε) can be written as [29]:

$$\rho[\frac{\partial}{\partial x_j}(ku_j)] = \frac{\partial}{\partial x_j}[(\mu + \frac{\mu_t}{\sigma_k})\frac{\partial k}{\partial x_j}] + G_k - \rho \varepsilon \tag{17}$$

$$\rho[\frac{\partial}{\partial x_j}(\varepsilon u_j)] = \frac{\partial}{\partial x_j}[(\mu + \frac{\mu_t}{\sigma_\varepsilon})\frac{\partial \varepsilon}{\partial x_j}] + \rho C_1 S \varepsilon - \rho C_2 \frac{\varepsilon^2}{k + \sqrt{\nu \varepsilon}} \tag{18}$$

where $C_1 = \max[0.43; \frac{\eta}{\eta+5}]$; $C_2 = 1.9$; $\sigma_k = 1$; $\sigma_\varepsilon = 1.2$; η is the time scale ratio defined as $= S\frac{k}{\varepsilon}$ with S is the modulus of the mean rate of strain tensor defined by $S = \sqrt{2S_{ij}S_{ij}}$; G_k represents the production of the turbulence kinetic energy.

It is worth mentioning that the authors have opted for the standard values of the model constants in the present study since the required model outputs (pressure loss and superficial velocity) are found to be in judicious agreement with the experimental ones. However, the standard values of the model constants may not always provide satisfactory output especially while modelling with no or little turbulence. In such cases, some low Reynolds number closure [32–35] may be used.

6.2. Model Description

The 3D geometric model with an inlet section, a tank, a parallel flow permeameter, and an outlet section of similar dimensions as the experimental set-up is created using the geometry feature in the ANSYS FLUENT workbench. Owing to the high degree of randomness and the fact that no two porous media shows similar structures at the pore level, it is unnecessary to construct the whole porous media at the fine level as it may not affect the overall physical property of the media. Therefore, the permeameter section is identified as the porous zone with uniform porosity similar to that achieved during experimentation. The porous media is numerically accounted as a momentum sink or a resistance in the momentum equation composed of viscous and inertial terms as given in Equation (19) [29]:

$$S_i = -(\sum_{j=1}^{3} D_{ij}\mu v_j + \sum_{j=1}^{3} C_{ij}\frac{1}{2}\rho|v|v_j) \qquad (19)$$

where S_i is the source term for i^{th} (x, y or z) momentum equation; $|v|$ is the magnitude of the velocity; D and C are the prescribed matrices. For a homogeneous media:

$$S_i = -(\frac{\mu}{\alpha}v_i + C_2\frac{1}{2}\rho|v|v_i) \qquad (20)$$

where α is the permeability; C_2 is the inertial resistance factor simply specifying D and C as diagonal matrices; and μ and ρ are the dynamic viscosity and the density of the fluid, respectively. The viscous and inertial resistance are measured based on the pressure loss observed from the experimental setup [29,36]. Furthermore, the model is solved by the finite volume method with the velocity pressure coupling done by the SIMPLE algorithm. The governing equations are discretized using the second order upwind differencing scheme. Convergence is attained when the scaled residuals are less than 10^{-4} times of their initial values.

6.3. Meshing

By creating the mesh, a continuous domain describing the flow with partial differential equations is replaced with a finite number of volumes (meshes). The discretization is performed using structured tetrahedral meshes. Nine different grid sizes are selected arbitrarily having element number ranging from 8867 to 6,463,169. Iterative convergence is obtained for each of these grid sizes whilst keeping all other input conditions identical. The output parameters, hydraulic gradient, and velocity are compared for all these meshes which have different numbers of elements. Figure 5 shows that the output parameters become constants as the grid size is refined. Furthermore, a grid sensitivity analysis is performed using the Grid Convergence Index (GCI) method to authenticate the precision of the numerical model [37].

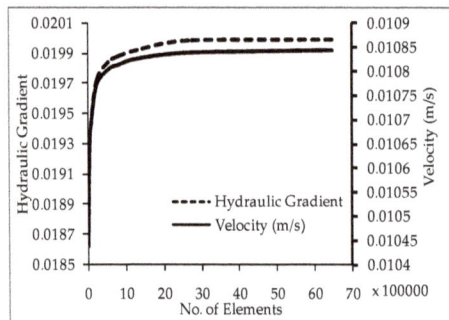

Figure 5. The variation of the pressure drop with different mesh sizes.

The GCI is the representation of the discretization error between the numerical solutions obtained from the finer grid and the coarser one. For three different selected grids (h_1, h_2, h_3) simulations are carried out to estimate the key variables (ϕ); in the present study these variables are Hydraulic gradient, Velocity (m/s) at 0.285 m and Velocity (m/s) at 1.11 m. The apparent order (p) of the method is then calculated using the following equations:

$$p = \frac{1}{\ln(r_{21})}|\ln|\varepsilon_{32}/\varepsilon_{21}| + q(p)| \qquad (21)$$

$$q(p) = \ln(\frac{r_{21}^p - [1.\text{sgn}(\varepsilon_{32}/\varepsilon_{21})]}{32 - [1.\text{sgn}(\varepsilon_{32}/\varepsilon_{21})]}) \tag{22}$$

where $\varepsilon_{32} = \phi_3 - \phi_2$, $\varepsilon_{21} = \phi_2 - \phi_1$, and $r_{21} = h_2/h_1$. The equations are solved using fixed-point iteration using the first term as an initial guess [37]. The relative error (e_a^{21}, e_a^{32}) and the extrapolated relative error (e_{ext}^{21}, e_{ext}^{32}) are calculated using Equations (23) and (24):

$$e_a^{21} = \frac{\phi_1 - \phi_2}{\phi_1} \tag{23}$$

$$e_{ext}^{21} = \frac{\phi_{ext}^{21} - \phi_2}{\phi_{ext}^{21}} \text{ where } \phi_{ext}^{21} = (r_{21}^p \phi_1 - \phi_2)/(r_{21}^p - 1) \tag{24}$$

Finally, the fine Grid Convergence Index (GCI) is calculated using the following equation:

$$GCI_{fine}^{21} = \frac{1.25 e_a^{21}}{r_{21}^p - 1} \tag{25}$$

Table 4 represents values of these parameters for three selected grids (227,414; 72,709; and 32,301). The output parameters are mentioned as ϕ_1, ϕ_2, ϕ_3 or grid sizes h_1, h_2 and h_3 in Table 4. A similar analysis is performed for other selected grid sizes. Numerical uncertainty in the fine grid solution is observed to be 1.09%, 0.56%, and 0.02% respectively for the hydraulic gradient, the velocity at a vertical distance of 0.285 m from the entrance of the permeameter, and the velocity at a vertical distance of 1.11 m from the entrance of the permeameter (Table 4). The GCI values indicate a negligible discretization error and therefore a grid independent solution for the selected grid. Considering the simulation time and precision, a grid system of 1,288,576 elements is adopted for the model.

Table 4. Sample calculations of numerical uncertainty using the GCI method [37].

Parameter	Hydraulic Gradient	Velocity (m/s) at 0.285 m	Velocity (m/s) at 1.11 m
h_1	0.005216	0.005216	0.005216
h_2	0.006937	0.006937	0.006937
h_3	0.009226	0.009226	0.009226
r_{21}	1.33	1.33	1.33
r_{32}	1.33	1.33	1.33
ϕ_1	0.01992	0.01083	0.01156
ϕ_2	0.01984	0.01080	0.01159
ϕ_3	0.01973	0.01077	0.01096
P	1.32772	1.30788	10.5924
ϕ_{ext}^{21}	0.02010	0.01088	0.01156
ϕ_{ext}^{32}	0.02010	0.01088	0.01163
e_a^{21}	0.40%	0.20%	0.27%
e_a^{32}	0.59%	0.29%	5.45%
e_{ext}^{21}	0.86%	0.45%	0.01%
e_{ext}^{32}	1.26%	0.65%	0.28%
GCI_{fine}^{21}	1.09%	0.56%	0.02%

6.4. Boundary Conditions

The boundary conditions are very important for the numerical solution of the problem. The type and the numerical values of the boundary conditions are carefully chosen. The inlet velocity is calculated by dividing the experimental discharge with the area of the inlet pipe. The turbulent quantities (k and ε) at the inlet are calculated from the Equations (26) and (28) using the turbulent intensity (I); turbulent length scale (l),which depends on the hydraulic diameter (D_h); and the inlet velocity(V) [38]. Equation (26) is used to calculate the turbulent intensity at the core of a fully developed duct [29]:

$$I = 0.16(\text{Re})^{-\left(\frac{1}{8}\right)} \tag{26}$$

The value of k is then calculated using the turbulent intensity from the following equation:

$$k = \frac{3}{2}(V \cdot I)^2 \tag{27}$$

where Re is the Reynolds number for a pipe flow defined as $\frac{\rho V d}{\mu}$.

$$\varepsilon = C_\mu^{\frac{3}{4}} \frac{k^{\frac{3}{2}}}{l} \text{ with } l = 0.07 D_h \tag{28}$$

The outlet is kept as an outflow boundary, assuming that the flow is completely developed; thereby causing the diffusion flux for all flow variables in the exit direction to be zero [29].

7. Results and Discussion

The values of the pressure and velocity along the direction of the flow (length of the permeameter) for the media sizes of 29.8, 34.78 and 41.59 mm are presented in Figure 6a,b, Figure 7a,b and Figure 8a,b, respectively. It can be observed that the total pressure decreases as the flow passes through the length of the permeameter, whereas the velocity plots illustrate a constant superficial velocity throughout the length of permeameter. The pressures and velocities at the tank section (-0.3 m to 0 m) are not used for further analysis since only the permeameter section (0 m to $+1.1$ m) is identified as a porous zone and subjected to the study. It is worth mentioning that, during the simulation, the operating pressure is set to zero in order to cut down the rounding errors, resulting in negative values in the pressure distribution diagram. However, it does not affect the simulation because of two main reasons; (a) the flow is assumed to be incompressible and (b) the pressure difference is taken into account by the Navier–Stokes equation which drives the flow. The validation of the results and the importance of the model are discussed further in the following sections.

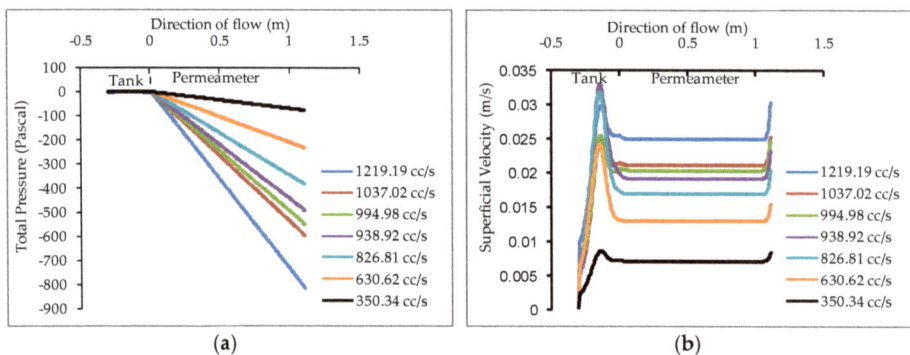

Figure 6. The values of (a) the total pressure and (b) the velocity in the direction of the flow subjected to different discharges for 29.8 mm media packed with 43.34% porosity.

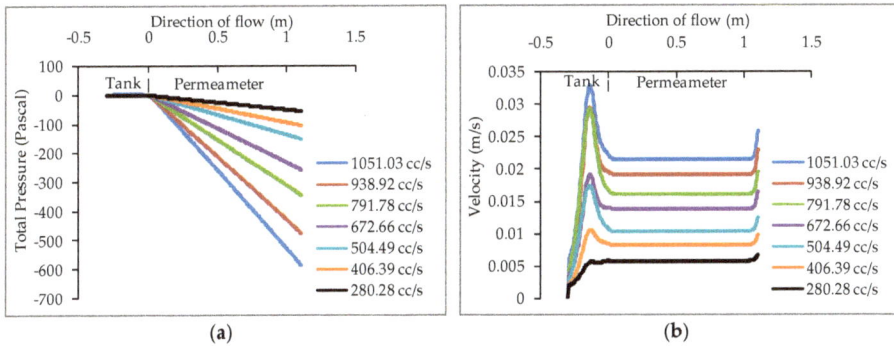

Figure 7. The values of (**a**) the total pressure and (**b**) the velocity in the direction of the flow subjected to different discharges for 34.78 mm media packed with 44.70% porosity.

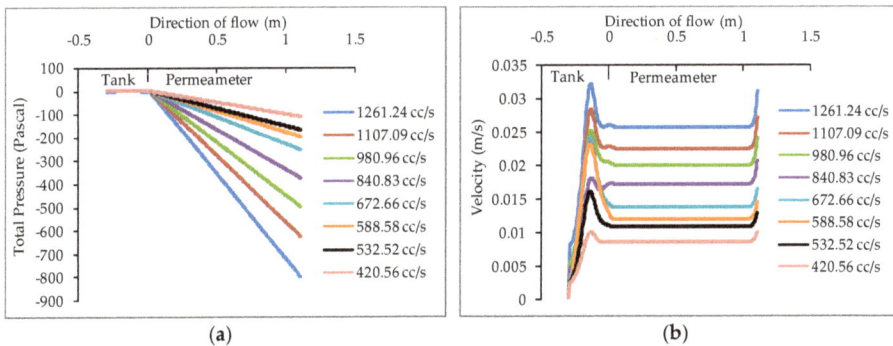

Figure 8. The values of (**a**) the total pressure and (**b**) the velocity in the direction of the flow subjected to different discharges for 41.59 mm media packed with 43.62% porosity.

7.1. Comparison between the Experimental and Simulation Data and Statistical Validation of the Simulation

Experimentally and numerically calculated hydraulic gradients are plotted as exponential function (Equation (10)) of the respective velocities in Figures 9a–c, 10a–c and 11a–c. The plots suggest an acceptable correlation between the experimental results and the simulation results for the range of the data tested.

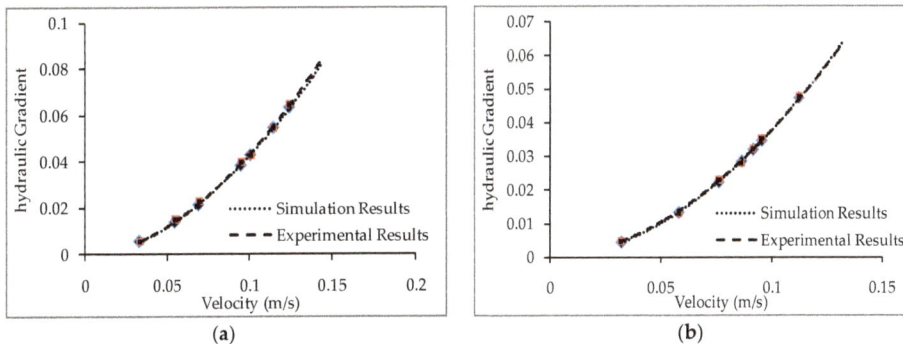

Figure 9. *Cont.*

(c)

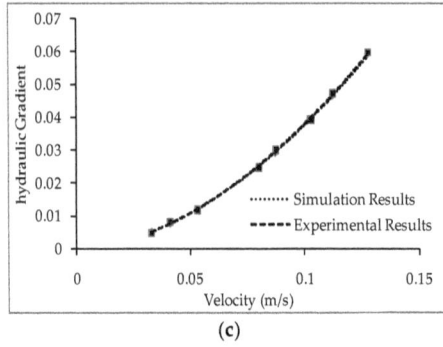

Figure 9. Comparison of simulation and experimental results for 29.8 mm media packed with (**a**) 40.59%; (**b**) 43.34%; and (**c**) 45.69% porosities.

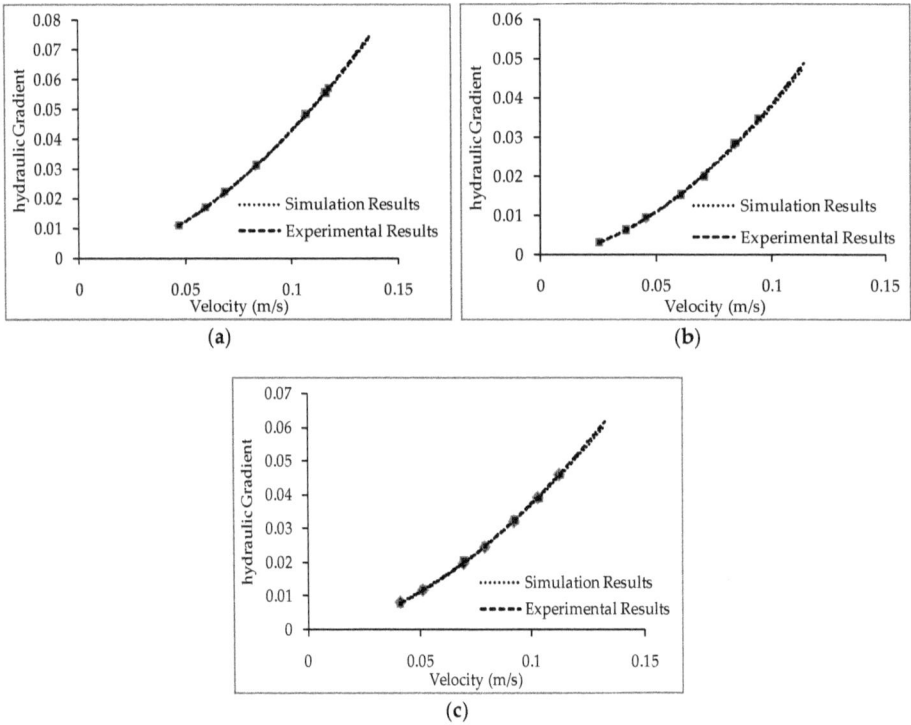

(a)

(b)

(c)

Figure 10. Comparison of simulation and experimental results for 34.78 mm media packed with (**a**) 41.72%; (**b**) 44.70%; and (**c**) 46.34% porosities.

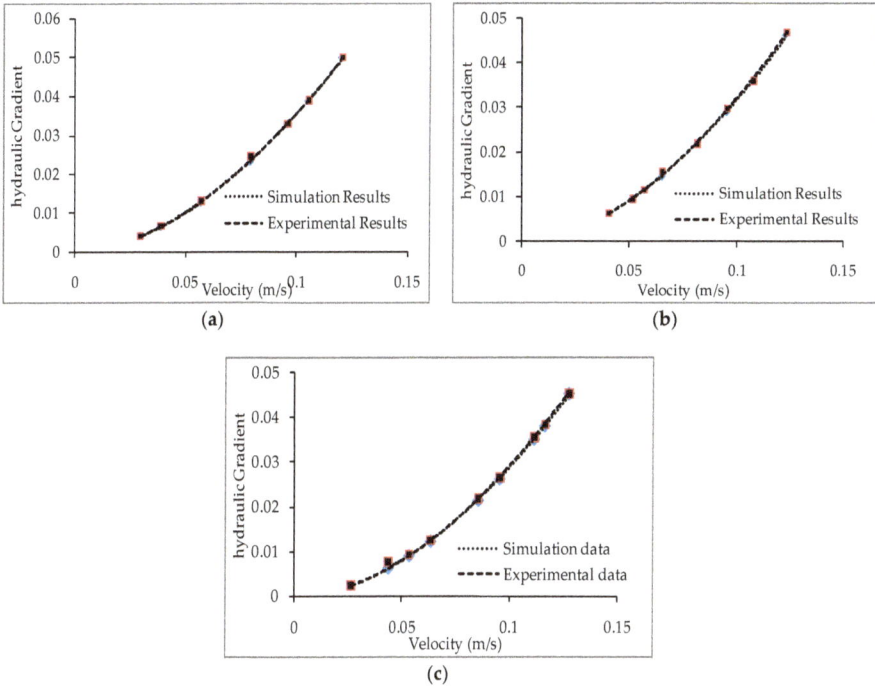

Figure 11. Comparison of simulation and experimental results for 41.59 mm media packed with (a) 41.03%; (b) 43.62%; and (c) 46.15% porosities.

Figure 12a–c represent the percentage deviations between experimental and simulation results applied to the different velocities for all media sizes and porosities. Simulation results are found to be in good agreement with the experimental results for the range of velocities presented. However, the deviation between the simulation and experimental results is not constant. The deviation reduces to a minimum value at a certain velocity and then further increases (Figure 12). This is due to the change in the flow regime with the variation of the velocity. With the change in velocity, the magnitudes of the resistive forces affecting the flow differ, causing a variation in the total resistance thereafter in the velocity-flow resistance relationship. This finally results in a great deviation between the simulation and experimental results. Furthermore, the simulation results have been validated statistically with the experimental results using the standard "Z-test" [39,40], which is explained further in the next section.

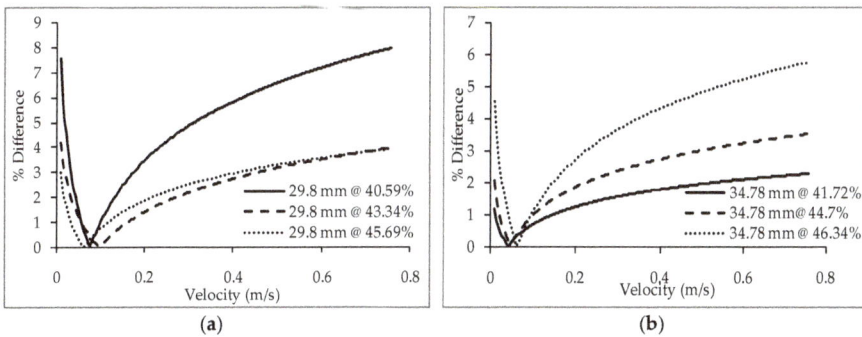

Figure 12. *Cont.*

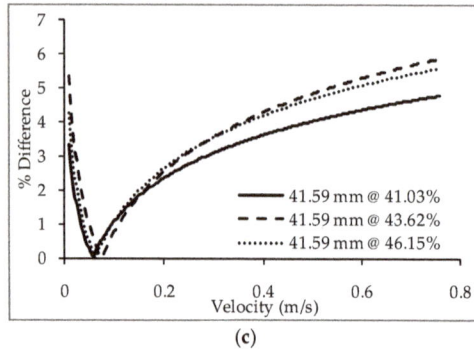

(c)

Figure 12. The variation of the percentage deviation between the experimental and simulation results with flow velocities for (**a**) 29.8 mm; (**b**) 34.78 mm; and (**c**) 41.59 mm media.

7.2. Validation of the Simulated Data Using the Z-Test

In order to test the validity of the simulation result, first the hypothesis is introduced [39–41] as follows:

$$H_0 : \mu_1 = \mu_2$$
$$H_1 : \mu_1 \neq \mu_2$$

(29)

where μ_1 and μ_2 are the population mean of the experimental and simulation result. In order to validate the simulated result, the null hypothesis, given by H_0, must be accepted.

The level of significance (α) is taken to be 5%, which implies that a confidence level of 95% is considered [41]. As the present study deals with two-tailed hypothesis, $\alpha = 0.05/2 = 0.025$. The corresponding critical value of Z for $\alpha = 0.025$ is ± 1.96 (from the Z table). Therefore, for the hypothesis given as Equation (29) to be accepted, the calculated Z value should be within the interval of $[-1.96, 1.96]$. The calculated Z values are obtained by employing the Z-test, as given by Equation (30):

$$Z_{calculated} = \frac{(\overline{x_1} - \overline{x_2}) - (\mu_1 - \mu_2)}{\sqrt{\frac{\sigma_1^2}{N_D} + \frac{\sigma_2^2}{N_D}}} \text{ for } N_D \geq 30$$

(30)

where $\overline{x_1}$, $\overline{x_2}$ and μ_1, μ_2 are the sample and population means for the experimental and simulation results. N_D is the total number of data points (sample size) and σ_1, σ_2 are the standard deviations for the experimental and simulation results.

The results of the Z-test for the three different media sizes are presented in Table 5. The calculated values of Z for all three media sizes are found to be within the acceptable range $[-1.96, 1.96]$ (Figure 13a–c). Therefore, the null hypothesis H_0 is accepted, validating the simulation results with the experimental results. Result from the statistical analysis along with the percentage deviation between the simulation and experimental data signify the accuracy of the CFD model used.

This type of CFD model can be very useful to predict the nature of the flow for any hydraulic structure. For a given media size and porosity, this type of simulation can provide a complete velocity and pressure loss profile throughout any given hydraulic structure which can immensely aid the designers and engineers. Furthermore, the obtained velocity and pressure loss data from such models can also be used to analyse the behaviour of different post-laminar flow equations used in porous media flow including the Wilkins equation.

Table 5. Values from the Z-test for different media sizes and porosities.

Media Size (mm)	Porosity (%)	Range of Velocity (m/s)	$\bar{x}_1$	$\bar{x}_2$	σ_1	σ_2	N_D	Z Value
	40.59	0.01–0.757	0.643	0.599	0.540	0.359	250	−0.940
29.80	43.34	0.01–0.757	0.586	0.606	0.497	0.517	250	0.429
	45.69	0.01–0.757	0.554	0.535	0.464	0.446	250	−0.459
	41.72	0.01–0.757	0.565	0.554	0.463	0.452	250	−0.276
34.78	44.70	0.01–0.757	0.527	0.510	0.435	0.420	250	−0.425
	46.34	0.01–0.757	0.496	0.471	0.407	0.383	250	−0.694
	41.03	0.01–0.757	0.511	0.489	0.422	0.406	250	−0.574
41.59	43.62	0.01–0.757	0.464	0.440	0.391	0.368	250	−0.684
	46.15	0.01–0.757	0.439	0.418	0.369	0.349	250	−0.661

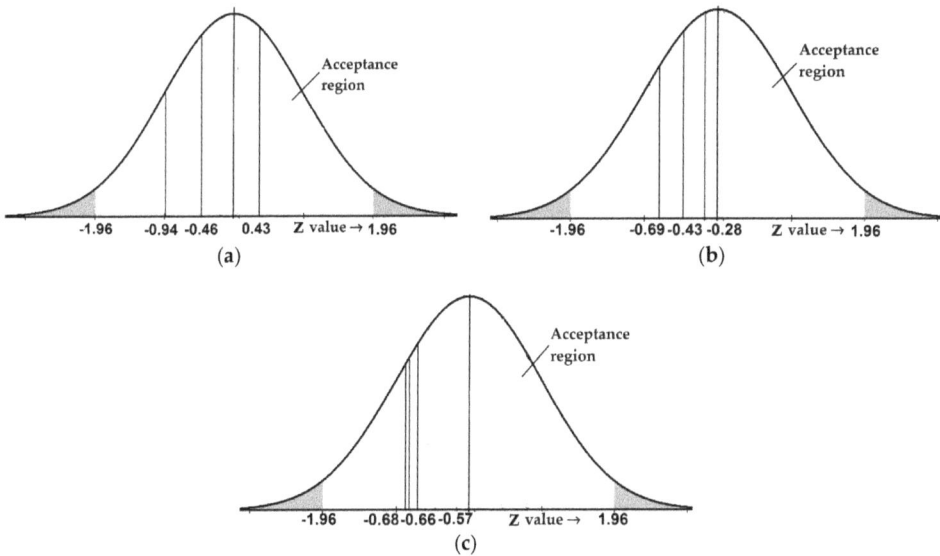

Figure 13. The $Z_{calculated}$ values on the binomial distribution curve for (**a**) 29.8 mm; (**b**) 34.78 mm; and (**c**) 41.59 mm media sizes.

8. Conclusions

A number of non-linear equations were proposed to predict post-laminar flow through porous media at a macroscopic level. However, the complex variation pattern of their respective coefficients limits their applicability in field conditions. The Wilkins equation is studied both theoretically and experimentally with respect to parallel flow. The obtained results are presented as follows:

1. The Wilkins equation can be satisfactorily used to represent post-laminar flow through porous media.

2. The Wilkins coefficients are found to have a non-deviating nature with varying hydraulic radius. The obtained results from the present study are similar to the results reported in the literature.

3. When subjected to variation in media size, the coefficients of the Wilkins equation are constant, given that the porosity is constant. However, variations in the porosity result in small variations of the coefficient *W*.

4. The flow condition inside the experimental set up is simulated with a CFD model in the ANSYS FLUENT software. Trends similar to the experimental ones are obtained from the simulation

results. The percentage deviation between the simulation and experimental results are within the acceptable range.

5. For further validation, the experimental results are statistically compared with the simulation results using the standard Z-test. The values of Z calculated are found to be within the acceptable region for all the experimental results.

Finally, the experimental and simulation results in the present study conclude that the Wilkins equation can represent non-linear flow through porous media with a satisfactory accuracy, compared to Darcy's equation. The coefficients of the Wilkins equations are found to be constant for variations of porosity and media size, unlike the Forchheimer type equations. Hence, this equation can be regarded as a convenient tool for designing and measuring the discharge for parallel flow through porous media in the non-laminar regimes such as rock fill dams, water filters, aquifers, water, oil and gas wells, and so on.

Acknowledgments: The authors of the present manuscript are thankful to the faculty of IIT (ISM) Dhanbad for extending their facilities during the study. The authors would also like to acknowledge the funding received from IIT(ISM), Dhanbad under the FRS (Faculty Research Scheme) with vide No.: FRS(62)/2013-2014/CE for the fabrication of the permeameter experimental set up utilised in the study.

Author Contributions: Ashes Banerjee, Srinivas Pasupuleti, and G.N. Pradeep Kumar conceived and designed the research. Ashes Banerjee conducted experiments, processed and evaluated the data under the supervision of Srinivas Pasupuleti. Mritunjay Kumar Singh and Srinivas Pasupuleti supervised the data analysis. Ashes Banerjee wrote the manuscript. Srinivas Pasupuleti and Mritunjay Kumar Singh edited the manuscript and also helped in improving the technical content. G.N. Pradeep Kumar outlined the framework of this paper and made key suggestions in improving the method, validation and results.

Conflicts of Interest: The authors declare no conflict of interest. Furthermore, with reference to the received funding from the institute mentioned in the Acknowledgements, there are no conflicts of interest regarding the publication of this manuscript.

References

1. Kovacs, G. Seepage through saturated and unsaturated layers. *Hydrol. Sci. J.* **1971**, *16*, 27–40. [CrossRef]
2. Kovacs, G. *Seepage Hydraulics*; ESPC: New York, NY, USA, 1981; ISBN 0-444-99755-5.
3. Vincent, M.C.; Pearson, C.M.; Kullman, J. Non-Darcy and multiphase flow in propped fractures: Case studies illustrate the dramatic effect on well productivity. In Proceedings of the SPE Gas Technology Symposium, Calgary, AB, Canada, 30 April–2 May 2002; pp. 71–84.
4. Parkins, A.K.; Trollope, D.H.; Lawson, J.D. Rockfill structures subject to water flow. *J. Soil Mech. Found. Div.* **1966**, *92*, 135–151.
5. Curtis, R.P.; Lawson, J.D. Flow over and through rockfill banks. *J. Hydraul. Div.* **1967**, *93*, 1–22.
6. Thiruvengadam, M.; Pradip Kumar, G.N. Validity of Forchheimer equation in radial flow through coarse granular media. *J. Eng. Mech.* **1997**, *123*, 696–704. [CrossRef]
7. Ergun, S. Fluid flow through packed columns. *Chem. Eng. Prog.* **1952**, *48*, 89–94.
8. Nasser, M.S.S. Radial Non-Linear Flow through Porous Media. Master's Thesis, University of Windsor, Windsor, ON, Canada, 1970.
9. Niranjan, H.S. Non-Darcy Flow through Porous Media. Master's Thesis, Indian Institute of Technology, Kanpur, India, 1973.
10. Venkataraman, P.; Rao, P.R.M. Darcian, transitional, and turbulent flow through porous media. *J. Hydraul. Eng.* **1998**, *124*, 840–846. [CrossRef]
11. Venkataraman, P.; Rao, P.R.M. Validation of Forchheimer's law for flow through porous media with converging boundaries. *J. Hydraul. Eng.* **2000**, *126*, 63–71. [CrossRef]
12. Kumar, G.N.P.; Thiruvengadam, M.; Murali, T. A further study on Forchheimer coefficients as applied in seepage flow. *ISH J. Hydraul. Eng.* **2004**, *10*, 1–13. [CrossRef]
13. Reddy, N.B.P.; Rao, P.R.M. Convergence effect on the flow resistance in porous media. *Inst. Eng. (I) J.* **2004**, *85*, 36–43.
14. Sadeghian, J.; Kholghi, M.K.; Horfar, A.; Bazargan, J. Comparison of Binomial and Power Equations in Radial Non-Darcy Flows in Coarse Porous Media. *J. Water Sci. Res.* **2013**, *5*, 65–75.

15. Bu, S.; Yang, J.; Dong, Q.; Wang, Q. Experimental study of flow transitions in structured packed beds of spheres with electrochemical technique. *Exp. Therm. Fluid Sci.* **2015**, *60*, 106–114. [CrossRef]
16. Muljadi, B.P.; Blunt, M.J.; Raeini, A.Q.; Bijeljic, B. The impact of porous media heterogeneity on non-Darcy flow behaviour from pore-scale simulation. *Adv. Water Resour.* **2016**, *95*, 329–340. [CrossRef]
17. Li, Z.; Wan, J.; Huang, K.; Chang, W.; He, Y. Effects of particle diameter on flow characteristics in sand columns. *Int. J. Heat Mass Transf.* **2017**, *104*, 533–536. [CrossRef]
18. Dukhan, N.; Bağcı, Ö.; Özdemir, M. Experimental flow in various porous media and reconciliation of Forchheimer and Ergun relations. *Exp. Therm. Fluid Sci.* **2014**, *57*, 425–433. [CrossRef]
19. Hellström, G.; Lundström, S. Flow through porous media at moderate Reynolds number. In Proceedings of the International Scientific Colloquium Modelling for Material Processing, Riga, Latvia, 8–9 June 2006; pp. 129–134.
20. Kumar, G.N.P.; Venkataraman, P. Non-Darcy converging flow through coarse granular media. *J. Inst. Eng. (India) Civ. Eng. Div.* **1995**, *76*, 6–11.
21. Wilkins, J.K. Flow of water through rockfill and its application to the design of dams. *N. Zeal. Eng.* **1955**, *10*, 382–387.
22. Garga, V.K.; Hansen, D.; Townsend, R.D. Considerations on the design of flow through rockfill drains. In Proceedings of the 14th Annual British Columbia Mine Reclamation Symposium, Cranbrook, BC, Canada, 14–15 May 1990.
23. Banerjee, A.; Pasupuleti, S.; Singh, M.K.; Kumar, G.N.P. A study on the Wilkins and Forchheimer equations used in coarse granular media flow. *Acta Geophys.* **2017**, 1–11. [CrossRef]
24. Scheidegger, A.E. *The Physics of Flow through Porous Media*; University of Toronto Press: Toronto, ON, Canada; London, UK, 1958.
25. Yu, B.M.; Li, J.H. A geometry model for tortuosity of flow path in porous media. *Chin. Phys. Lett.* **2004**, *21*, 1569–1571. [CrossRef]
26. Rose, H.E.; Rizk, A.M.A. Further researches in fluid flow through beds of granular material. *Proc. Inst. Mech. Eng.* **1949**, *160*, 493–511. [CrossRef]
27. Sedghi-Asl, M.; Rahimi, H.; Salehi, R. Non-Darcy Flow of Water Through a Packed Column Test. *Transp. Porous Media* **2014**, *101*, 215–227. [CrossRef]
28. Kumar, G.N.P. Radial Non-Darcy Flow through Coarse Granular Media. Ph.D. Thesis, Sri Venkateswara University, Tirupati, India, 1994, unpublished.
29. ANSYS Inc. *Ansys Fluent 15.0: Users Guide*; ANSYS Inc.: Canonsburg, PA, USA, November 2013.
30. Shih, T.H.; Liou, W.W.; Shabbir, A.; Yang, Z.; Zhu, J. A new k–ε eddy viscosity model for high Reynolds number turbulent flows. *Comput. Fluids* **1995**, *24*, 227–238. [CrossRef]
31. Kim, S.E.; Choudhury, D.; Patel, B. *Computations of Complex Turbulent Flows Using the Commercial Code Fluent*; ICASE/LaRC Interdisciplinary Series in Science and Engineering; Springer: Dordrecht, The Netherlands, 1997; Volume 7, ISBN 978-94-011-4724-8.
32. Jones, W.P.; Launder, B. The prediction of laminarization with a two-equation model of turbulence. *Int. J. Heat Mass Transf.* **1972**, *15*, 301–314. [CrossRef]
33. Chien, K.Y. Predictions of channel and boundary-layer flows with a low-Reynolds-number turbulence model. *AIAA J.* **1982**, *20*, 33–38. [CrossRef]
34. Myong, H.K.; Kasagi, N. A new approach to the improvement of k-ε turbulence model for wall-bounded shear flows. *JSME Int. J. Ser. 2 Fluids Eng. Heat Transf. Power Combust. Thermophys. Prop.* **1990**, *33*, 63–72. [CrossRef]
35. Crowe, C.T. On models for turbulence modulation in fluid-particle flows. *Int. J. Multiph. Flow* **2000**, *26*, 719–727. [CrossRef]
36. Banerjee, A.; Pasupuleti, S.; Kumar, G.N.P.; Dutta, S.C. A Three-Dimensional CFD Simulation for the Nonlinear Parallel Flow Phenomena through Coarse Granular Porous Media. In *Lecture Notes in Mechanical Engineering*; Springer: Berlin/Heidelberg, Germany, 2018; pp. 469–480. [CrossRef]
37. Celik, I.B.; Ghia, U.; Roache, P.J.; Freitas, C.J.; Coleman, H.; Raad, P.E. Procedure for estimation and reporting of uncertainty due to discretization in CFD applications. *J. Fluids Eng.* **2008**, *130*, 0780011–0780014. [CrossRef]
38. Safer, N.; Woloszyn, M.; Roux, J.J. Three-dimensional simulation with a CFD tool of the airflow phenomena in single floor double-skin facade equipped with a venetian blind. *Sol. Energy* **2005**, *79*, 193–203. [CrossRef]

39. Camblor, P.M. On correlated z-values distribution in hypothesis testing. *Comput. Stat. Data Anal.* **2014**, *79*, 30–43. [CrossRef]
40. Chen, Z.; Nadarajah, S. On the optimally weighted z-test for combining probabilities from independent studies. *Comput. Stat. Data Anal.* **2014**, *70*, 387–394. [CrossRef]
41. Mann, P.S.; Lacke, C.J. *Introductory Statistics*; John Wiley and Sons Inc.: Delhi, India, 2010; ISBN 10:812652734X.

energies

MDPI

Article

The Effect of Oil Properties on the Supercritical CO$_2$ Diffusion Coefficient under Tight Reservoir Conditions

Chao Zhang [†], **Chenyu Qiao** [†], **Songyan Li** * and **Zhaomin Li**

College of Petroleum Engineering, China University of Petroleum, Qingdao 266580, China;
zhangc@upc.edu.cn (C.Z.); S16020339@s.upc.edu.cn (C.Q.); lizhm@upc.edu.cn (Z.L.)
* Correspondence: lsyupc@163.com; Tel.: +86-532-8698-1717
† These authors contributed equally to this work.

Received: 18 May 2018; Accepted: 6 June 2018; Published: 8 June 2018

Abstract: In this paper, a generalized methodology has been developed to determine the diffusion coefficient of supercritical CO$_2$ in cores that are saturated with different oil samples, under reservoir conditions. In theory, a mathematical model that combines Fick's diffusion equation and the Peng-Robinson equation of state has been established to describe the mass transfer process. In experiments, the pressure decay method has been employed, and the CO$_2$ diffusion coefficient can be determined once the experimental data match the computational result of the theoretical model. Six oil samples with different compositions (oil samples A to F) are introduced in this study, and the results show that the supercritical CO$_2$ diffusion coefficient decreases gradually from oil samples A to F. The changing properties of oil can account for the decrease in the CO$_2$ diffusion coefficient in two aspects. First, the increasing viscosity of oil slows down the speed of the mass transfer process. Second, the increase in the proportion of heavy components in oil enlarges the mass transfer resistance. According to the results of this work, a lower viscosity and lighter components of oil can facilitate the mass transfer process.

Keywords: oil properties; diffusion coefficient; supercritical CO$_2$; Peng-Robinson equation of state (PR EOS)

---

1. Introduction

Insufficient oil and gas supplies and global warming have aroused interest in enhanced oil recovery (EOR) with CO$_2$ and geological CO$_2$ storage [1–3]. CO$_2$ is usually injected into geological formations to improve oil recovery and to store and sequester the greenhouse gas in the atmosphere through the interaction of CO$_2$ with the crude oil, which restricts CO$_2$ molecules in the pores of the rock, and the reaction of CO$_2$ molecules with mineral grains. The main purpose of CO$_2$ EOR is to produce more hydrocarbons from oil reservoirs [4–8]. Therefore, a full understanding of the behavior of CO$_2$ under reservoir conditions has always been one of the main interests of researchers and the petroleum industry. There are several processes involved in the CO$_2$ EOR, i.e., diffusion of CO$_2$ into the crude oil in the porous rocks [9–19], the chemical reaction of CO$_2$ with formation minerals [20,21], and rock mechanics caused by pore pressure changes [22]. Only the first process of CO$_2$ diffusion in porous media is considered in this study, which has theoretical importance and meaning for applications in the petroleum industry.

Diffusion is a spontaneously dispersing process of molecules or ions [23], which is caused by the concentration difference in solutions or dispersions. The quantitative description of the rate of the diffusion process is usually expressed by the diffusion coefficient [24–27]. There have been several methods for the quantitative determination of the gas (N$_2$, CO$_2$, CH$_4$ [28], C$_2$H$_6$, C$_3$H$_8$, or a mixture gas) diffusion coefficient in crude oil. These methods include the pressure decay method, X-ray computer-assisted tomography (CAT) method, magnetic resonance imaging (MRI) method [29,30], dynamic drop volume analysis (DPDVA), and pore-scale network modeling method. All of these methods have advantages and flaws:

Direct testing method. Hill and Lacey [31] tested the diffusion coefficient of CH_4 in isopentane with a constant pressure method in a Pressure-Volume-Temperature (PVT) system. Sigmund [32] also used this method to study the binary dense gas diffusion coefficients under reservoir conditions. Islas-Juarez et al. [33] set up an experimental device to determine the effective diffusion coefficient of N_2 in a sandpack model. They devised a special setting to sample crude oil that was dissolved with N_2 in their porous model. The samples were then analyzed by gas chromatograph, and the gas concentration in the oil phase was determined. The corresponding diffusion coefficient was determined by matching the mathematical diffusion model with the experimentally measured concentration curve. The main shortcomings of the direct testing method for the solvent diffusion coefficient are that it is both time and labor consuming, which results in an expensive process.

Pressure decay method. Riazi [34] proposed a technique known as the pressure decay method based on the fact that the pressure of the gas phase in a closed diffusion cell decreases with the migration of the gas phase into the liquid phase. One of the remarkable features of this method is that it does not require costly component measurements. In the following two decades, many scholars adopted this method to test the diffusion coefficient for different gas and oil systems and modified the basic method for even more complex conditions and higher accuracy [35–39].

CAT and MRI methods. Wen et al. [40] and Afsahi [41] utilized the nuclear magnetic resonance (NMR) method to estimate the diffusion coefficients in bitumen and solvent mixtures. Guerrero-Aconcha et al. [42] used CAT to obtain the density profiles and then back-calculated the concentration-dependent diffusion coefficients. Wen et al. [43] used the two nondestructive methods of NMR and CAT to determine the solubility profile inside a solvent and heavy oil system. Wang et al. [44] estimated the diffusion coefficients in porous media with an X-ray and investigated the effect of the thickness of a diffusion interface on convection flow. The advantage of NMR and CAT is that they are nondestructive and do not affect the distribution of the solvent in the liquid and diffusion process. However, the flaw is the requirement of expensive equipment.

DPDVA method. Yang and Gu [45] studied the mass transfer behavior of CO_2–crude oil systems with DPDVA technology and measured solvent diffusivity in heavy oil under reservoir pressure and temperature. The DPDSA is a special method that correlates the oil swelling effect due to the mass transfer from the solvent into crude oil to the interfacial tension reduction between the oil and the solvent phases.

Pore-scale network modeling method. Recently, pore-scale network modeling has aroused the research interests of scholars in petroleum engineering [46–48]. The method can now be used as a platform to analyze various phenomena in porous media, including wettability alternation, multiphase flow, mass transfer process, etc. Garmeh et al. [49] solved the single-phase flow in a pore-scale network model to study the dispersion and mixing phenomena in porous media. Taheri et al. [50] predicted the solvent diffusion coefficient in heavy oil and bitumen with the sub-pore-scale modeling method and drew the conclusion that it produced similar results to classic experimental measurements.

The CO_2 diffusion coefficients in bulk crude oils and in cores that are saturated with formation fluids have had wide concern recently due to the popularity of CO_2 EOR. Many significant studies have been conducted in this field, while few attempts have been made to analyze the effect of crude oil properties on the CO_2 diffusion process. In this paper, a generalized methodology has been developed to determine the diffusion coefficient of supercritical CO_2 for cores that are saturated with different crude oils, under reservoir conditions. In theory, a mathematical model that describes the mass transfer process of CO_2 in crude-oil-saturated cores has been established and combines the Peng-Robinson equation of state (PR EOS) and Fick's diffusion equation. Experimentally, the pressure decay method has been employed to determine the diffusion coefficient. The pressure of the CO_2 phase in the annular space of the diffusion cell was monitored and recorded during the mass transfer process of CO_2 into the oil-saturated low-permeable cores. Once the difference between the tested and calculated pressure decay curves reached a minimum value, the diffusion coefficient of CO_2 could be determined. In addition, the influence of oil properties on the CO_2 diffusion coefficient has also been analyzed.

2. Experimental Section

2.1. Materials

The oil samples used in the present work were obtained by mixing kerosene and crude oil in various ratios to form a series of oil samples with different viscosities and component proportions. The kerosene was purchased from the China University of Petroleum (East China) and has a density and viscosity of 802.1 kg/m^3 and 1.34 mPa·s under 50 °C and atmospheric pressure, respectively. The crude oil used in this work was collected from Changji Oilfield in Xinjiang Province (China). The crude oil sample is dead oil without sands and brine, whose density measured with a densitometer (DMA 4200M, Anton Paar, Graz, Austria) was 936.8 kg/m^3 under 50 °C and atmospheric pressure. The viscosity, which was measured by a rheometer (MCR 302, Anton Paar), is 1770.91 mPa·s under the same conditions. The component distributions of both the kerosene and crude oil were determined with gas chromatography (GC), and the result is depicted in Figure 1. The CO_2 used in this work was purchased from Tianyuan Co., Ltd. (Qingdao, China) and has a purity higher than 99.99 mol %. Artificial cores with a permeability that ranged from 0.096 to 0.103 mD were used as porous media in this work and were compressed with sand particles. The diameters of the sand particles in the cores were restricted to a tiny range, which can ensure the homogeneity and isotropy of the cores. The parameters of the cores used in this work are tabulated in Table 1.

Real reservoirs are heterogeneous and anisotropic, which makes the mass transfer process in reservoirs complicated [51–53]; however, this paper concentrates on the effect of oil properties on the CO_2 diffusion coefficients. To simplify the model and analysis, artificial cores are introduced to eliminate the influence of this factor. Therefore, the effects of heterogeneity and anisotropy are ignored in this study.

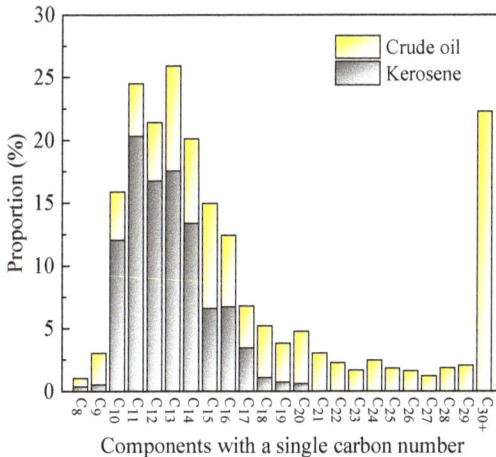

Figure 1. The component distributions of kerosene and the crude oil.

Table 1. Detailed parameters of the artificial cores.

Test No.	Core Diameter (mm)	Core Length (mm)	Permeability (mD)	Porosity (%)	Initial Pressure (MPa)	Temperature (°C)
1	38.16	90.42	0.102	3.92	15.29	70
2	38.18	90.20	0.099	3.83	15.36	70
3	38.14	89.66	0.096	4.95	15.35	70
4	38.20	89.12	0.103	3.76	15.33	70
5	38.18	89.74	0.100	4.16	15.36	70
6	38.16	89.40	0.102	3.75	15.32	70

According to Table 1, the characteristics (porosity and permeability) of the cores are not exactly the same. The artificial cores are compressed with sand particles and crosslinker; therefore, two factors contribute to the variable characteristics of the cores. First, the sand particles are not identical; thus, the inner structure of the cores is not totally regular. Second, the distribution of crosslinker may influence the properties of the cores; for instance, the layer of crosslinker on the particle surface can decrease the diameter of the pores. However, the properties of the cores that are employed in this work are so similar that the influence of different cores can be ignored.

2.2. Apparatus

The schematic diagram of the apparatus used in this paper was illustrated in our previously published work and again in Figure 2 [54,55]. The CO_2 diffusion experiments are conducted in a diffusion cell located in a water bath to keep a constant temperature for all the experiments. The core is placed in the center of the diffusion cell, and CO_2 is introduced into the cylinder. The pressure of the CO_2 decreases due to the diffusion of CO_2 into the core. The pressure decay of the CO_2 in the diffusion cell during each experiment is monitored and recorded by a pressure transducer, which can be used for diffusion coefficient calculation.

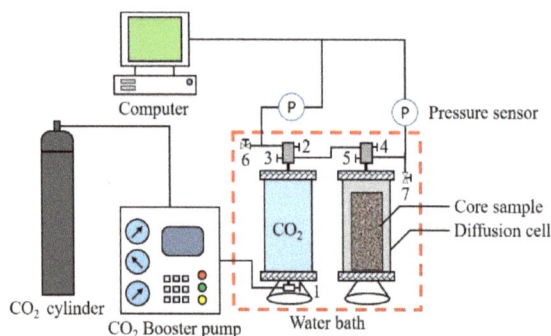

Figure 2. The schematic diagram of the apparatus used in the CO_2 diffusion experiment.

2.3. Experimental Procedures

The experimental procedures used in this study to test the CO_2 diffusion coefficients in the oil-saturated cores are described as follows:

(a) Clean and dry the core for the experiment, put it into an intermediate container and vacuum for 10.0 h. Then, the oil sample is injected into the intermediate container at room temperature until the pressure of the oil sample reaches 15.0 MPa; maintain this pressure for 48.0 h to ensure that the core pores are completely saturated with crude oil.

(b) Seal the two ends of the oil-saturated core with epoxy resin and aluminum foil to ensure that CO_2 can diffuse only through the side surface of the core.

(c) Connect the apparatus that is required for the diffusion experiment according to Figure 2. After testing the air tightness of the diffusion cell, place the core in it. Replace the air in the diffusion cell with low pressure CO_2.

(d) Put the diffusion cell and CO_2 container into the water bath at the required temperature for 4.0 h, and open valve 5 to monitor the pressure in the CO_2 container.

(e) When the pressure inside the CO_2 container is stable, open valves 2, 3 and 4 to inject CO_2 into the diffusion cell. Close valves 3 and 4 quickly after the pressure in the diffusion cell and CO_2 container reach a balance, and record the pressure decay in the diffusion cell.

(f) When the pressure in the diffusion cell does not change, finish the diffusion experiment. Slowly open all valves, release the fluid in the diffusion cell, and clean the equipment for the next set of experiments.

3. Mathematical Model

3.1. Diffusion Model in Porous Media

The physical model used in this study is shown in Figure 3 [54,55]. The two end faces of the core sample are sealed with aluminum foil with epoxy resin, just as we did in our previous work [55], which makes the end faces impermeable. Thus, CO_2 can diffuse only in the radial direction of the core. The mathematical model used in this study includes Fick's diffusion equation and PR EOS to describe the diffusion process from the CO_2 phase to the cores being saturated with different oil samples. The mass transfer of CO_2 is considered by the diffusion equation, while the phase behavior between CO_2 and the oil sample is described by PR EOS. Several assumptions that were adopted for this mathematical model are elaborated below:

(1) Cores are homogenous and isotropic, i.e., the influences of different cores are ignored.

(2) Oil saturates all pores in the cores, i.e., oil saturation is 100%.

(3) CO_2 concentration at the side surface of the core is constant during the diffusion process.

(4) The extraction effect of CO_2 on oil is negligible. (The literature has reported that the extraction effect of CO_2 on hydrocarbons that are heavier than C_9 is tiny [56], and Figure 1 shows that it is reasonable to ignore extraction in this study.)

(5) The convection flow caused by the density difference is ignored.

(6) The CO_2 transfer process occurs only in the radial direction.

(7) There is no heat exchange during the diffusion process.

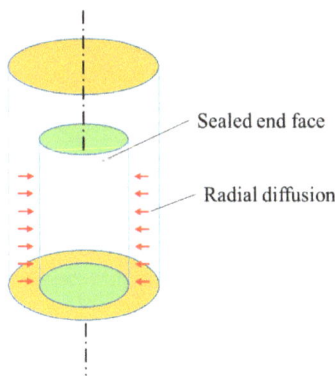

Figure 3. The diagram of the physical model for CO_2 diffusion.

In the cylindrical coordinates system, Equation (1) is used to describe the radial diffusion process of CO_2 in oil-saturated porous media. A velocity item is introduced into the equation to consider the influence of oil swelling due to CO_2 dissolution:

$$\frac{\partial c}{\partial t} = D\frac{\partial^2 c}{\partial r^2} - c\frac{\partial u}{\partial r} - \frac{cu}{r} - (u - \frac{D}{r})\frac{\partial c}{\partial r} \tag{1}$$

where c is the concentration of CO_2 in the oil sample, mol/m^3, t is diffusion time, s, u is the flow velocity that is generated by the volume expansion of the oil sample, m/s, r is the distance to the central axis of the core, m, and D is the effective diffusion coefficient of CO_2 in the oil-saturated core, m^2/s.

To facilitate the solution of the mathematical model, Equation (2) is used to nondimensionalize the diffusion equation, which can be expressed as Equation (3):

$$\begin{cases} \bar{r} = \frac{r}{r_0} \quad \bar{c} = \frac{c}{c_0} \\ \tau = \frac{tD}{r_0^2} \quad \bar{u} = \frac{ur_0}{D} \\ \lambda = \bar{u} - \frac{1}{\bar{r}} \end{cases} \tag{2}$$

$$\frac{\partial \bar{c}}{\partial \tau} = \frac{\partial^2 \bar{c}}{\partial \bar{r}^2} - \lambda\frac{\partial \bar{c}}{\partial \bar{r}} - \bar{c}\frac{\partial \bar{u}}{\partial \bar{r}} - \frac{\bar{c}\bar{u}}{\bar{r}} \tag{3}$$

where $\bar{r}$ is the dimensionless distance, r_0 is the radius of the core, m, $\bar{c}$ is the dimensionless concentration, c_0 is the saturation concentration of CO_2 in crude oil under experimental conditions, mol/m^3, τ is dimensionless time and $\bar{u}$ is the dimensionless velocity.

According to the characteristics of diffusion, the side surface of the core can be considered the Dirichlet boundary, and the central axis of the core can be considered a closed boundary. The boundary conditions and initial conditions of the mathematical model can be represented by Equations (4) and (5), respectively. The full implicit finite difference method is employed to solve the above model. In the discrete process, the derivative of the concentration and velocity with space is a second-order central difference scheme, and the derivative of concentration with time is a first-order forward difference scheme:

$$\begin{cases} \bar{c} = 1 & (\bar{r} = 1, \tau > 0) \\ \bar{u} = 0, \frac{\partial \bar{c}}{\partial \bar{r}} = 0 & (\bar{r} = 0, \tau \geq 0) \end{cases}, \tag{4}$$

$$\begin{cases} \bar{u} = 0, \bar{c} = 1 & (\bar{r} = 1, \tau = 0) \\ \bar{u} = 0, \bar{c} = 0 & (\bar{r} < 1, \tau = 0) \end{cases}, \tag{5}$$

3.2. Peng-Robinson Equation of State (PR EOS)

The CO_2 pressure of the diffusion cell is recorded for the pressure decay method, and the pressure prediction thus affects the accuracy of the CO_2 diffusion coefficient. PR EOS is introduced in this paper, and the interaction between CO_2 and the oil sample is considered, which increases the calculation accuracy of the annular pressure and diffusion coefficient. PR EOS is a third-order equation with two constants that were proposed by Peng and Robinson [57]. It is a semi-empirical model that is widely applied in the petrochemical industry, and it describes the phase behavior and phase equilibrium of multicomponent systems. PR EOS requires the specific parameters of the components in the system and the binary interaction parameters (BIPs) between each component. PR EOS can be expressed by Equations (6) and (7):

$$P = \frac{RT}{V - b} - \frac{a}{V(V + b) + b(V - b)} \tag{6}$$

$$\begin{cases} a = a_c \alpha(T_r, \omega) \\ a_c = \frac{0.457235 R^2 T_c^2}{P_c} \\ b = \frac{0.0777969 R T_c}{P_c} \end{cases} \tag{7}$$

where P is the system pressure, Pa, R is the general gas constant, 8.314 J/mol/K, T is the system temperature, K, V is the molar volume, m^3/mol, T_c is the critical temperature, K, and P_c is the critical pressure, Pa. α is a function of the relative temperature and acentric factor.

The composition of the oil sample is complex. Although PR EOS sets no restriction on the number of components in the mixture, the computing load sharply increases with the number of components. Thus, it is important to simplify the calculation process with the prerequisite of ensuring a high accuracy of calculation. In this study, several pseudo-components are introduced as substitutes for all hydrocarbon components in the oil sample, which can efficiently decrease the calculation load of PR EOS without influencing the computing accuracy [56–58]. Specifically, in this work, hydrocarbons with a single carbon number (SCN) are lumped into three pseudo-components, which have been tested as having a good result on the oil/CO_2 system [59,60]. The parameters of the pseudo-components were determined with a series of empirical models [61–68], and the specific data are listed in the results section.

3.3. Determination of the Diffusion Coefficients

The CO_2 diffusion coefficient is determined by fitting the experimental and calculated data with particle swarm optimization (PSO). The calculated pressure-time (*p-t*) curve can be adjusted to minimize the error with a measured *p-t* curve by optimizing the value of the diffusion coefficient, according to Equation (2). The CO_2 diffusion coefficient is determined once the value of the objective function reaches a minimum, which is shown in Equation (8):

$$\text{Error} = \frac{1}{PN} \sum_{i=1}^{PN} \sqrt{(t_{Ei} - t_{Ci})^2} \tag{8}$$

where PN is the data number of the *p-t* curve, and t_{Ei} and t_{Ci} are the experimental and calculated times, respectively. For the calculated *p-t* curve, the pressure values can be determined with the following steps:

(1) Determine CO_2 concentration distribution in the core.
(2) Calculate the mole composition in the annular space of the diffusion cell according to the amount of swelled oil and dissolved CO_2.
(3) Determine the pressure value by solving PR EOS with data that are obtained in step 2.

Moreover, global regression and piecewise regression are employed to determine the CO_2 diffusion coefficients. In a global regression, all experimental data points are used, and a constant coefficient is obtained to describe the average rate of mass transfer. In the early stage and later stage regressions, part of the data points is employed to calculate the diffusion coefficients, and two diffusion coefficients are obtained to describe the rate of mass transfer in the early and later stages, respectively. The relevant contents will be elaborated in the Results section.

4. Results and Discussion

4.1. Characterization of the Oil Samples

Six groups of oil samples, which were prepared by mixing kerosene and crude oil under different volume ratios, are used in the diffusion experiments to study the influence of oil properties on the CO_2 diffusion coefficient. The viscosity-temperature curves of the oil samples are presented in Figure 4, and the viscosity of each oil sample at the experimental temperature is marked in the figure (the pentagram symbols). The carbon distributions of the oil samples that were determined with the method in the literatures [55,59] are illustrated in Figure 5. It is notable that the carbon distributions of kerosene and crude oil are determined with the above method, and the carbon distribution of mixed oil is determined by combining the carbon distribution of the two oils according to their mixing ratio. The detailed process can be described with Equation (9):

$$z_i = \frac{z_i^c n^c + z_i^k n^k}{n^c + n^k} \tag{9}$$

where z_i, z_i^c and z_i^k are the mole friction of component i in the mixed oil, crude oil and kerosene samples, respectively, and n^c and n^k are the mole numbers of crude oil and kerosene, respectively.

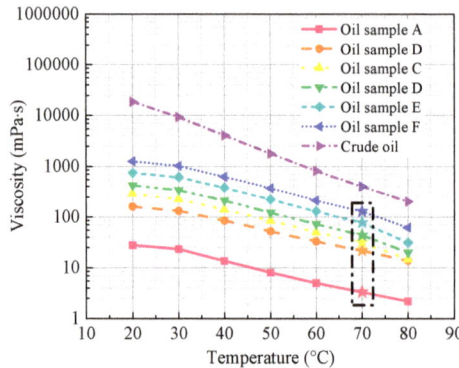

Figure 4. Viscosity-temperature curves of the oil samples (the pentagram symbols are the viscosities of the oil samples under experimental conditions.).

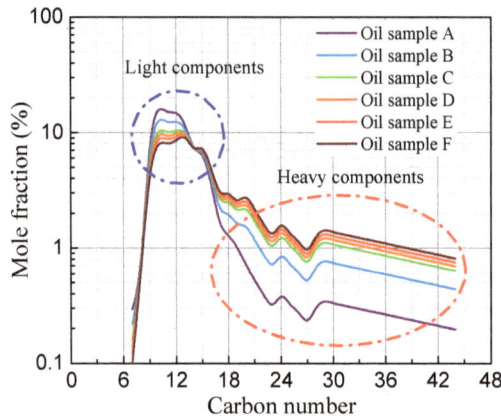

Figure 5. The composition analysis of the oil samples.

Figure 5 shows that the components lighter than C_9 account for a tiny part of those oil samples (<1%), which means that there is no obvious extraction effect during the mass transfer process [56,59,60]. The components of the CO_2 phase were analyzed by a GC method after the diffusion experiments, and no hydrocarbons were found in the CO_2 phase. The GC analysis result agrees with our previous work [55] and proves that neglecting the extraction effect is reasonable. According to the figure, from oil samples A to E, the number of light components (C_9–C_{15}) decreases, while the amount of heavy components (heavier than C_{18}) increases gradually. Multiple pseudo-components are introduced in this work to replace the entire carbon distribution to characterize the oil samples, which greatly reduces the computational cost [60]. The pseudo-component parameters are tabulated in Table 2. The tendencies and values of the data in Table 2 are analogous to our previous work [55], which proves the reliability of the parameters. The binary interaction parameter (BIP) matrix is listed in Table 3 and describes the interactions among the components in the system. BIPs between hydrocarbons are set at 0 according to [61,69]. Moreover, only the parameters of oil sample A are given in Tables 2 and 3; the parameters of the other oil samples are presented in the Appendix A.

Table 2. Parameters of the pseudo-components of oil sample A.

Oil No.	Pseudo-Component	Z (mol %)	MW (g/mol)	SG	T_b (K)	T_c (K)	P_c (kPa)	ω
	A1	43.302	143.19	0.813	454.768	644.008	2587.222	0.449
A	A2	34.479	180.762	0.841	509.554	697.369	2168.917	0.559
	A3	22.220	315.116	0.893	632.112	806.063	1527.015	0.832

Table 3. BIP matrix for the CO_2 and pseudo-components.

Oil No.	Component	A1	A2	A3	CO_2
	A1	0	0	0	1.027×10^{-4}
	A2	0	0	0	7.162×10^{-5}
A	A3	0	0	0	2.057×10^{-3}
	CO_2	1.027×10^{-4}	7.162×10^{-5}	2.057×10^{-3}	0

4.2. Solution of the Diffusion Model in the Oil-Saturated Cores

The features of the CO_2 diffusion process in the oil-saturated porous media can be characterized by concentration and velocity profiles, which are obtained by solving the diffusion mathematical model that is proposed in Section 3.1 [54,55]. The CO_2 concentration profile during the diffusion process is presented in Figure 6, where black curves identify the CO_2 concentrations at different space and time grids.

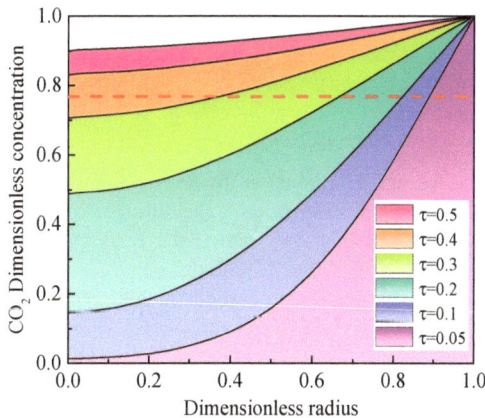

Figure 6. CO_2 concentration profile of the cores in the diffusion process (the red line reflects the average CO_2 dimensionless concentration in the core at $\tau = 0.2$.).

The area between a curve and the x-axis characterizes the total CO_2 amount in the core at a specific time point, and the area between two curves can characterize the increment of CO_2 during this period of time. According to Figure 6, the CO_2 concentration at the central axis ($\bar{r} = 0$) almost reaches 0.5 when dimensionless time is 0.2, and the area between the curve and the x-axis show that the average CO_2 concentration was above 0.7 at this moment (the red dashed line in Figure 6). The tendency shows that most of the CO_2 diffuses into the oil-saturated core at the early stage of the diffusion process. Moreover, the area of the colored region between two curves decreases with the increase in dimensionless time, which shows that the increase in CO_2 in the core gradually slows down, i.e., the speed of mass transfer slows down. Therefore, the diffusion process is divided into two stages, namely, an early stage with a high diffusion rate and a later stage with a lower diffusion rate, which is identical to our previous work [55].

The velocity profile caused by oil volume swelling is illustrated in Figure 7. The trend of Figure 7 shows that the velocity of oil is relatively high at the early stage ($\tau < 0.2$), which means that the fast volume expansion of oil, i.e., the diffusion rate of CO_2, is relatively high. The following decrease in velocity characterizes the relatively slow diffusion rate in the later stage. Thus, Figures 6 and 7 complement and confirm one another. It should be noted that the velocity difference between the outer part and inner part of the core is obvious at an early stage; then, the difference decreases gradually, which shows CO_2 diffusing from the outer part to the inner part of the core. More details have already been elaborated in our previous work.

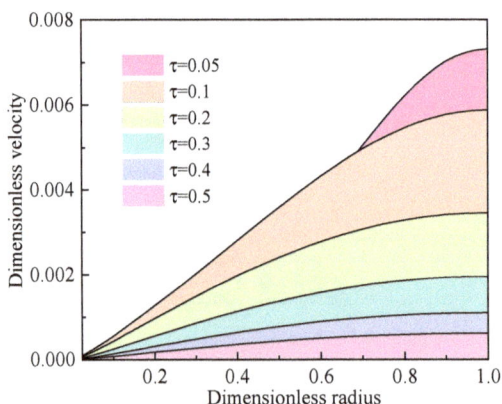

Figure 7. Velocity of oil in the cores caused by volume swelling.

4.3. Effect of the Oil Properties on the Diffusion Coefficient

A pressure decay experiment is employed in this paper to study the CO_2 diffusion process in oil-saturated tight cores under reservoir conditions. The diffusion coefficient of CO_2 is determined with the experimental pressure recording and mathematical model that is listed in Section 3, and the pressure curves of the diffusion experiments with different oil samples are depicted in Figure 8. Except for the experimental data, the global fitting (red line) with a constant diffusion coefficient and the piecewise fitting (blue and violet dashed lines) with variable diffusion coefficients are also given in Figure 8. Although there are some differences between the results of the global regression of the different experiments (see Figure 8a,b), which can be attributed to experimental and calculation errors, the value of the goodness-of-fit in each set of experiments shows that both regression methods have acceptable accuracy. This value also shows that the piecewise regression has a better result than the global regression. The highly precise result of the piecewise fitting agrees with the literature [70,71] that suggests that the diffusion coefficient is a variable during the diffusion process, and it also proves the reliability of the conclusions in Section 4.2.

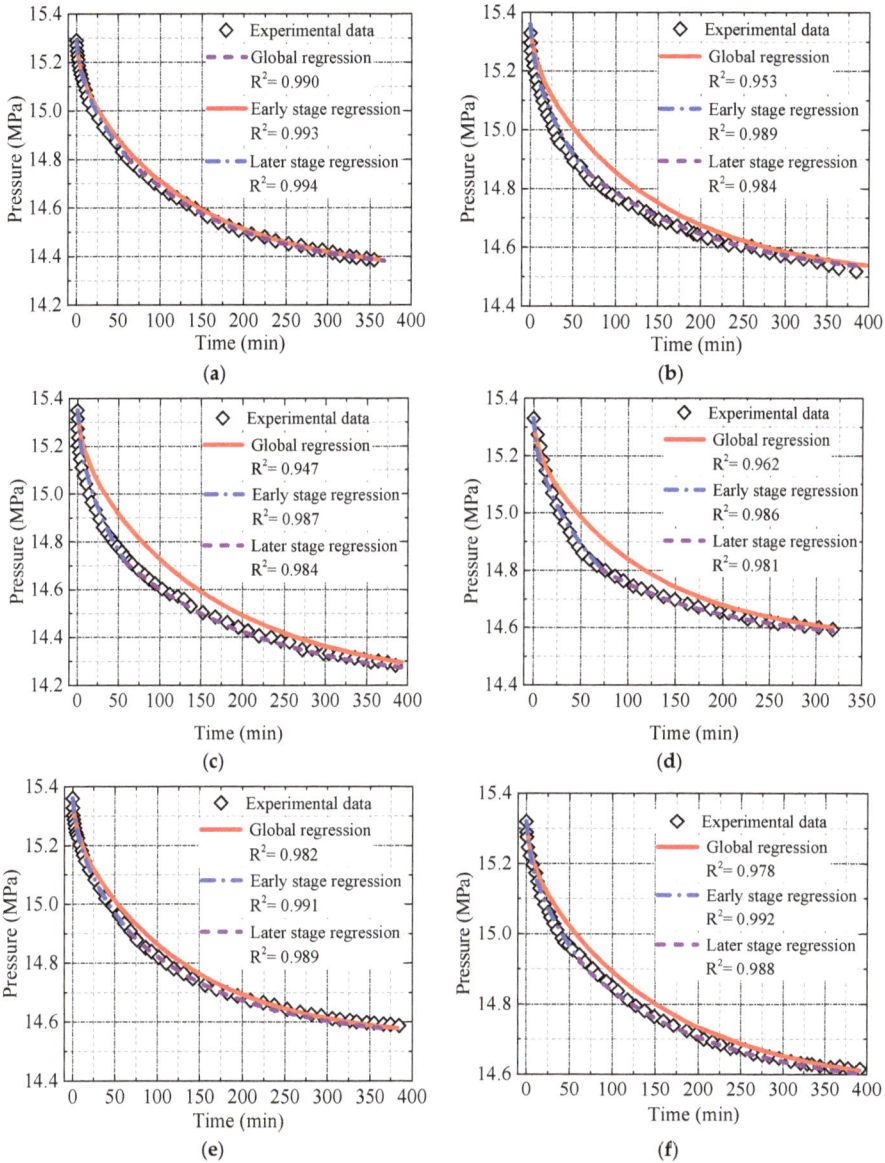

Figure 8. Pressure-time curves of the CO_2 diffusion experiments. (**a**) Oil sample A, 3.30 mPa·s; (**b**) Oil sample B, 21.57 mPa·s; (**c**) Oil sample C, 31.09 mPa·s; (**d**) Oil sample D, 43.50 mPa·s; (**e**) Oil sample E, 76.81 mPa·s; (**f**) Oil sample F, 127.47 mPa·s.

The CO_2 diffusion coefficients in tight porous media that is saturated with different oil samples, at 70 °C and 15.29–15.36 MPa, are illustrated in Figure 9. In the experimental range of this study, the diffusion coefficients obtained with the global regression range from 55.325×10^{-10} to 107.886×10^{-10} m^2/s. For the piecewise regression, the diffusion coefficients range from 74.975×10^{-10} to 128.925×10^{-10} m^2/s at the early stage and from 39.389×10^{-10} to 89.462×10^{-10} m^2/s at the later stage. The figure also shows that the CO_2 diffusion coefficient decreases gradually from oil samples A to F, and the decreasing tendency can be attributed to the following two factors:

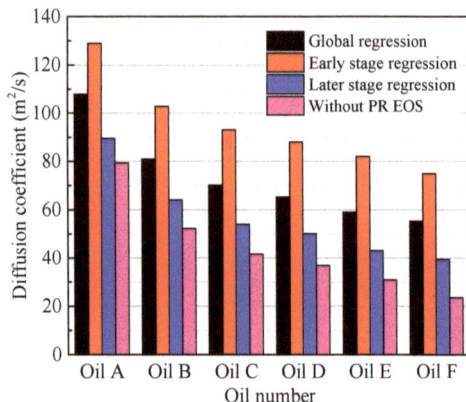

Figure 9. The CO_2 diffusion coefficient in tight cores saturated with different oil samples (70 °C, 15.29–15.36 MPa).

(1) *The increase in the viscosity of the oil samples.* Numerous scholars have indicated that the viscosity of the liquid phase has an obvious influence on the diffusion coefficient [59,72–75]. The effect of oil viscosity on the CO_2 diffusion coefficient is depicted in Figure 10. The figure shows that a lower viscosity can facilitate the CO_2 diffusion process, which agrees with the negative relationship between the two parameters that are proposed in the literature [59,74–76]. Moreover, Figures 9 and 10 also show the obvious tendency of the diffusion coefficients at different stages. The diffusion coefficient at the early stage is always higher than the average level (global regression), while the diffusion coefficient at the later stage is always lower than the average level. The coefficient that is determined without PR EOS is minimal because it ignores the interaction of oil and CO_2. It is notable that the experimental data form a good linear trend in the semilogarithmic coordinate system.

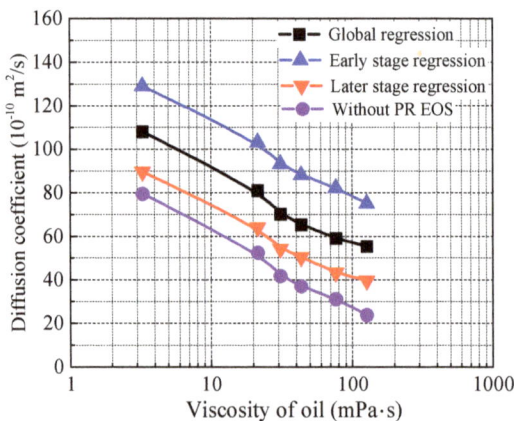

Figure 10. The effect of oil viscosity on the CO_2 diffusion coefficient.

Hayduk and Cheng [77] tested a large amount of the experimental data in the literature to reveal the relationship between the diffusion coefficient and the viscosity of the solvent. They indicated that any diffusing substance has a specific exponential correlation between its diffusion coefficient and the viscosity of the solvent (or pure liquid phase), which is irrelevant to the components of the solvent. Their conclusion is summarized as equation 10, where μ is the viscosity of the solvent or pure liquid phase:

$$D = A\mu^B \tag{10}$$

Equation (10) is employed to fit the diffusion coefficient data in Figure 10, which are obtained from the global regression. Moreover, a linear fitting method in a semi-logarithmic coordinate system is also introduced, because of the obvious linear relationship between the diffusivity and viscosity in Figure 10. The experimental data and two fitting lines are illustrated in Figure 11. As seen, both fitting methods have a satisfactory goodness-of-fit. The similar trend of experimental data and the classic exponential model (the blue dash-and-dotted line in Figure 11) proves that the data in this paper are reliable and that the viscosity of the liquid phase greatly influences the diffusion process. However, the linear fitting of D and $\ln\mu$ shows a better result than the exponential model, and the deviation between the two fitting lines increases gradually as the viscosity increases. This finding may reveal that for a complex liquid such as petroleum, viscosity is not the exclusive factor that affects the diffusion process.

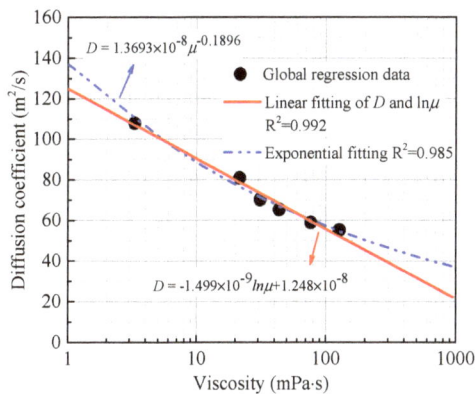

Figure 11. The effect of oil viscosity on the CO_2 diffusion coefficient.

(2) *The increasing proportion of heavy components in the oil samples.* As shown in Figure 5, from oil samples A to F, the proportion of heavy components increases gradually. Several scholars have suggested that the components of oil also directly influence the diffusion process of CO_2 [73,75].

The solubility of CO_2 in each oil sample, which is determined with a two-phase equilibrium calculation [78–80], is presented in Figure 12 and reflects the capacity of oil to accommodate CO_2. Furthermore, the CO_2 solubility of each pseudo-component at 15.3 MPa and 70 °C is depicted in Figure 13. Figures 12 and 13 show that the CO_2 solubility decreases with the increase in the proportion of heavy components in oil, i.e., the resistance for CO_2 diffusing into oil increases [55]. Thus, the increase in the proportion of heavy components restricts the diffusion process from two aspects. First, it improves the viscosity of oil, which indirectly decreases the diffusion coefficient [81]. Second, heavy components directly increase the mass transfer resistance. Moreover, the resistance effect of heavy components also accounts for the deviation between the experimental data and the exponential model in Figure 11.

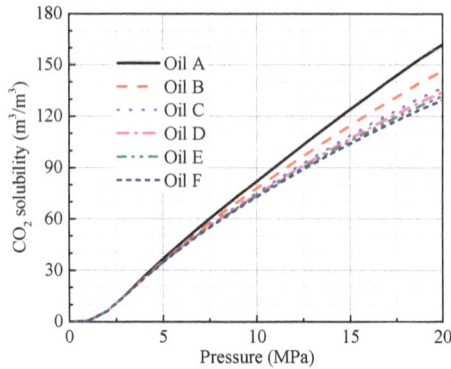

Figure 12. CO_2 solubility of the oil samples at 70 °C.

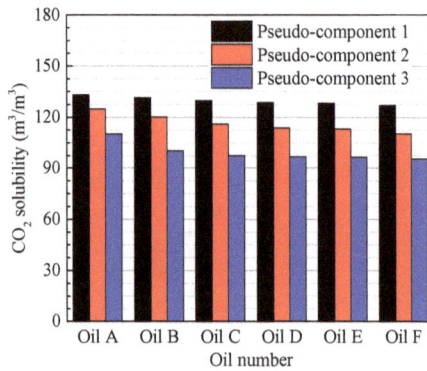

Figure 13. CO_2 solubility of the pseudo-components (70 °C, 15.3 MPa).

4.4. Comparison

The data of the CO_2 diffusion coefficient in this paper are compared with the data in the literature, and both are listed in Table 4.

According to the data in the table, the CO_2 diffusivity that was obtained in this paper is somewhat larger than the CO_2 diffusivity in the literature. However, it is still within a reasonable range, and the differences in the data in the different papers can be attributed to several reasons. First, the experimental pressure and temperature in this work are higher than the experimental pressure and temperature in the literature [59,60,75,76]. Thus, the larger diffusion coefficient in this study agrees with the theory that an increase in pressure or temperature facilitates the diffusion process. Second, the viscosities of the oil samples used in this work are lower than the viscosities of the heavy oils used in the literature, which contributes to the high diffusion rate. Third, the proportion of heavy components in the oil samples used in this paper is smaller (by comparing Figure 1 and other works [59,60]); thus, the resistance for CO_2 diffusing into oil is small [55]. Moreover, the interaction between CO_2 and the oil samples is characterized by PR EOS in this work, and the data that were obtained with this model are more precise.

Table 4. Data in the literature of the CO_2 diffusion coefficient in porous media saturated with oil.

Fluid	Viscosity (mPa·s)	Pressure (MPa)	Temperature (K)	Permeability (mD)	Diffusion Coefficient (10^{-10} m²/s)		Sources
Mixed oil samples	3.30–127.47 @343.15K	15.29–15.36	343.15	0.096–0.103	Early stage 74.97–128.92	Later stage 39.38–89.46	This study
					Global regression 55.33–107.89		
					Without PR EOS 23.63–79.37		
Changji light oil	7.26 @323.15K	14.56–14.89	298.15–358.15	0.058–0.192	Early stage 67.04–164.38	Later stage 33.82–100.37	Li et al. [54]
N-hexadecan	2.14 @313.15	2.28–6.03	313.15	80.67–227.74	5.98–8.01		Li et al. [24]
Lloydminster heavy oil	12,854.00 @294.55K	3.74–3.37	294.55	/	4.30		Zheng et al. [58]
Lloydminster heavy oil	12,854.00 @294.55K	5.40	317.65	/	14.97		Zheng et al. [59]
Athabasca bitumen	821,000.00 @298.15K	4.00–8.00	323.15	/	2.20–8.90		Upreti [74]
Lloydminster heavy oil	23,000.00 @297.15K	2.00–6.00	297.15	/	2.00–5.50		Yang [45]
Athabasca bitumen	106,000.00 @313.15K	3.24	348.15	/	5.03		Rasmussen et al. [75]

5. Conclusions

A generalized methodology was developed in this paper to determine the diffusion coefficient of supercritical CO_2 in cores saturated with different oil samples, under reservoir conditions. A mathematical model that describes the mass transfer process of CO_2 in oil-saturated cores was established. The pressure decay method was used to determine the diffusion process. The results show that the supercritical CO_2 diffusion coefficient decreases gradually from oil samples A to F. It decreases from 128.92×10^{-10} to 74.97×10^{-10} m²/s at the early stage, decreases from 89.46×10^{-10} to 39.38×10^{-10} m²/s at the later stage, and decreases from 107.89×10^{-10} to 55.33×10^{-10} m²/s with the global regression method. The changing properties of oil can account for the decrease in the CO_2 diffusion coefficient in two aspects. First, the increasing viscosity of oil slows down the speed of the mass transfer process. Second, the increase in the proportion of heavy components in oil enlarges the mass transfer resistance. These findings can provide direction in predicting CO_2 storage potential in reservoirs and the effect of CO_2 EOR. Moreover, these findings can also help to optimize engineering techniques in oil fields.

Author Contributions: Data curation, C.Z.; Investigation, C.Z. and C.Q.; Methodology, C.Q.; Supervision, S.L. and Z.L.; Writing—original draft, C.Z. and C.Q.; Writing—review & editing, S.L.

Acknowledgments: This work was supported by the National Key Basic Research Program of China (2015CB250904), the National Natural Science Foundation of China (No. 51774306), the National Science and Technology Major Project of China (2017ZX05072005-004), the National Key Scientific, Technological Project for the Oil & Gas Field and Coalbed Methane of China (2016ZX05031002-004-002), the Natural Science Foundation of Shandong Province, China (Grant ZR2017BEE059), the China Postdoctoral Science Foundation (Grant 2016M600572), and Fundamental Research Funds for the Central Universities (14CX02185A and 18CX02160A). The authors sincerely thank their colleagues at the Foam Fluid Research Center at the China University of Petroleum (East China) for helping us with the experiments.

Conflicts of Interest: The authors declare no conflict of interest.

Appendix A. Parameters in PR EOS of Oil Samples B to F

Table A1. Properties of the pseudo-components of oil samples B to F.

Oil No.	Pseudo-Component	Z (mol %)	MW (g/mol)	SG	T_b (K)	T_c (K)	P_c (kPa)	ω
B	B1	47.706	150.741	0.819	466.154	655.221	2497.765	0.472
	B2	35.239	209.78	0.857	544.786	729.824	1942.549	0.636
	B3	17.353	439.464	0.930	726.081	886.938	1122.854	1.043
C	C1	49.459	158.515	0.824	477.370	666.098	2412.984	0.494
	C2	33.400	244.079	0.872	580.821	762.058	1739.855	0.719
	C3	17.143	480.992	0.941	753.344	909.992	1023.217	1.101
D	D1	53.065	164.282	0.828	485.358	673.746	2354.775	0.511
	D2	30.215	263.996	0.880	600.216	779.188	1638.055	0.763
	D3	16.721	494.173	0.944	760.992	916.415	998.275	1.119
E	E1	50.339	164.845	0.829	486.152	674.510	2348.895	0.512
	E2	32.411	268.563	0.882	604.173	782.621	1619.513	0.771
	E3	17.253	498.801	0.945	763.573	918.579	990.160	1.124
F	F1	55.032	173.010	0.834	497.042	684.824	2272.200	0.535
	F2	28.996	298.023	0.892	630.406	805.497	1492.491	0.831
	F3	16.736	519.753	0.949	774.237	927.510	958.930	1.148

Table A2. BIP matrix of oil sample B.

Oil No.	Component	B1	B2	B3	CO_2
B	B1	0	0	0	3.652×10^{-5}
	B2	0	0	0	3.998×10^{-4}
	B3	0	0	0	4.415×10^{-3}
	CO_2	3.652×10^{-5}	3.998×10^{-4}	4.415×10^{-6}	0

Table A3. BIP matrix of oil sample C.

Oil No.	Component	C1	C2	C3	CO_2
C	C1	0	0	0	4.181×10^{-6}
	C2	0	0	0	9.641×10^{-4}
	C3	0	0	0	5.148×10^{-3}
	CO_2	4.181×10^{-6}	9.641×10^{-4}	5.148×10^{-3}	0

Table A4. BIP matrix of oil sample D.

Oil No.	Component	D1	D2	D3	CO_2
D	D1	0	0	0	6.155×10^{-7}
	D2	0	0	0	1.338×10^{-3}
	D3	0	0	0	5.353×10^{-3}
	CO_2	6.155×10^{-7}	1.338×10^{-3}	5.353×10^{-3}	0

Table A5. BIP matrix of oil sample E.

Oil No.	Component	E1	E2	E3	CO_2
E	E1	0	0	0	1.117×10^{-6}
	E2	0	0	0	1.421×10^{-3}
	E3	0	0	0	5.422×10^{-3}
	CO_2	1.117×10^{-6}	1.421×10^{-3}	5.422×10^{-3}	0

Table A6. BIP matrix of oil sample F.

Oil No.	Component	F1	F2	F3	CO_2
	F1	0	0	0	2.365×10^{-5}
F	F2	0	0	0	1.998×10^{-3}
	F3	0	0	0	5.704×10^{-3}
	CO_2	2.365×10^{-5}	1.998×10^{-3}	5.704×10^{-3}	0

References

1. Yang, Q.; Zhong, C.; Chen, J. Computational study of CO_2 storage in metal–organic frameworks. *J. Phys. Chem. C* **2011**, *112*, 1562–1569. [CrossRef]
2. Lydonrochelle, M.T. Amine ccrubbing for CO_2 capture. *Science* **2009**, *325*, 1652–1654.
3. Rutqvist, J. The geomechanics of CO_2 storage in deep sedimentary formations. *Geotech. Geol. Eng.* **2012**, *30*, 525–551. [CrossRef]
4. Holm, L.W.; Josendal, V.A. Mechanisms of oil displacement by carbon dioxide. *J. Pet. Technol.* **1974**, *26*, 1427–1438. [CrossRef]
5. Monger, T.G.; Ramos, J.C.; Thomas, J. Light oil recovery from cyclic CO_2 injection: Influence of low pressure impure CO_2 and reservoir gas. *SPE Reserv. Eng.* **1991**, *6*, 25–32. [CrossRef]
6. Moberg, R. The Wyburn CO_2 monitoring and storage project. *Greenh. Issues* **2001**, *57*, 2–3.
7. Baines, S.J.; Worden, R.H. Geological storage of carbon dioxide. *Rudarsko-Geološko-Naftni Zbornik.* **2004**, *28*, 9–22. [CrossRef]
8. Wang, S.; Feng, Q.; Javadpour, F.; Xia, T.; Li, Z. Oil Adsorption in shale nanopores and its effect on recoverable oil-in-place. *Int. J. Coal Geol.* **2015**, *147–148*, 9–24. [CrossRef]
9. Zhang, Z.; Chen, F.; Rezakazemi, M.; Zhang, W.; Lu, C.; Chang, H.; Quan, X. Modeling of a CO_2-piperazine-membrane absorption system. *Chem. Eng. Res. Des.* **2018**, *131*, 375–384. [CrossRef]
10. Zhang, Z.; Cai, J.; Chen, F.; Li, H.; Zhang, W.; Qi, W. Progress in enhancement of CO_2 absorption by nanofluids: A mini review of mechanisms and current status. *Renew. Energy* **2018**, *118*, 527–535. [CrossRef]
11. Yang, Y.; Qiu, L.; Cao, Y.; Chen, C.; Lei, D.; Wan, M. Reservoir quality and diagenesis of the Permian Lucaogou Formation tight ccarbonates in Jimsar Sag, Junggar Basin, West China. *J. Earth Sci.* **2017**, *28*, 1032–1046. [CrossRef]
12. Cao, M.; Gu, Y. Temperature effects on the phase behaviour, mutual interactions and oil recovery of a light crude oil–CO_2 system. *Fluid Phase Equilib.* **2013**, *356*, 78–89. [CrossRef]
13. Du, F. An Experimental Study of Carbon Dioxide Dissolution into a Light Crude Oil. Regina. Master's Thesis, University of Regina, Regina, SK, Canada, 2016.
14. Luo, P.; Yang, C.; Gu, Y. Enhanced solvent dissolution into in-situ upgraded heavy oil under different pressures. *Fluid Phase Equilib.* **2007**, *252*, 143–151. [CrossRef]
15. Yang, D.; Gu, Y. Visualization of interfacial interactions of crude Oil-CO_2 systems under reservoir conditions. In Proceedings of the 14th Symposium on Improved Oil Recovery, Tulsa, OK, USA, 17–21 April 2004.
16. Cui, G.; Zhang, L.; Tan, C.; Ren, S.; Zhuang, Y.; Enechukwu, C. Injection of supercritical CO_2 for geothermal exploitation from sandstone and carbonate reservoirs: CO_2–water–rock Interactions and their Effects. *J. CO_2 Util.* **2017**, *20*, 113–128. [CrossRef]
17. Zhang, L.; Li, X.; Zhang, Y.; Cui, G.; Tan, C.; Ren, S. CO_2 Injection for geothermal development associated with EGR and geological storage in depleted high-temperature gas reservoirs. *Energy* **2017**, *123*, 139–148. [CrossRef]
18. Ghasemi, M.; Astutik, W.; Alavian, S.A.; Whitson, C.H.; Sigalas, L.; Olsen, D. Determining diffusion coefficients for carbon dioxide injection in oil-saturated chalk by use of a constant-volume-diffusion method. *SPE J.* **2017**, *22*, 505–520. [CrossRef]
19. Zhang, K.; Gu, Y. New Qualitative and quantitative technical criteria for determining the minimum miscibility pressures (MMPs) with the rising-bubble apparatus (RBA). *Fuel* **2016**, *175*, 172–181. [CrossRef]
20. Izgec, O.; Demiral, B.; Bertin, H. CO_2 Injection into saline carbonate aquifer formations I: Laboratory investigation. *Transp. Porous Media* **2008**, *72*, 1–24. [CrossRef]

21. Sayegh, S.G.; Rao, D.N.; Kokal, S.; Najman, J. Phase behavior and physical properties of lindbergh heavy oil/CO₂ mixtures. *J. Can. Pet. Technol.* **1990**, *29*, 31–39. [CrossRef]
22. Comerlati, A.; Ferronato, M.; Gambolati, G.; Putti, M.; Teatini, P. Fluid-dynamic and gmechanical effects of CO₂ sequestration below the venice lagoon. *Environ. Eng. Geosci.* **2006**, *12*, 211–226. [CrossRef]
23. Zhang, X.; Trinh, T.T.; Santen, R.A.V. Mechanism of the initial stage of silicate oligomerization. *J. Am. Chem. Soc.* **2011**, *133*, 6613–6625. [CrossRef] [PubMed]
24. Li, Z.; Dong, M. Experimental study of carbon dioxide diffusion in oil-saturated porous media under reservoir conditions. *Ind. Eng. Chem. Res.* **2009**, *48*, 9307–9317. [CrossRef]
25. Jia, Y.; Bian H, B.; Duveau, G.; Shao, J. Numerical analysis of the thermo-hydromechanical behaviour of underground storages in hard rock. In Proceedings of the Geoshanghai International Conference, Shanghai, China, 3–5 June 2010; pp. 198–205.
26. Li, Z.; Dong, M.; Li, S.; Dai, L. A New method for gas effective diffusion coefficient measurement in water-saturated porous rocks under high pressures. *J. Porous Media* **2006**, *9*, 445–461. [CrossRef]
27. Hou, S.; Liu, F.; Wang, S.; Bian, H. Coupled heat and moisture transfer in hollow Concrete block wall filled with compressed straw bricks. *Energy Build.* **2017**, *135*, 74–84. [CrossRef]
28. Wang, S.; Feng, Q.; Zha, M.; Javadpour, F.; Hu, Q. Supercritical methane diffusion in shale nanopores: Effects of pressure, mineral types, and moisture content. *Energy Fuels* **2017**, *32*, 169–180. [CrossRef]
29. Zhao, P.; Wang, Z.; Sun, Z.; Cai, J.; Wang, L. Investigation on the pore structure and multifractal characteristics of tight oil reservoirs using NMR measurements: Permian Lucaogou Formation in Jimusaer Sag, Junggar Basin. *Mar. Petrol. Geol.* **2017**, *86*, 1067–1081. [CrossRef]
30. Wang, F.; Yang, K.; Cai, J. Fractal characterization of tight oil reservoir pore structure using nuclear magnetic resonance and mercury intrusion porosimetry. *Fractals* **2018**, *2*, 1840017. [CrossRef]
31. Hill, E.S.; Lacey, W.N. Hate of solution of propane in quiescent liquid hydrocarbons. *Ind. Eng. Chem.* **1934**, *25*, 1014–1019.
32. Sigmund, P.M. Prediction of molecular diffusion at reservoir conditions. Part I—Measurement and prediction of binary dense gas diffusion coefficients. *J. Can. Pet. Technol.* **1976**, *15*, 48–57. [CrossRef]
33. Islas-Juarez, R.; Samanego, V.F.; Luna, E.; Perez-Rosales, C.; Cruz, J. Experimental study of effective diffusion in porous media. In Proceedings of the SPE International Petroleum Conference in Mexico, Puebla, Mexico, 7–9 November 2004; pp. 781–787.
34. Riazi, M.R. A new method for experimental measurement of diffusion coefficients in reservoir fluids. *J. Pet. Sci. Eng.* **1996**, *14*, 235–250. [CrossRef]
35. Upreti, S.R.; Mehrotra, A.K. Experimental measurement of gas diffusivity in bitumen: Results of carbon dioxide. *Ind. Eng. Chem. Res.* **2000**, *39*, 1080–1087. [CrossRef]
36. Zhang, Y.; Hyndman, C.L.; Maini, B.B. Measurement of gas diffusivity in heavy oils. *J. Pet. Sci. Eng.* **1999**, *25*, 37–47. [CrossRef]
37. Tharanivasan, A.K.; Yang, C.; Gu, Y. Measurements of molecular diffusion coefficients of carbon dioxide, methane, and propane in heavy oil under reservoir conditions. *Energy Fuels* **2006**, *20*, 2509–2517. [CrossRef]
38. El-Haj, R.; Lohi, A.; Upreti, S.R. Experimental determination of butane dispersion in vapour extraction of heavy oil and bitumen. *J. Pet. Sci. Eng.* **2009**, *67*, 41–47. [CrossRef]
39. Okazawa, T. Impact of concentration—Dependence of diffusion coefficient on VAPEX drainage rates. *J. Can. Pet. Technol.* **2009**, *48*, 47–53. [CrossRef]
40. Wen, Y.; Bryan, J.; Kantzas, A. Estimation of diffusion coefficients in bitumen solvent mixtures as derived from low field NMR spectra. *J. Can. Pet. Technol.* **2005**, *44*, 29–35. [CrossRef]
41. Afsahi, B. Advanced in Diffusivity and Viscosity Measurements of Hydrocarbon Solvents in Heavy Oil and Bitumen. Master's Thesis, University of Calgary, Calgary, AB, Canada, 2006.
42. Guerrero-Aconcha, U.; Salama, D.; Kantzas, A. Diffusion of n-alkanes in heavy oil. In Proceedings of the SPE Annual Technical Conference and Exhibition, Denver, CO, USA, 21–24 September 2008.
43. Wen, Y.; Kantzas, A.; Wang, G. Estimation of diffusion coefficients in bitumen solvent mixtures using X-ray CAT scanning and low field NMR. In Proceedings of the Canadian International Petroleum Conference, Calgary, AB, Canada, 8–10 June 2004.
44. Wang, L.; Nakanishi, Y.; Teston, A.D.; Suekane, T. Effect of diffusing layer thickness on the density-driven natural convection of miscible fluids in porous media: Modeling of mass transport. *J. Fluid Sci. Technol.* **2018**, *13*, 1–20. [CrossRef]

45. Yang, C.; Gu, Y. A new method for measuring solvent diffusivity in heavy oil by dynamic pendant drop shape analysis (DPDSA). *SPE J.* **2006**, *11*, 48–57. [CrossRef]

46. Oren, P.E.; Bakke, S.; Arntzen, O.J. Extending predictive capabilities to network models. *SPE J.* **1998**, *3*, 324–336. [CrossRef]

47. Blunt, M.J.; Piri, M.; Valvatne, P. Detailed physics, predictive capabilities and upscaling for pore-scale models of multiphase flow. *Adv. Water Resour.* **2002**, *25*, 1069–1089. [CrossRef]

48. Piri, M.; Blunt, M.J. Pore-scale modeling of three-phase flow in mixed wet systems. In Proceedings of the SPE Annual Technical Conference and Exhibition, San Antonio, TX, USA, 29 September–2 October 2002.

49. Garmeh, G.; Johns, R.T.; Lake, L.W. Pore-scale simulation of dispersion in porous media. In Proceedings of the SPE Annual Technical Conference and Exhibition, Anaheim, CA, USA, 11–14 November 2007.

50. Taheri, S.; Kantzas, A.; Abedi, J. Mass diffusion into bitumen: A sub-pore scale modeling approach. In Proceedings of the Canadian Unconventional Resources & International Petroleum Conference, Calgary, AB, Canada, 19–21 October 2010.

51. De Paoli, M.; Zonta, F.; Soldati, A. Dissolution in anisotropic porous media: Modelling convection regimes from onset to shutdown. *Phys. Fluids* **2017**, *29*, 026601. [CrossRef]

52. Xu, X.; Chen, S.; Zhang, D. Convective stability analysis of the long-term storage of carbon dioxide in deep saline aquifers. *Adv. Water Resour.* **2006**, *29*, 397–407. [CrossRef]

53. De Paoli, M.; Zonta, F.; Soldati, A. Influence of anisotropic permeability on convection in Porous media: Implications for geological CO_2 sequestration. *Phys. Fluids* **2016**, *28*, 367–370. [CrossRef]

54. Li, S.; Li, Z.; Dong, Q. Diffusion coefficients of supercritical CO_2 in oil-saturated cores under low permeability reservoir conditions. *J. CO_2 Util.* **2016**, *14*, 47–60. [CrossRef]

55. Li, S.; Qiao, C.; Zhang, C.; Li, Z. Determination of diffusion coefficients of supercritical CO_2 under tight oil reservoir conditions with pressure-decay method. *J. CO_2 Util.* **2018**, *24*, 430–443. [CrossRef]

56. Li, H.; Yang, D. Determination of individual diffusion coefficients of solvent/CO_2 mixture in heavy oil with pressure-decay method. *SPE J.* **2015**, *21*, 131–143. [CrossRef]

57. Peng, D.; Robinson, D.B. A new two-constant equation of state. *Ind. Eng. Chem. Fundam.* **1976**, *15*, 92–94. [CrossRef]

58. Zuo, J.; Zhang, D. Plus fraction characterization and PVT data regression for reservoir fluids near critical conditions. In Proceedings of the SPE Asia Pacific Oil and Gas Conference and Exhibition, Brisbane, QLD, Australia, 16–18 October 2000.

59. Zheng, S.; Li, H.; Sun, H.; Yang, D. Determination of diffusion coefficient for alkane solvent–CO_2 mixtures in heavy oil with consideration of swelling effect. *Ind. Eng. Chem. Res.* **2016**, *55*, 1533–1549. [CrossRef]

60. Zheng, S.; Yang, D. Determination of individual diffusion coefficients of C_3H_8/n-C_4H_{10}/CO_2/heavy-oil systems at high pressures and elevated temperatures by dynamic volume analysis. In Proceedings of the SPE Improved Oil Recovery Conference, Tulsa, OK, USA, 11–13 April 2016.

61. Fateen, S.E.K.; Khalil, M.M.; Elnabawy, A.O. Semi-empirical correlation for binary interaction parameters of the Peng–Robinson equation of state with the van der Waals mixing rules for the prediction of high-pressure vapor–liquid equilibrium. *J. Adv. Res.* **2013**, *4*, 137–145. [CrossRef] [PubMed]

62. Elsharkawy, A.M. An empirical model for estimating the saturation pressures of crude oils. *J. Pet. Sci. Eng.* **2003**, *38*, 57–77. [CrossRef]

63. Pedersen, K.S.; Thomassen, P.; Fredenslund, A. SRK-EOS calculation for Crude OILS. *Fluid Phase Equilib.* **1983**, *14*, 209–218. [CrossRef]

64. Twu Chorng, H. Prediction of thermodynamic properties of normal paraffins using only normal boiling point. *Fluid Phase Equilib.* **1983**, *11*, 65–81.

65. Twu Chorng, H. An internally consistent correlation for predicting the critical properties and molecular weights of petroleum and coal-tar liquids. *Fluid Phase Equilib.* **1984**, *16*, 137–150.

66. Kesler, M.G.; Lee, B.I. Improve prediction of enthalpy fractions. *Hydrocarb. Process.* **1976**, *55*, 153–158.

67. Danesh, A.; Xu, D.; Todd, A.C. A Grouping method to optimize oil description for compositional simulation of gas-injection processes. *SPE Reserv. Eng.* **1992**, *7*, 343–348. [CrossRef]

68. Renner, T.A. Measurement and correlation of diffusion coefficients for CO_2 and rich-gas applications. *SPE Reserv. Eng.* **1988**, *3*, 517–523. [CrossRef]

69. Moysan, J.M.; Paradowski, H.; Vidal, J. Prediction of phase behaviour of gas-containing systems with cubic equations of state. *Chem. Eng. Sci.* **1986**, *41*, 2069–2074. [CrossRef]

70. Crank, J. *The Mathematics of Diffusion*; Oxford University Press: Oxford, UK, 1979.

71. Zhao, R.; Ao, W.; Xiao, A.; Yan, W.; Yu, Z.; Xiao, X. Diffusion law and measurement of variable diffusion coefficient of CO_2 in oil. *J. China Univ. Pet.* **2016**, *40*, 136–142.

72. Kavousi, A.; Torabi, F.; Chan, C.W.; Shirif, E. Experimental measurement and parametric study of CO_2 solubility and molecular diffusivity in heavy crude oil systems. *Fluid Phase Equilib.* **2014**, *371*, 57–66. [CrossRef]

73. Behzadfar, E.; Hatzikiriakos, S.G. Diffusivity of CO_2 in bitumen: Pressure–decay measurements coupled with rheometry. *Energy Fuels* **2014**, *28*, 1304–1311. [CrossRef]

74. Umesi, N.O.; Danner, R.P. Predicting diffusion coefficients in nonpolar solvents. *Ind. Eng. Chem. Proc. Des. Dev.* **1981**, *20*, 662–665. [CrossRef]

75. Upreti, S.R.; Mehrotra, A.K. Diffusivity of CO_2, CH_4, C_2H_6 and N_2 in athabasca bitumen. *Can. J. Chem. Eng.* **2002**, *80*, 116–125. [CrossRef]

76. Rasmussen, M.L.; Civan, F. Parameters of gas dissolution in liquids obtained by isothermal pressure decay. *AICHE J.* **2009**, *55*, 9–23. [CrossRef]

77. Hayduk, W.; Cheng, S.C. Review of relation between diffusivity and solvent viscosity in dilute liquid solutions. *Chem. Eng. Sci.* **1971**, *26*, 635–646. [CrossRef]

78. Dan, V.N.; Graciaa, A. A new reduction method for phase equilibrium calculations. *Fluid Phase Equilib.* **2011**, *302*, 226–233.

79. Leibovici, C.F.; Neoschil, J. A solution of Rachford-Rice equations for multiphase systems. *Fluid Phase Equilib.* **1995**, *112*, 217–221. [CrossRef]

80. Okuno, R.; Johns, R.T.; Sepehrnoori, K. Application of a reduced method in compositional simulation. *SPE J.* **2010**, *15*, 39–49. [CrossRef]

81. Shu, W.R. Viscosity correlation for mixtures of heavy oil, bitumen, and petroleum fractions. *SPE J.* **1984**, *24*, 277–282. [CrossRef]

![energies logo] *energies*

MDPI

Brief Report

Data Report: Molecular and Isotopic Compositions of the Extracted Gas from China's First Offshore Natural Gas Hydrate Production Test in South China Sea

Jianliang Ye, Xuwen Qin, Haijun Qiu, Wenwei Xie, Hongfeng Lu *, Cheng Lu, Jianhou Zhou, Jiyong Liu, Tianbang Yang, Jun Cao and Rina Sa

Guangzhou Marine Geological Survey, China Geological Survey, Guangzhou 510760, China; jianliangye@hydz.cn (J.Y.); qinxuwen@hydz.cn (X.Q.); qiuhaijun@hydz.cn (H.Q.); xiewenwei@hydz.cn (W.X.); jaluch@126.com (C.L.); zhoujianhou1989@163.com (J.Z.); ljy1986512@163.com (J.L.); yangtianbang5@163.com (T.Y.); caojun_031051@126.com (J.C.); sarina_gmgs@163.com (R.S.)
* Correspondence: luhongfeng@hydz.cn

Received: 2 September 2018; Accepted: 14 October 2018; Published: 17 October 2018

Abstract: Three hundred gas samples recovered from SHSC-4 during China's first gas hydrate production test in the South China Sea were examined for gas component and isotopic composition. According to the gas chromatography analysis, all the gas samples from SHSC-4 are predominated by CH_4, with minor $N_2 + O_2$, as well as trace amounts of CO_2, C_2H_6, and C_3H_8. No H_2S was detected. The molecular and isotopic data of the gas samples fall into the region of "mixed origin" on the plot of $C_1/(C_2 + C_3) - \delta^{13}C_1$, which is close to the microbial origin. The discrimination diagram of $\delta^{13}C_1 - \delta D_{CH4}$ shows that the methane in all of the samples is of microbial origin, and is derived from the CO_2 reduction.

Keywords: isotopic composition; methane; gas hydrate; South China Sea

---

1. Introduction

In 2017, the first offshore natural gas hydrate (NGH) production test in the South China Sea (SCS) was conducted by the China Geological Survey (CGS). The production test site, SHSC-4, is located in the middle of the continental slope of southeast Shenhu area, northern SCS, about 300 km southeastward away from Hong Kong (Figure 1). The water depth of SHSC-4 is about 1263.5 m, according to the ROV (Remote Operated Vehicle) in situ survey. The production test lasted for 60 days, from 10 May to 9 July. The cumulative gas volume produced during the sixty-day test was approximately 309,000 m^3 (under the atmospheric pressure). The gas production rate was approximately 5454 m^3/day, with a maximum of 35,000 m^3/day. The gas hydrate production test in SCS holds the new record for the longest production time and the highest cumulative volume in the world's offshore gas hydrate production test, which is essential to the future global energy supply.

The core data from site SHSC-4 showed that gas hydrate occurred over a 50-meter interval from 201 mbsf (meter blow the seafloor) to 251 mbsf [1]. The characteristics of gas hydrate-bearing sediment is clayey silt, which accounts for more than 90% of the global NGH [2]. The gas hydrate saturation from both the pore-water freshening and pressure core mass balance is variable throughout the interval, with mean values of 35%. The mean effective porosity of the gas hydrate reservoir is about 34%, and the mean permeability is 2.9 mD [1].

In this document, we report on the shore-based analyses of the produced gas component and isotopic composition from the SHSC-4 site. The objective of this report is to present the gas characteristics of China's first gas hydrate production test in SCS, based on the molecular and isotopic analysis.

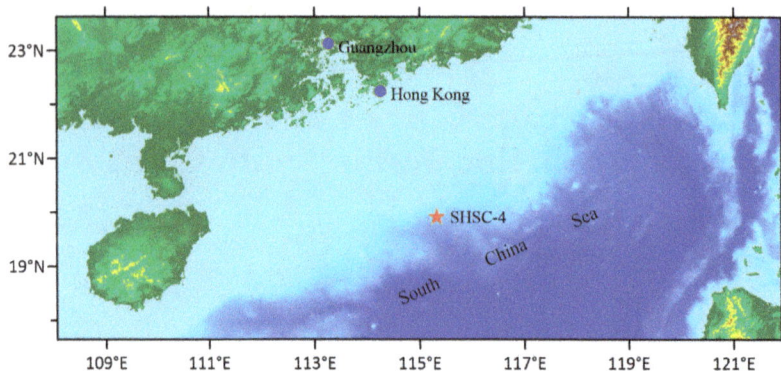

Figure 1. Location of China's first offshore gas hydrate production test in the South China Sea (SCS). The production test site named SHSC-4 (red solid star), which water depth is about 1263.5 m.

2. Samples and Analytical Methods

All of the gas samples were collected from the flow line on the rig, which was extracted from the gas hydrate interval between 201 mbsf and 251 mbsf with depressurization. All of the samples had no hydrogen sulfide smell. They were sealed in sample bags and were taken back for onshore analysis immediately. Three hundred gas samples were examined for geochemistry by gas chromatography (GC) and isotope ratio mass spectrometer.

An Agilent 7890A GC equipped with a 30 m × 0.53 mm HP-PLOT/Q column was used to measure the hydrocarbon gases from C1 to C5. The Agilent 7890 A is configured with a 1-mL, valve-actuated, sample loop for injection, and a thermal conductivity detector (TCD) for gas detection. The samples were introduced by syringes under atmospheric pressure, and a minimum of 10 mL of gas was used to flush the injection loop. The experiments were initially performed under 50 °C, with an increasing temperature rate of 20 °C/min to 200 °C. Helium was used as the carrier gas at a constant flow rate of 5 mL/min. The TCD temperature was kept under 200 °C. The GC was calibrated using the hydrocarbon gas standards of known concentrations.

Isotopic ratios of carbon and hydrogen in the hydrocarbon gases were measured with a Thermo Fisher Trace GC ultra-gas chromatography coupled to a MAT-253 isotope ratio mass spectrometer via a GC-isolink interface, at Guangzhou Marine Geological Survey Laboratory. The gas components were separated on a 30 m × 0.3 mm ID HP-PLOT/Q column using helium as the carrier gas. The flow rate was 1.5 mL/min. The GC oven was set under 50 °C for 2 min, then programmed to increase to 140 °C at a rate of 30 °C/min, and was maintained under 140 °C for 4 min. The GC injection port was set under 100 °C and then the split ratio was 50:1. The combustion oven temperature was 1000 °C in order to convert all of the hydrocarbons to CO_2, and the high temperature cracked oven temperature was 1420 °C in order to determine the hydrogen isotope. The $\delta^{13}C$ values were reported in the unit of per mil (‰), relative to the VPDB (Vienna Pee Dee Belemnite) standard. The carbon isotope data are more accurate for the hydrocarbon gases (accuracy within ±0.15‰). The δD values were also reported in the unit of ‰ (accuracy within ±1.0‰), relative to VSMOW (Vienna Standard Mean Ocean Water). The carbon and hydrogen isotopic ratio for both methane and ethane were measured, and other hydrocarbon gases were not detected because of the instrument limit.

3. Results

All of the 300 gas samples show similar GC curves, which exhibit a large C1 peak (Methane), but a weak peak of the other components (Figure 2). The GC analysis shows that all of the gas samples are predominated by CH_4 (91.35–97.69%, average = 95.87%) with minor $N_2 + O_2$ (1.79–8.18%, average = 3.63%), and trace CO_2 (0.007–0.142%, average = 0.013%), C_2H_6 (0.294–0.521%) and C_3H_8 (0.062–0.108%). No H_2S was detected. Among all of the hydrocarbon gases, CH_4 consists 99.38~99.60%, with an average of 99.50% (Figure 3).

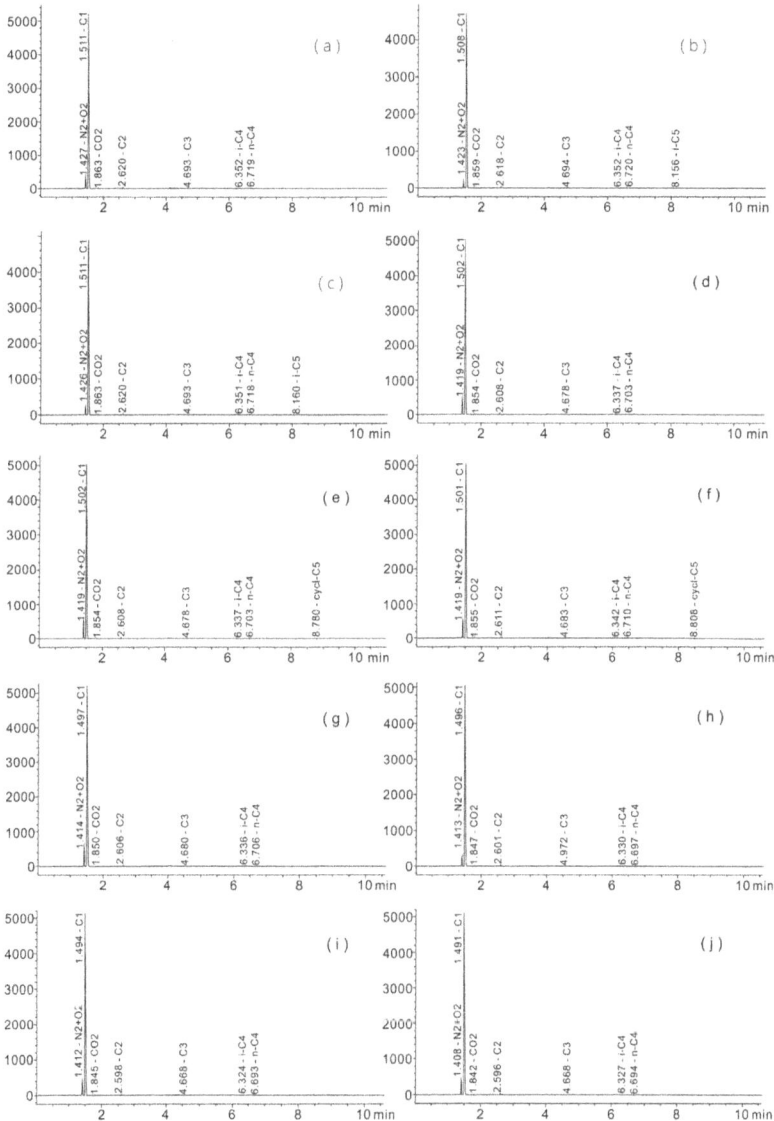

Figure 2. The representative GC curves of the gas samples from China's first offshore gas hydrate production test in SCS. (**a,b**) stand for the GC curves of gas extracted from the early depressurization; (**c–h**) are the curves of the gas samples from the middle production; (**i,j**) are the samples from the late production. All of the samples exhibit a large C1 peak and other weak peaks, which show that all of the gas samples are predominated by CH_4.

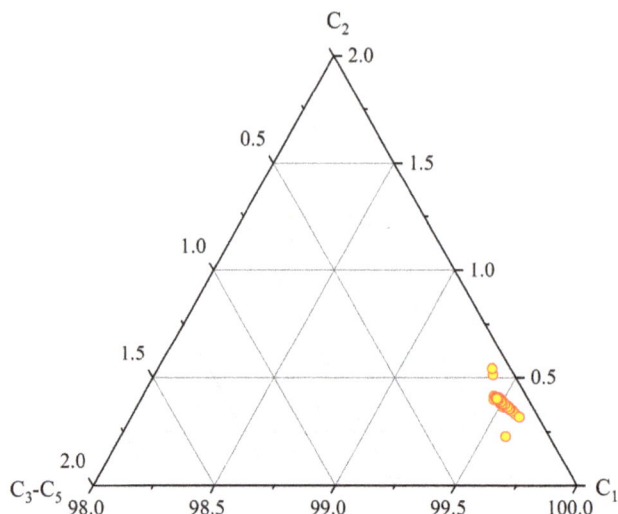

Figure 3. Ternary diagram of hydrocarbons from China's first offshore gas hydrate production test in SCS. All of the data points fall into the C1 end member, which indicate that methane is the main component of gas.

The molecular $\{C_1/(C_2 + C_3)\}$ compositions of the 300 gas samples collected from the production test range from 165 to 259 (ave. = 209). The $\delta^{13}C$ and δD in methane are within the range of -66.09–$-63.14‰$ (average = $-64.88‰$) and -195.8–$-186.0‰$ (average = $-191.3‰$), respectively, which are similar to those of the methane experimentally released from the natural gas hydrate samples collected from the Shenhu Area ($\delta^{13}C_1 = -64.38$–$-61.57‰$, $\delta D_{CH4} = -220.00$–$-191.00‰$ [3]).

The relationship between the C_1/C_{2+} ratios and the $\delta^{13}C$ values of methane is a good way to distinguish the origin of hydrocarbon. High C_1/C_{2+} ratios ($C_1/C_{2+} \geq 1000$) and low methane $\delta^{13}C$ values ($\delta^{13}C \leq -55‰$) are characteristic of a microbial origin, while low C_1/C_{2+} ratios ($C_1/C_{2+} \leq 100$) and high methane $\delta^{13}C$ values ($\delta^{13}C \geq -50‰$) are characteristic of a thermogenic origin [4–10]. On the plot of $C_1/(C_2 + C_3) - \delta^{13}C_1$, all of the gas samples from the production test fall into the "mixed origin" region close to the microbial origin (Figure 4), which is different from the gas hydrate from Blake Ridge, Hydrate Ridge, Mexico gulf, and a Japan gas hydrate prodution test site in Nankai Trough, but close to the Ulleung basin of Korea. Moreover, the discrimination diagram of $\delta^{13}C_1 - \delta D_{CH4}$ (Figure 5) shows that the methane in all of the samples is of microbial origin and is derived from a CO_2 reduction, which is different from the nearby LW3-1-1 gas well with methane thermogenic origin in SCS, but is the same as Blake Ridge, Hydrate Ridge, and the Ulleung basin of Korea. The information from Figures 4 and 5 indicate that the gas extracted from China's first gas hydrate production test possibly originate from bacterial and thermogenic gas mixtures, but have more of a bacterial component.

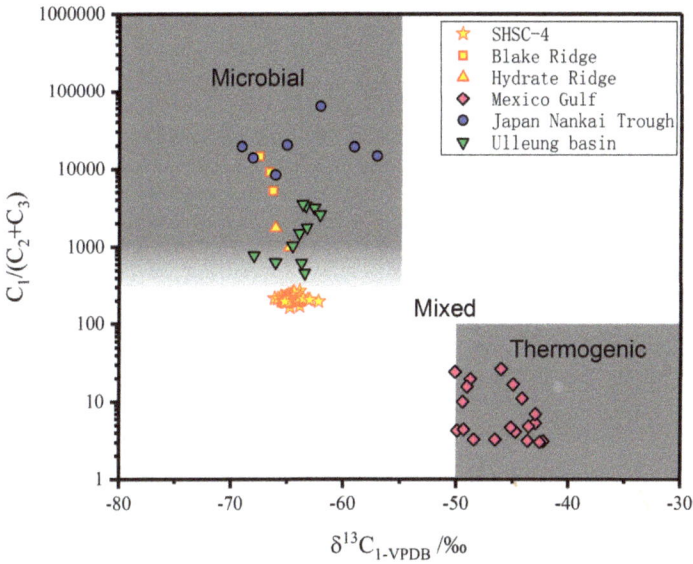

Figure 4. Relationship between $C_1/(C_2 + C_3)$ and $\delta^{13}C\text{-}C_1$ of hydrocarbon from China's first offshore gas hydrate production test in SCS (adapted from Whiticar et al. [6]). The data from the Blake Ridge [11], the Hydrate Ridge [7], the Mexico Gulf [12], the Japan Nankai Trough [13], and the Ulleung Basin [14] were plotted for comparison.

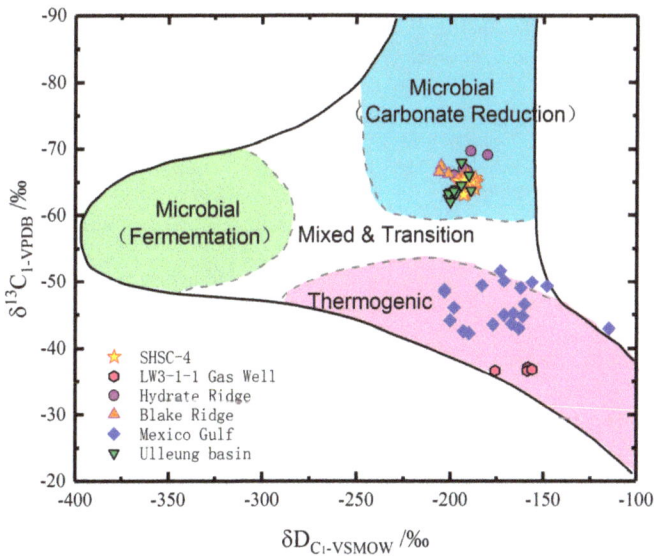

Figure 5. Relationship between $\delta^{13}C$ and δD of CH_4 from China's first offshore gas hydrate production test in SCS (base map from Whiticar et al. [6]). The data from the LW3-1-1 gas well in SCS [15], the Hydrate Ridge [16], the Blake Ridge [11], the Mexico Gulf [12], and the Ulleung Basin [14] were plotted for comparison.

Energies **2018**, *11*, 2793

4. Summary and Conclusions

The extracted gas from China's first natural gas hydrate production test in SCS are predominantly by methane, which accounts for 99.38~99.60% of all of the hydrocarbon gases. The δ^{13}C and δD in methane are within the range of -66.09–-63.14‰ and -195.8–-186.0‰, respectively. Different from those of the nearby LW3-1-1 gas well with a thermogenic origin, these values imply that the methane of the natural gas hydrate in the Shenhu area was mainly derived from the bacterial reduction of CO_2. Nevertheless, the contribution of the thermogenic origin cannot be excluded in this study. It is possible that the bacterial gas is the main gas origin for gas hydrate formation in SHSC-4.

Author Contributions: J.Y. made the substantial contributions of research conception and design. X.Q. contributed to made the critical revision. H.Q. contributed to data collection. W.X. made the contribution of data interpretation. H.L. drafted the article and made final approval of the article. C.L., J.Z. and J.L. analysed the component of samples. T.Y. collected the gas samples. J.C. and R.S. made the isotope analysis.

Funding: This research was funded by the GH Programs of China Geological Survey grant number DD20189320 and DD20160221.

Acknowledgments: This study was financially supported by the GH Programs of China Geological Survey (No. DD20189320 and No. DD20160221). We thank all of the scientists and staff who participated in China's natural gas hydrate production test in SCS. The authors are grateful to Kuang Zenggui, Zhang Ruwei, Li Zhanzhao, Wang Tinghui, Wei Jiangong, and Liang Qianyong for their sampling and measurements in the field. We also thank Jiang Xuexiao and Deng Yinan for helping the carbon and oxygen isotope analyses in the laboratory.

Conflicts of Interest: The authors declare no conflicts of interest.

References

1. Li, J.; Ye, J.; Qin, X.; Qiu, H.; Wu, N.; Lu, H.; Xie, W.; Lu, J.; Peng, F.; Xu, Z.; et al. The first offshore natural gas hydrate production test in South China Sea. *China Geol.* **2018**, *1*, 1–12. [CrossRef]
2. Johnson, H. Global resource potential of gas hydrate—A new calculation. *Fire Ice Methane Hydrate Newslett.* **2011**, *11*, 1–4.
3. Liu, C.; Meng, Q.; Li, C.; Sun, J.; He, X.; Yang, S.; Liang, J. Characterization of natural gas hydrate and its deposits recovered from the northern slope of the South China Sea. *Earth Sci. Front.* **2017**, *24*, 41–50, (In Chinese with English abstract).
4. Bernard, B.B.; Brooks, J.M.; Sackett, W.M. Light-hydrocarbons in recent Texas continental-shelf and slope sediments. *J. Geophys. Res.-Oceans* **1978**, *83*, 4053–4061. [CrossRef]
5. Claypool, G.E.; Kvenvolden, K.A. Methane and other hydrocarbon gases in marine sediments. *Earth Planet. Sci.* **1983**, *11*, 299–327. [CrossRef]
6. Whiticar, M.J. Carbon and hydrogen isotope systematics of bacterial formation and oxidation of methane. *Chem. Geol.* **1999**, *161*, 291–314. [CrossRef]
7. Milkov, A.V.; Claypool, G.E.; Lee, Y.-J.; Sassen, R. Gas hydrate systems at Hydrate Ridge offshore Oregon inferred from molecular and isotopic properties of hydrate-bound and void gases. *Geochim. Cosmochim. Acta* **2005**, *69*, 1007–1026. [CrossRef]
8. Pape, T.; Bahr, A.; Rethemeyer, J.; Kessler, J.D.; Sahling, H.; Hinrichs, K.-U.; Klapp, S.A.; Reeburgh, W.S.; Bohrmann, G. Molecular and isotopic partitioning of low molecular-weight hydrocarbons during migration and gas hydrate precipitation in deposits of a high-flux seepage site. *Chem. Geol.* **2010**, *269*, 350–363. [CrossRef]
9. Kim, J.-H.; Park, M.H.; Chun, J.H.; Lee, J.Y. Molecular and isotopic signatures in sediments and gas hydrate of the central/southwestern Ulleung Basin: High alkalinity escape fuelled by biogenically sourced methane. *Geo-Mar. Lett.* **2011**, *31*, 37–49. [CrossRef]
10. Kim, J.-H.; Torres, M.E.; Choi, J.Y.; Bahk, J.J.; Park, M.H.; Hong, W.L. Inferences on gas transport based onmolecular and isotopic signatures of gases at acoustic chimneys and background sites in the Ulleung Basin. *Org. Geochem.* **2012**, *43*, 26–38. [CrossRef]
11. Matsumoto, R.; Takashi, U.; Waseda, A.; Uchida, T.; Takeya, S.; Hirano, T.; Yamada, K.; Maeda, Y.; Okui, T. Occurrence, structure, and composition of natural gas hydrate recovered from the Blake Ridge, Northwest Atlantic. *Proc. Ocean Drill. Program Sci. Results* **2000**, *164*, 13–28.

12. Sassen, R.; Sweet, S.T.; Milkov, A.V.; DeFreitas, D.A.; Kennicutt, M.C.; Roberts, H.H. Stability of thermogenic gas hydrate in the Gulf of Mexico: Constraints on models of climate change. In *Natural Gas Hydrates: Occurrence, Distribution, and Detection*; Paull, C.K., Dillon, W.P., Eds.; American Geophysical Union: Washington, DC, USA, 2001; pp. 131–143.

13. Kida, M.; Jin, Y.; Watanabe, M.; Konno, Y.; Yoneda, J.; Egawa, K.; Ito, T.; Nakatsuka, Y.; Suzuki, K.; Fujii, T.; et al. Chemical and crystallographic characterizations of natural gas hydrates recovered from a production test site in the eastern Nankai Trough. *Mar. Pet. Geol.* **2015**, *66*, 396–403. [CrossRef]

14. Choi, J.; Kim, J.-H.; Torres, M.E.; Hong, W.-L.; Lee, J.-W.; Yi, B.Y.; Bahk, J.J.; Lee, K.E. Gas origin and migration in the Ulleung Basin, East Sea: Results from the Second Ulleung Basin Gas Hydrate Drilling Expedition (UBGH2). *Mar. Pet. Geol.* **2013**, *47*, 113–124. [CrossRef]

15. Zhu, J.; Shi, H.; He, M.; Pang, X.; Yang, S.; Li, Z. Origins and Geochemical Characteristics of Gases in LW3-1-1 Well in the Deep Sea Region of Baiyun Sag, Pearl River Mouth Basin. *Nat. Gas Geosci.* **2008**, *19*, 229–233, (In Chinese with English abstract).

16. Winckler, G.; Werner, A.H.; Holocher, J.; Kipfer, R.; Levin, I.; Poss, C.; Rehder, G.; Suess, E.; Schlosser, P. Noble gas and radiocarbon in natural gas hydrates. *Geophys. Res. Lett.* **2002**, *29*, 1–4. [CrossRef]

energies

MDPI

Review

Prevention of Potential Hazards Associated with Marine Gas Hydrate Exploitation: A Review

Fangtian Wang [1,2], Bin Zhao [1,2,*] and Gang Li [1,2]

[1] State Key Laboratory of Coal Resources and Safe Mining, China University of Mining and Technology, Xuzhou 221116, China; wangfangtian111@cumt.edu.cn (F.W.); Lgang_27@cumt.edu.cn (G.L.)
[2] School of Mines, Key Laboratory of Deep Coal Resource Mining, Ministry of Education of China, China University of Mining and Technology, Xuzhou 221116, China
* Correspondence: 01120120@cumt.edu.cn

Received: 20 August 2018; Accepted: 7 September 2018; Published: 10 September 2018

Abstract: Marine gas hydrates (MGHs), which have great potential for exploitation and utilization, account for around 99% of all global natural gas hydrate resources under current prospecting technique. However, there are several potential hazards associated with their production and development. These are classified into four categories by this paper: marine geohazards, greenhouse gas emissions, marine ecological hazards, and marine engineering hazards. In order to prevent these risks from occurring, the concept of "lifecycle management of hazards prevention" during the development and production from MGHs is proposed and divided into three stages: preparation, production control, and post-production protection. Of these stages, economic evaluation of the resource is the foundation; gas production methods are the key; with monitoring, assessment, and early warning as the guarantee. A production test in the Shenhu area of the South China Sea shows that MGH exploration and development can be planned using the "three-steps" methodology: commercializing and developing research ideas in the short term, maintaining economic levels of production in the medium term, and forming a global forum to discuss effective MGH development in the long term. When increasing MGH development is combined with the lifecycle management of hazards prevention system, and technological innovations are combined with global cooperation to solve the risks associated with MGH development, then safe access to a new source of clean energy may be obtained.

Keywords: marine gas hydrate; submarine landslide; greenhouse gas emission; lifecycle management; hazard prevention

1. Introduction

As the global economy develops, the demand for energy is increasing and with the resultant rise in consumption of fossil fuels, there is a need to find alternative forms of energy to maintain sustainable development [1,2]. As a non-traditional fossil fuel, natural gas hydrates (NGHs) are the subject of increasing research since their discovery in the 1960s [3,4], because of their high calorific value and potential utilization. NGHs (combustible ice) are non-stoichiometric crystalline compounds that form ice-like solid structures of gas (i.e., usually methane, ethane, propane and lower order hydrocarbons [5]) and water in a low-temperature (2–18 °C) or high-pressure (3.5–14.5 MPa) environment [6–8]. These conditions occur in near-surface, deep-water marine sediments and in terrestrial permafrost areas that are widely distributed around the world (Figure 1). Currently, NGH reserves are believed to be in the order of 3.0×10^{15} m^3 (3.0×10^{12} t oil equivalent), which is approximately twice the world's known supply of fossil fuels (coal, oil, natural gas) [9]. Of the NGH reserves, marine gas hydrates (MGHs) account for more than 99% of these reserves, which if developed could provide many years of production [10,11]. As a result, research on the development of MGHs has become an important research issue.

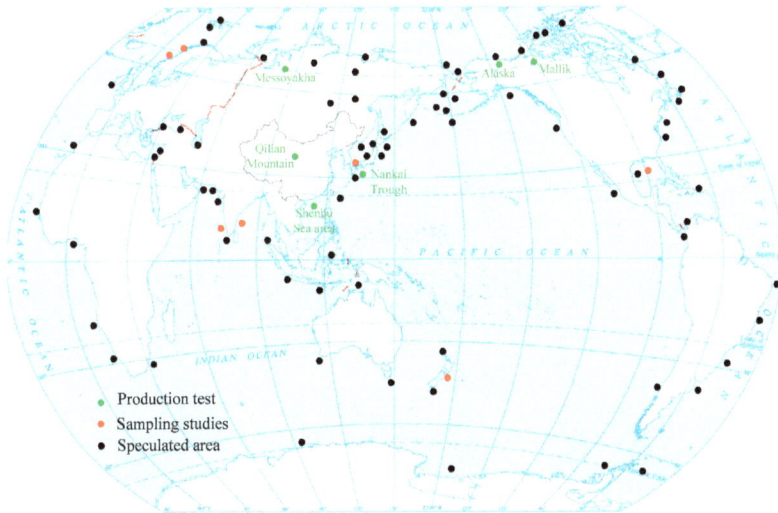

Figure 1. Distribution of NGH deposits on Earth [8]. (1) "Production test" represent the places where have been successfully tested NGH production; (2) "Sampling studies" represent the places just where the hydrate samples are taken but not production tested; (3) "Speculated area" represent the places where there may be hydrates under current prospecting technique.

NGHs are only stable within a specific range of temperature and pressure conditions (Figure 2) and understanding these is crucial to the development of NGH reserves. At present, there are five NGH production methods: depressurization, thermal activation, chemical agent injection, CO_2 replacement and solid fluidization [12,13].

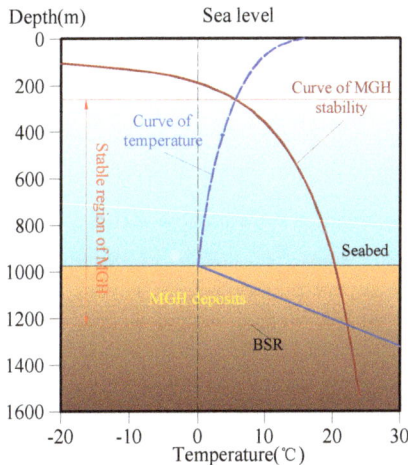

Figure 2. Pressure-temperature equilibrium curves for MGHs [2].

There have been three important production tests of MGHs in the 21st century [14], namely within the Nankai Trough, Japan (2013, 2017) [15,16] and in the Shenhu area of the South China Sea (2017) [10]. Production tests of NGHs in terrestrial permafrost have been undertaken in the Mackenzie Delta, Canada (2002, 2007, 2008) [17] and the North Slope of Alaska, United States of America (2012) [18]; other countries such as Germany, India and South Korea have also conducted sampling studies on MGHs [19–21]. The results of the production tests and sampling studies have found that the crystal structure, sediment morphology and occurrence characteristics of NGHs show great diversity (Figure 3). This difference means that development of these resources will be complicated [6]. There is the potential for submarine landslides, climate warming, marine ecological damage and other hazards, if MGHs are not developed carefully. Current levels of technological development still face many technical and environmental challenges before the economic benefits of MGHs can be realized [10,14].

Figure 3. Occurrence characteristics of NGHs [6]. (**a**) Contained in the pore spaces of coarse-grained sand; (**b**) Massive lenses and nodules in muds; (**c**) Disseminated in muds; (**d**) Contained in the pore spaces of fine-grained marine sand; (**e**) Thin veins in muds; (**f**) Massive mounds on sea-floor.

This paper classifies and summarizes the different types of potential hazards in the development of MGHs based upon research to date, and proposes a comprehensive prevention and control strategy for these hazards, based on the concept of "lifecycle management". Additionally, the key challenges and lessons learned from a production test in the Shenhu area of the South China Sea are presented.

2. Classification and Causes of Potential Hazards from MGHs

The development of MGH reserves is controlled by the environment and the geology of the sediments at the location, so development of these reserves is complex and the production methodology is location specific [14]. Four categories of potential hazards have been identified which may affect the development of MGHs: marine geohazards, greenhouse gas emissions, marine ecological hazards, and marine engineering hazards.

2.1. Marine Geohazards

2.1.1. Submarine Landslide

Submarine landslides are the most important type of marine geohazard that may be encountered during the development of MGH reserves. As the pressure-temperature conditions change during development of MGH reserves, methane is released, and the filling and cementation of the reservoirs is reduced [22,23]. This results in a decrease of effective stress and an increase in pore pressure, which reduces the shear strength and bearing capacity of the sediments and can lead to reduced slope stability (Figure 4). Geohazards, such as sediment deformation, slumping, and even debris flows may occur. The submarine landslides of Storrega (Norway), at Cape Fear (USA) and in the Beaufort Sea (Canada) may have been related to the decomposition of MGHs [24–26]. Three criteria for the potential occurrence of submarine landslide have been identified [26,27]: (1) hydrates are widely distributed within the landslide areas; (2) the initial position of the landslide zone must be located at the phase boundary of the pressure-temperature field, (3) there is a low-permeability deposit under the hydrates which can maintain high pore pressure. The criteria suggest that submarine landslides caused by MGH decomposition are more likely to appear on shallow submarine slopes.

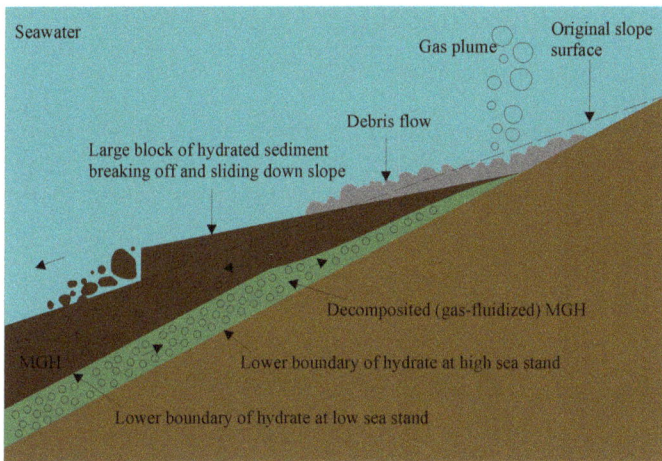

Figure 4. Schematic depiction of the mechanism that causes submarine landslides [23].

2.1.2. Earthquakes and Other Geohazards

The development of MGH reserves may cause other geohazards (Figure 5), such as earthquakes, active faulting, mud diapirism, and turbidity currents. Rapid venting of MGH reservoirs may cause the development of active faults which could provide a further conduit for methane escape, which would lead to a further decrease of reservoir pressure and thereby increase the rate of methane production [28]. Once the reservoir is drained, secondary hazards such as earthquakes may occur as a result of sediment settlement into the produced voids. Mud diapirism caused by plastic sediment flow around over-pressurized sand layers may also occur [22,29,30].

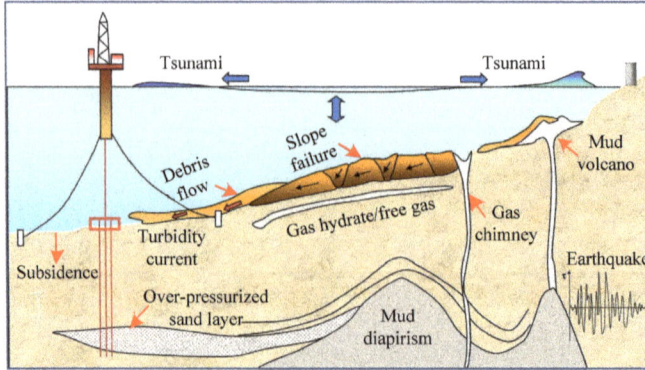

Figure 5. Sketch map of marine geohazards [22].

It should be noted that the release of MGHs and the associated marine geohazards triggered by this release are a normal process under natural conditions; the commercial development of MGHs could alter the balance and trigger these marine geohazards as an unnatural response [14]. Further studies are required to understand the causation of marine geohazards due to MGH development to ensure safe access to these resources.

2.2. Greenhouse Gas Emissions

One of the main impacts on the Earth's climate of increased greenhouse gases emissions is global warming. Methane is a greenhouse gas whose global warming potential index is twenty-five times that of carbon dioxide by unit mass and is an accelerator for environmental change [31,32]. Both the Paleocene-Eocene Thermal Maximum (PETM), with a global temperature rise of 4–8 °C that occurred 55.5 million years ago, and the global warming during the Quaternary interglacial periods were possibly caused by the large-scale decomposition of MGHs [23,33,34]. It suggested that during the Quaternary there was a cyclical link between the decomposition of MGHs (formation of methane) and global warming resulting in the glacial/intra-glacial cycle. As Figure 6 shows, (1) global cooling in the glacial epoch and sea-level decline leads to lower hydrostatic pressure, which results in the decomposition of MGHs as a result of the loss of stable pressure-temperature conditions; (2) the resulting methane enters the atmosphere which causes global warming, and an interglacial period ensues. As the glaciers melt as a response to the warmer conditions (interglacial period), the subsequent sea-level rise leads to increased hydrostatic pressure, and the MGHs restabilize until the next interglacial [35–37].

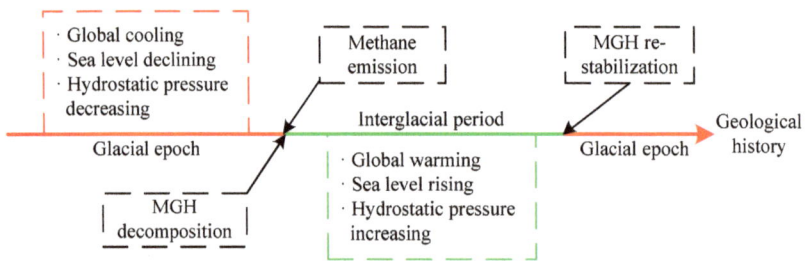

Figure 6. Relationship between gas hydrate decomposition and climate change.

At present, there is extensive research into establishing links between sudden releases of methane from MGH decomposition and specific changes in the global climate at certain times in the geological past, as well as the impact that these events had on geological history. One of the major problems with determining these effects is the fact that MGH-derived methane is dissolved and oxidized by seawater, so the amount entering the atmosphere is not representative of the release event or period [36]. Hence, the greenhouse gas emissions effect of MGH decomposition under normal conditions needs further observation and research [37].

2.3. Marine Ecological Hazards

In this paper, marine ecological hazards refer to the adverse effects on marine organisms and other components of the marine environment caused by the MGH decomposition. Under normal conditions, the gases from MGH decomposition reach the surface in a cold spring via a variety of means, including migration up fault planes and pore space expansion, where they form autotrophic chemosynthetic communities that consume methane, hydrogen sulfide, and other substances to provide the basic driving force for the entire marine community [38]. Inorganic carbon is also formed when MGH decompose and react with seawater to form carbonate minerals, providing a habitat for marine plankton and other biological communities (Figure 7). Uncontrolled releases from MGH may lead to faulting, eruptions at the seabed, and the collapse of the carbonate deposits and chemosynthetic communities, which may adversely affect the health of the marine environment. Meanwhile, some of the methane derived from the MGH will be regenerated as new hydrates and return to the seabed while the rest will react with the dissolved oxygen in the seawater to form CO_2. If excessive amounts of methane enter the seawater, large volumes of oxygen are consumed to form CO_2 so that the growth and evolution of marine animals and plants are prevented which could lead to extinction in extreme cases [39–41]. Marine faunal extinctions at the end of the Permian, Triassic, and Cretaceous may have been caused by the release of methane and subsequent changes to the marine environment [42,43].

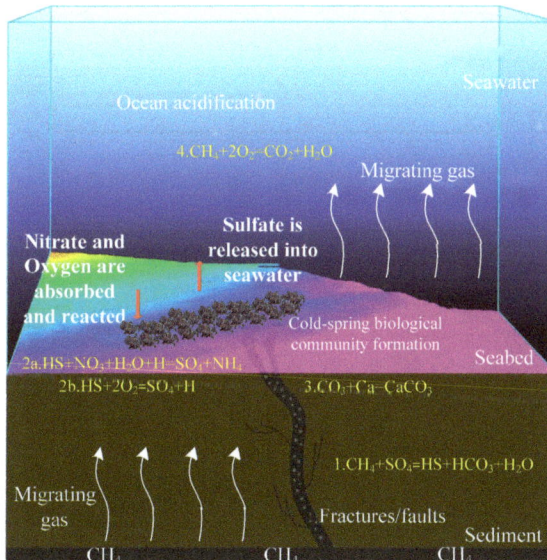

Figure 7. Chemical reaction after MGH decomposition [39].

Methane release by decomposition of MGH is a natural process and is part of the marine environment. Further research is needed into the effects on the marine environment that which might happen if MGH are developed as a commercial resource.

2.4. Marine Engineering Hazards

Two types of marine engineering hazards can be caused by the development of MGH: (1) MGH instability induced during drilling, (2) the risks associated with drilling or installing submarine structures in MGH areas. These hazards are inter related and the links and differences are discussed below.

2.4.1. MGH Stability Hazards

The MGH stability zone shown in Figure 2 is restricted to a limited temperature and pressure regime. Consequently, drilling through MGH is challenging as it involves changing the pressure regime and increasing the temperature profile of the sediments immediately surrounding the well bore (Figure 8). Furthermore, the use of organic alcohol hydrate thermodynamic inhibitors and inorganic salts in the drilling fluid can enhance the production of methane and barite precipitates, respectively [44]. On drilling into an MGH, large amounts of methane can be released that infiltrate into the drill pipe, resulting in a sharp increase in the mud pressure. As the methane rises and cools, it can reform as hydrate crystals that clog the drill pipe, potentially causing well abandonment. As drilling continues into the underlying free-gas zone, formation pressure can rise, causing increased release of methane and the buildup of large amounts of high-pressure gas in the drill pipe, which could lead to a blowout [45]. With significant releases of methane, soil formation stability is reduced through the development of numerous voids, which could impact borehole stability if they collapse. Furthermore, poorly consolidated sediments can result in significant sand production, which can affect the operation of safety equipment such as blowout preventers [46].

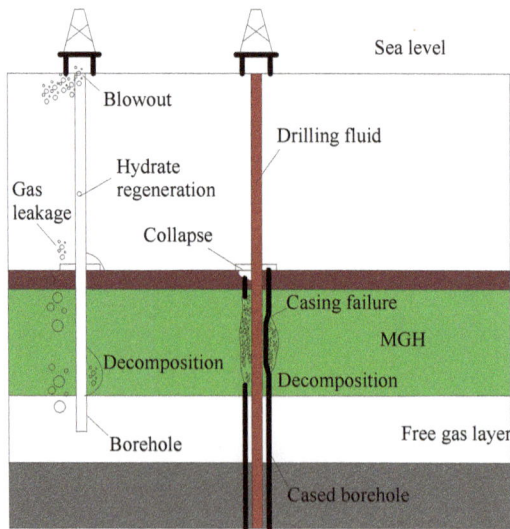

Figure 8. Sketch of drilling engineering hazards [38].

2.4.2. Risks Associated with Drilling or Installing Structures in MGH Areas

Exploration and development drilling for hydrocarbons in deep water can involve drilling through MGH deposits, where decomposition and regeneration of MGH may occur [47]. When drilling through these deposits, the formation properties will change [48]. The risks such as changes in drilling fluid properties, borehole stability issues, well cleaning, and cementation problems will follow. However, these issues can be successfully mitigated by appropriate drilling techniques just like we can adjust the drilling fluid according to the formation properties change. Secondary generation of MGH within blowout preventers, as well as changes in the rheology of the drilling fluid through the formation of barite scale and the subsequent blockage of pipework within the blowout preventer, may occur, but these effects can be reduced by appropriate composition of the drilling muds [49]. If foundations have to be piled into or laid on top of these deposits, then the seabed stability can be reduced if the pressure–temperature regime is disrupted with the subsequent damaged to the structures, associated pipelines, and communication cables [50].

3. Control and Prevention of Potential Hazards in MGH Development

Many methods have been proposed to reduce the potential hazards associated with the development of MGH as a resource. A concept of "lifecycle management of hazards prevention" (LMHP) in the development of MGH is proposed to cover the different stages of MGH development. These stages are: preparation; control during development; and post production protection. Each stage requires research into specific issues (Figure 9) and these are discussed below.

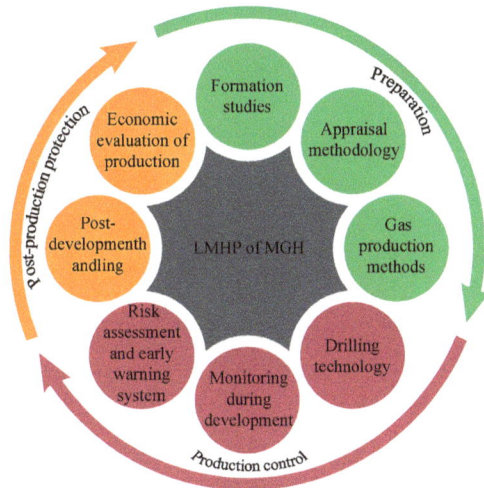

Figure 9. Lifecycle management schematic.

3.1. Preparation

The preparation phase is the primary step and the LMHP covers the mechanism of MGH formation, appraisal methodology and gas production methods.

3.1.1. MGH Formation Studies

A comprehensive study of the formation, migration, accumulation, and storage of methane within the MGH, in conjunction with expected pressure-temperature conditions, gas source, gas migration, and the presence of suitable reservoirs within the MGH accumulation are the main factors to be determined [51]. The characteristics of each MGH accumulation depend on the interaction of all these factors [52]. The degree of difficulty for the development of NGH reservoirs can be represented in pyramidal form (Figure 10). The difficulty of developing MGH reservoirs varies in a continuum from low in sandy reservoirs, then permeable clay reservoirs, then cold spring-related massive reservoirs, to high in non-permeable clay reservoirs. Each reservoir-type can be subdivided into three types according to recoverable value: Class I, Class II and Class III [6,12]. Class I is that one hydrate bearing layer covers on a two-phase fluid zone with free gas which is considered to be the most suitable reservoir for natural gas production. Class II is that one hydrate bearing layer covers on a mobile water zone and Class III is that there is only one hydrate bearing layer, which are still not well defined as gas production targets [53]. The successful production test of high-saturation diffusion-sourced hydrates in a viscous siltstone reservoir in the Shenhu area of the South China Sea demonstrates that production can be successfully obtained from complex MGH reservoirs [54].

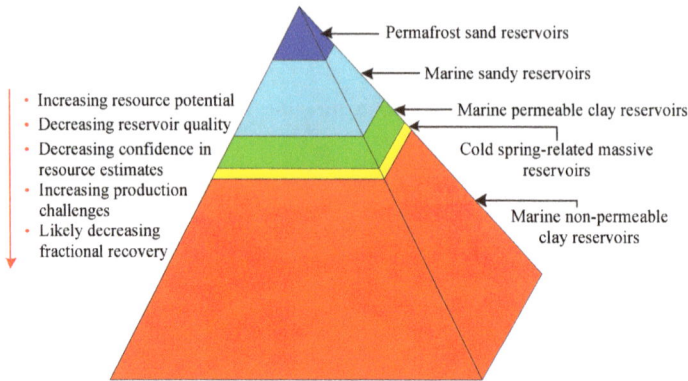

Figure 10. The pyramid of NGH exploitation difficulty [53].

3.1.2. Appraisal Methodology

An efficient appraisal methodology is a prerequisite for effective economic development of MGH reserves. The appraisal methodology involves the elimination of prospects with insufficient reserves, as well as reserves which are deemed risky or too complex to develop [14]. Resources can be defined as the total quantity of gas stored in the MGH reservoir (the sum of discovered and undiscovered gas as well as economically recoverable and non-economically recoverable gas), while reserves can be defined as the amount of gas that can be recovered at a reasonable level of economic return [55,56]. Currently, appraisal methodology involves four methods for volumetric estimation: (1) area/depth method, (2) volumetric method, (3) probability statistics, and (4) material balance method, of which the volumetric method which most widely used [57]. Figure 11 shows that as geological certainty and economic cut-offs improve, the reserves gradually decrease. At present, the concept of "natural gas hydrate petroleum system" [58] based on volumetric method has been proposed by researchers, which combines the accumulation mechanism with the actual occurrence conditions, and has a higher feasibility in the management of MGH resource appraisal system in the future [59].

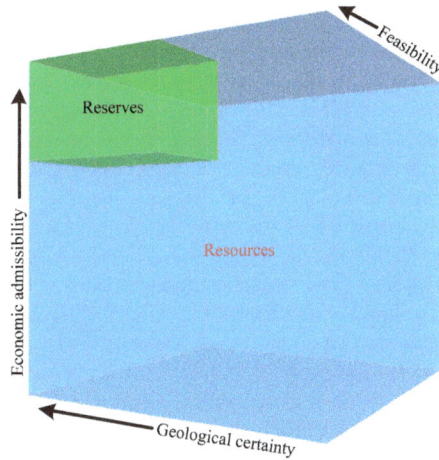

Figure 11. Volumetric method in resource appraisal [56]. (a) the "blue cube" represents "Resources"; (b) the "green cuboid" represents "Reserves"; (c) the "arrows" indicates an increase in each element.

3.1.3. Gas Production Methods

Choosing the most effective method for gas production from MGHs involves assessing a variety of factors to achieve the simplest, most practical, and environmentally safe method of production. Five methods are mentioned in the Introduction, their production benefits and risks are presented here as follows [6,60]:

(1) Chemical agent injection, used in permafrost NGH developments, is expensive and toxic and application in MGH development will damage marine ecological environments.

(2) Depressurization operations of MGHs do not require continuous operation so have lower costs, and are appropriate for large-scale production.

(3) Thermal activation methods can be used in situations of complex geology; however, these methods require high levels of energy consumption as well as having low heat exchange efficiency.

(4) CO_2 replacement methods are environmentally friendly, but are technically difficult and require a steady supply of CO_2.

(5) Solid fluidization is only applicable to deep-sea reservoirs at shallow depths, with limited-diagenesis and poor cementation. It has only been applied in the Liwan area of the northern South China Sea [35].

Production tests using a combination of depressurization and thermal activation methods have been successfully applied at the highest ranked offshore prospects (the Shenhu area of the South China Sea and the Nankai Trough of Japan) [10,15,16], and the CO_2 replacement method is the most effective onshore method to avoid potential hazards in the future (production test of permafrost in the North Slope of Alaska, USA) [18,61].

3.2. Production Control

Production control during MGH development is the key priority for LMHP and involves using appropriate drilling technology, hazard monitoring during drilling and production, and suitable monitoring systems to provide early warning of systems failure.

3.2.1. Drilling Technology Management

Drilling technology management is one of the core parts of MGH development which can avoid potential hazards if applied correctly. There are three aspects involved: drilling, cementing, and completion [62]. The drilling system includes drilling technology, fluid, and equipment, and their effective integration to ensure efficient drilling. The preferred method of drilling is selected to inhibit decomposition of MGHs by maintaining pressure or using casing. Cased drilling has been used on all the production tests because it is quicker, cheaper, and less prone to failure. Well cementation provides protection and support to the wellbore casing, particularly across the reservoir section. Well completion provides further stability to the wellbore and to prevent accidents caused by excessive sand production. Drilling has been done to date using semi-submersible drilling rigs or drill ships (Figure 12) and downhole gas-liquid-solid separation devices [63]. The Shenhu production test used the "Blue Whale I" semi-submersible drilling rig and a Chinese gas-water-sand tri-phase separator. The Nankai Trough production tests used the "Chikyu (Earth)" deep ocean drilling vessel and a Japanese gas-liquid separator. Both these systems proved effective during the respective production tests [64,65]. During the drilling of these production wells, drilling fluids had significant quantities of thermodynamic inhibitors added to the drilling mud to suppress or retard the regeneration of hydrates. This also had a significant effect on the prevention and control of marine geohazards. Special cements were used which had the following properties: low temperature, low density, low hydration heat, high early strength, low filtrate loss, good densification, and anti-channeling in order to ensure efficient completion of cementing operations. Appropriate well completion measures, such as mechanical sieve tubes and gravel packing, were used during the production tests [66]. Research is needed to improve sand control methods during well completion to reduce sand abrasion and enhance abrasion resistance.

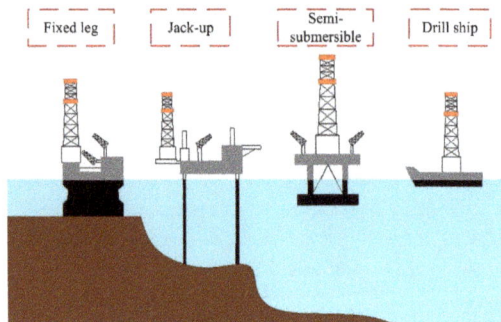

Figure 12. Different platforms for offshore drilling [10].

3.2.2. Monitoring during Development

Monitoring the changes in the submarine environment is an important part of the MGH development process (Figure 13). This ensures that potential hazards can be prevented during development, as well as establishing a baseline for monitoring changes in the marine environment during production and throughout development. Baseline monitoring is performed to detect changes in reservoir parameters, such as temperature, pressure, permeability, porosity, saturation, as well as in environmental parameters, such as seawater composition and submarine life that result from the development of MGHs. It is important to obtain baseline data before production commences in order to analyze the changes relating to production and development in the surrounding environment; these data can be collected along with other monitoring data in the planning stage. During production and development, it is important to have real time monitoring of seabed deformation, reservoir stability, and methane leakage; this was undertaken during the production test in the Nankai Trough of Japan and in the Shenhu area of the South China Sea [11,67,68]. Therefore, it is important to carry out long term, real-time, extensive, multi-parameter, in-situ monitoring to evaluate the impact that MGH production has on the marine environment during production, development, and afterwards.

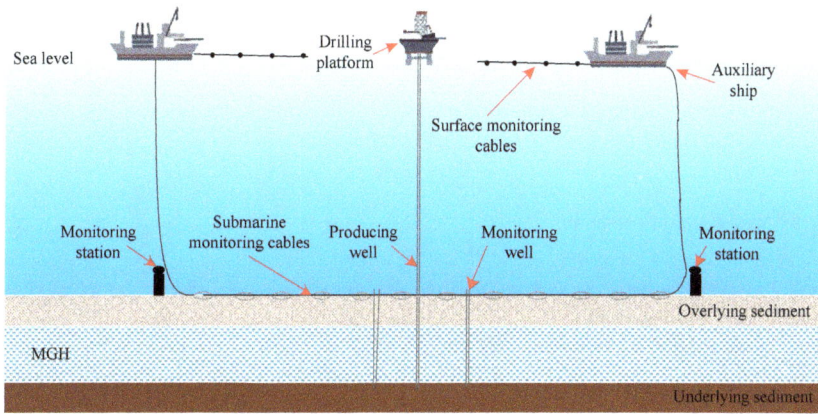

Figure 13. Submarine in-situ monitoring system [67].

3.2.3. Risk Assessment and Early Warning System

The purpose of a risk assessment system is to estimate and assess the scale and impact levels that potential hazards would have throughout the entire life cycle of a MGH development, and then establish a quantitative assessment and early warning system for potential hazards. This paper suggests a fuzzy comprehensive evaluation method [69] to quantitatively evaluate different types of hazards. First, a set of assessment object factors (U) is determined (Table 1), and then a comment set V = {High risk, medium risk, low risk, minimal risk} is determined.

Table 1. Assessment object factors set.

The First Index U_i (A_i)	The Second Index U_{ij} (A_{ij})	The Third Index U_{ijk} (A_{ijk})
Climate and environmental hazards U_1 (A_1)	Greenhouse gas U_{11} (A_{11}) Biocoenosis hazards U_{12} (A_{12})	—
Marine geohazards U_2 (A_2)	Submarine landslide U_{21} (A_{21}) Active fault U_{22} (A_{22}) Mud diaper U_{23} (A_{23}) Sea quake U_{24} (A_{24}) Turbidity current U_{25} (A_{25})	—
Marine engineering hazards U_3 (A_3)	Drilling engineering hazards U_{31} (A_{31})	Well blowout, leakage and borehole instability U_{311} (A_{311}) Sand production, well plugging and hydrate secondary generation U_{312} (A_{312}) Disused well U_{313} (A_{313})
	Deep-sea drilling hydrate crossing and other accidents U_{32} (A_{32})	submarine line accidents U_{321} (A_{321}) Oil-gas well accidents U_{322} (A_{322})
Note: the sum of A_i is 1	Note: the sum of A_{ij} is 1 (i is the same)	Note: the sum of A_{ijk} is 1 (i, j all are the same)

Next, a second set (A) of assessment object factors is derived using a variety of methods including expert estimation, Delphi method, and characteristic value, although they all involve subjective analysis. The fuzzy relation matrix R is determined using a single factor fuzzy assessment. The appropriate fuzzy synthesis operator M is selected to combine the weight set A and the matrix R to create a weight set B. Then by comparing the weight set B with the comment set V, the potential risk level for each potential hazard can be determined. As a result of this analysis of potential hazards, a quantified risk assessment for potential hazard levels during MGH development can be established and a set of corresponding early warning measures be established. Using appropriate technologies, personnel, and equipment in a prudent way to form a complete risk assessment and develop early warning system for hazards will prevent hazards occurring and provide a solution pathway when they occur.

3.3. Post-Production Protection

It is important to evaluate the post-production restoration cost of the environment surrounding an MGH development as part of the overall evaluation of the economic value of developing an MGH resource.

3.3.1. Post-Development Handling

The cessation of production at MGH developments involves a variety of issues such as well abandonment, reservoir protection, and monitoring of the surrounding environment. Well abandonment is similar to that of traditional deep-sea gas production and monitoring of the surrounding environment is concerned with whether there are anomalies in strata, seawater, or the atmosphere. These processes can be undertaken during the risk assessment process. Post production reservoir remediation is important to maintain seabed stability. During MGH production, large numbers of voids may appear in the reservoir, resulting in reduced sediment strength and problems relating to sediment collapse. A method to inject the produced voids with high-water content sediments under high-pressure and low-temperature is proposed, which would improve sediment stability and remediate the marine environment after production has ceased.

3.3.2. Economic Evaluation of Production

The economic evaluation of production considers the problem of coordination between input and output, technology and efficiency. One of the main purposes of developing MGHs is to maximize the economic benefit by using suitable technology in a safe environment. The two important parameters involved in the economic evaluation are the energy efficiency ratio (EER, ratio of combustion heat to decomposition heat in unit) and the energy return on energy invested (EROI, ratio of energy output to energy input during production) [70]. On the one hand, EER is affected by the production method, with the depressurization having the maximum energy efficiency; on the other hand, the reservoir type (Section 3.1.1) is also an important factor. The changes in reservoir energy efficiency and expected production cost trends are shown in Figure 14a. EROI is mainly influenced by the technical level and the amount of resources used during production. Generally speaking, the higher the resource value used, the higher the hydrate production efficiency that can be achieved. The continuous improvement in technology and constant consumption of resources indicates that the EROI has a peak value Q_{max} and then declines to the breakeven line (Figure 14b). Therefore, low-efficiency technology will inevitably consume large amounts of high-quality resources and shift the peak forward. One of the important uses of economic evaluation of the production costs associated with technological breakthrough and innovation is to ensure that these lie within a reasonable range on the EER and EROI graphs [71,72].

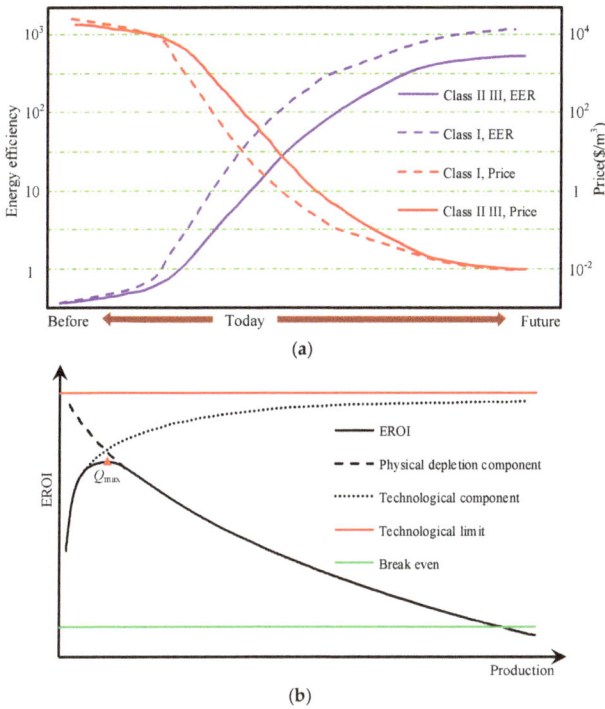

Figure 14. Economic evaluation indices of production [70]. (**a**) Changes of EER and production cost with time; (**b**) Change of EROI with production. The "Physical depletion component" means the trend of gas hydrate resource depletion; the "Technological component" means the trend of technology development; the "Technological limit" means the maximum that the technology can achieve; the "Break even" means that the EROI is maintained at a lower level and is no longer developed.

Based on the above analysis, all stages of LMHP are dynamic and interlinked processes. Effective integration of these will enable MGH development to be undertaken in a safe and effective fashion; ignoring small details may lead to significant problems during the project life cycle.

4. Key Challenges and Prospects

The development and exploitation of MGHs is undergoing rapid evolution. Between May to July 2017, China successfully tested in the Shenhu area of the South China Sea, where natural gas was extracted from silty clay reservoirs at a depth of 203 to 277 m below the seabed, in water depths of around 1266 m below sea level [10,35]. The production test was carried out by the "Blue Whale I" semi-submersible drilling rig. Firstly, the MGH in the silty clay reservoirs is decomposed by the depressurization, and then the natural gas developed by the hydrate sediment is taken out smoothly by using a gas-water-sand tri-phase separator which independently developed by China [10,73]. During the 60-day stable production test, the cumulative gas production exceeded 30.9×10^4 m^3, with an average daily output of 5151 m^3 and a maximum daily production of 3.5×10^4 m^3. The maximum methane content in the produced gas was 99.5% [10,72]. Several major technical breakthroughs were achieved during this production test, such as the longest sustained production time and the maximum volume of gas production, the duration of stable air flow and environmental safety, as well as establishing two new world records for the gas production time and volume. However, these production test successes are only an initial step in the process of MGH development where all involved countries, including China, are still facing great challenges and difficulties.

4.1. Key Challenges

The potential hazards and the implementation of the prevention and control measures within LMHP are all challenges. In this section, these are summarized into the "three-step" strategy (Figure 15) reflecting the importance that these problems have within the different development stages of MGH development.

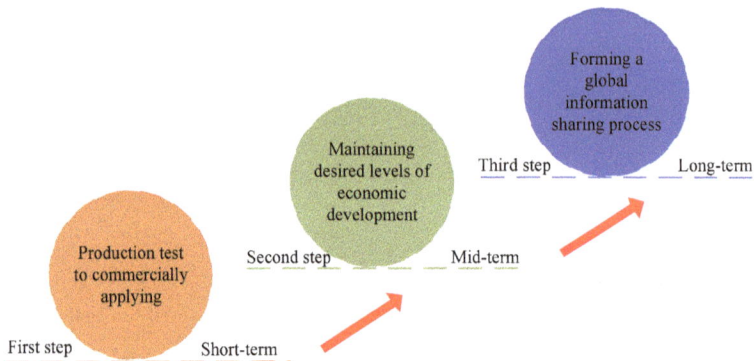

Figure 15. The "three-step" strategy of development.

The first step: short-term challenges. This stage contains most of the major problems that needs to be dealt with from production testing to development. These include target zone (s) selection, the mechanism and method of increasing production, flow security mechanisms, sand control methods, solid-gas-liquid multiphase flow monitoring, reservoir deformation and monitoring, in-situ bio-environmental analysis, risk assessment and establishment of an early warning system. This step is the bottleneck for the development of MGHs as large investments are required to drive technological innovation to increase production and development breakthroughs.

The second step: mid-term challenges. These challenges are more associated with ensuring that the economics of development are maintained to ensure successful development. Production levels can be maintained to ensure that costs associated with the exploration and development, production monitoring, technological innovation, and environmental monitoring and remediation are covered. This requires that a long-term comprehensive solution is found to guarantee the effective utilization of MGHs.

The third step: long-term challenges. A global resource management mechanism similar to that used in developing petroleum resources needs to be created at this stage. As a new source of energy, MGHs have the potential to change global consumption patterns. Their successful development and achieving maximum return on investment requires international cooperation, and exchange of ideas, as well as developing a scheme to allow global production and shared access to these resources.

4.2. Prospects

As world energy patterns change and environmental issues become more important, the identification and development of a new unconventional source of energy is becoming a global issue. MGHs are possibly a great source of potentially clean energy with large reserves, wide global distribution, and a high energy density. Production tests results from the Shenhu area of the South China Sea and the Nankai Trough of Japan demonstrate the accumulation and application of drilling technology to develop these reserves. The potential risks, the formation of MGHs, as well as the production and prevention hazards during development are the key issues to solve the economic utilization of MGHs, will be major research topics for scientists from around the world for many years.

5. Conclusions

(1) Four inter-related hazards which may occur during MGH exploitation were identified as follows: marine geohazards, greenhouse gas emissions, marine ecological hazards, and marine engineering hazards.

(2) Lifecycle management of potential hazards prevention in the exploitation of MGHs (LMHP) was proposed firstly. It has three stages: preparation, which involves investigating the accumulation mechanism, appraisal methodology and gas production methods; production control, which includes drilling techniques, monitoring during production risk assessment and early warning system; and post-production protection, including post-production remediation and economic evaluation of production. All these factors are inter-related and need to be systematically evaluated.

(3) A "three-step" strategy for the development of LMHP is proposed, which consists of commercially applying the results of MGH research in the short-term, maintaining desired levels of economic development in the mid-term and forming a global information sharing process associated with hydrate in the long-term. Understanding this "three step" strategy will allow the successful development of MGH resources.

(4) The production test in the Shenhu area of the South China Sea showed that the development of MGHs is a complex and constantly changing problem with difficult challenges. Safe and efficient development and production of MGHs can be achieved by innovation and breakthroughs in the use of technology as well as through extensive international cooperation and exchange of information around the world.

Author Contributions: The concept of LMHP and "three-step" strategy for MGHs development was proposed by F.W. and he also wrote the conclusions and abstract. The specific classifications of the two strategies were carried out by B.Z. and data collection during the revision of the paper was completed by G.L.

Funding: This research was funded by the [Independent Research Project of State Key Laboratory of Coal Resources and Safe Mining, CUMT] grant number [SKLCRSM18X012], the [Independent Innovation Project for Double World-class Construction, CUMT] grant number [2018ZZCX07] and the [China Scholarship Council] grant number [201802180034].

Acknowledgments: This work was supported by the Independent Research Project of State Key Laboratory of Coal Resources and Safe Mining of CUMT (SKLCRSM18X012), the Priority Academic Program Development of Jiangsu Higher Education Institutions, the Independent Innovation Project for Double World-class Construction of CUMT (No.2018ZZCX07) and the visiting scholarship funded by the China Scholarship Council (No.201802180034). The authors gratefully acknowledge financial support of the above-mentioned agencies.

Conflicts of Interest: The authors declare that they have no conflicts of interest.

References

1. US Energy Information Administration (EIA). *Annual Energy Outlook 2013*; US Energy Information Administration: Washington, DC, USA, 2013; pp. 60–62.

2. Chong, Z.R.; Yang, S.H.B.; Babu, P.; Linga, P.; Li, X.S. Review of natural gas hydrates as an energy resource: Prospects and challenges. *Appl. Energy* **2016**, *162*, 1633–1652. [CrossRef]

3. Makogon, Y.F.; Trebin, F.A.; Trofimuk, A.A.; Tsarev, V.P.; Cherskiy, N.V. Detection of a pool of natural gas in a solid (hydrate gas) state. *Dokl. Akad. Nauk. SSSR* **1972**, *196*, 197–200.

4. Trofimuk, A.A.; Chersky, N.V.; Tsaryov, V.P. The role of continental glaciation and hydrate formation on petroleum occurrence. In *The Future Supply of Nature-Made Petroleum and Gas Technical Reports*; Pergamon Press: Oxford, UK, 1977; pp. 919–926.

5. Sloan, E.D. Gas Hydrates: Review of Physical/Chemical Properties. *Energy Fuels* **1998**, *12*, 191–196. [CrossRef]

6. Xu, C.G.; Li, X.S. Research progress on methane production from natural gas hydrates. *RSC Adv.* **2015**, *5*, 54672–54699. [CrossRef]
7. Beaudoin, Y.C.; Waite, W.; Boswell, R.; Dallimore, S.R. *Frozen Heat: A UNEP Global Outlook on Methane Gas Hydrates*; United Nations Environment Programme, GRID-Arendal: Arendal, Norway, 2014; p. 1.
8. Makogon, Y.F. Natural gas hydrates—A promising source of energy. *J. Nat. Gas Sci. Eng.* **2010**, *2*, 49–59. [CrossRef]
9. Boswell, R.; Collett, T.S. Current perspectives on gas hydrate resources. *Energy Environ. Sci.* **2011**, *4*, 1206–1215. [CrossRef]
10. Wu, S.G.; Wang, J.L. On the China's successful gas production test from marine gas hydrate reservoirs. *Chin. Sci. Bull.* **2018**, *63*, 2–8.
11. Zhu, C.Q.; Zhang, M.S.; Liu, X.L.; Wang, Z.H.; Shen, Z.C.; Zhang, B.W.; Zhang, X.T.; Jia, Y.G. Gas hydrates: Production, geohazards and monitoring. *J. Catastrophol.* **2017**, *32*, 51–56.
12. Li, X.S.; Xu, C.G.; Zhang, Y.; Ruan, X.K.; Li, G.; Wang, Y. Investigation into gas production from natural gas hydrate: A review. *Appl. Energy* **2016**, *172*, 286–322. [CrossRef]
13. Zhou, S.W.; Chen, W.; Li, Q.P. The green solid fluidization development principle of natural gas hydrate stored in shallow layers of deep water. *China Offshore Oil Gas* **2014**, *26*, 1–7.
14. Wu, N.Y.; Huang, L.; Hu, G.W.; Li, Y.L.; Chen, Q.; Liu, C.L. Geological controlling factors and scientific challenges for offshore gas hydrate exploitation. *Mar. Geol. Quat. Geol.* **2017**, *37*, 1–11.
15. Fujii, T.; Suzuki, K.; Takayamaet, T.; Tamaki, M.; Komatsua, Y.; Konnoc, Y.; Yonedac, J.; Yamamotoa, K.; Nagaoc, J. Geological setting and characterization of a methane hydrate reservoir distributed at the first offshore production test site on the Daini-Atsumi Knoll in the eastern Nankai Trough, Japan. *Mar. Pet. Geol.* **2015**, *66*, 310–322. [CrossRef]
16. Yamamoto, K.; Terao, Y.; Fujii, T.; Ikawa, T.; Seki, M.; Matsuzawa, M.; Kanno, T. Operational overview of the first offshore production test of methane hydrates in the Eastern Nankai Trough. In Proceedings of the Offshore Technology Conference, Houston, TX, USA, 5–8 May 2014.
17. Collett, T.S.; Johnson, A.H.; Knapp, C.C.; Boswell, R. Natural gas hydrates: A review. *Browse Collections* **2009**, 146–219. [CrossRef]
18. Collett, T.S.; Lewis, R.E.; Winters, W.J.; Lee, M.W.; Rose, K.K.; Boswell, R.M. Downhole well log and core montages from the Mount Elbert gas hydrate stratigraphic test well, Alaska North Slope. *Mar. Pet. Geol.* **2011**, *28*, 561–577. [CrossRef]
19. Wang, L.F.; Fu, S.Y.; Liang, J.Q.; Shang, J.J.; Wang, J.L. A review on gas hydrate developments propped by worldwide national projects. *Geol. China* **2017**, *44*, 439–448.
20. Haeckel, M.; Bialas, J.; Klaucke, I.; Wallmann, K.; Bohrmann, G.; Schwalenberg, K. Gas hydrate occurrences in the Black Sea—New observations from the German SUGAR project. *Fire Ice Methane Hydr. Newslett.* **2015**, *15*, 6–9.
21. Beeskow-Strauch, B.; Schicks, J.; Zimmer, M. Evaluation of CH4 Gas Permeation Rates through Silicone Membranes and Its Possible Use as CH4-Extractor in Gas Hydrate Deposits. *Energies* **2015**, *8*, 5090–5106. [CrossRef]
22. Vanneste, M.; Sultan, N.; Garziglia, S.; Forsberg, C.F.; L'Heureux, J.S. Seafloor instabilities and sediment deformation processes: The need for integrated, multi-disciplinary investigations. *Mar. Geol.* **2014**, *352*, 183–214. [CrossRef]
23. Maslin, M.; Owen, M.; Betts, R.; Day, S.; Jones, T.D.; Ridgwell, A. Gas hydrates: Past and future geohazard? *Philos. Trans. R. Soc. Lond. A Math. Phys. Eng. Sci.* **2010**, *368*, 2369–2393. [CrossRef] [PubMed]
24. Kvalstad, T.J.; Andresen, L.; Forsberg, C.F.; Berg, K.; Bryn, P.; Wangen, M. The Storegga slide: Evaluation of triggering sources and slide mechanics. *Mar. Pet. Geol.* **2005**, *12*, 245–256. [CrossRef]
25. Hornbach, M.J.; Lavier, L.L.; Ruppel, C.D. Triggering mechanism and tsunamogenic potential of the Cape Fear Slide complex, US Atlantic margin. *Geochem. Geophys. Geosyst.* **2007**, *8*, Q12008. [CrossRef]
26. Qin, Z.L.; Wu, S.G.; Wang, Z.J.; Li, Q.P. Geohazards and risk of deepwater engineering induced by gas hydrate—A case study from oil leakage of deep water drilling well in GOM. *Prog. Geophys.* **2011**, *26*, 1279–1287.
27. Cai, J.C.; Wei, W.; Hu, X.Y.; Liu, R.C.; Wang, J.J. Fractal characterization of dynamic fracture network extension in porous media. *Fractals-Complex Geom. Patterns Scaling Nat. Soc.* **2017**, *25*, 1750023. [CrossRef]

28. Paull, C.K.; Iii, W.U.; Dallimore, S.R.; Blasco, S.M.; Lorenson, T.D.; Melling, H.; Medioli, B.E.; Nixon, F.M.; McLaughlin, F.A. Origin of pingo-like features on the Beaufort Sea shelf and their possible relationship to decomposing methane gas hydrates. *Geophys. Res. Lett.* **2007**, *34*, 223–234. [CrossRef]

29. Jia, Y.G.; Zhu, C.Q.; Liu, L.P.; Wang, D. Marine geohazards: Review and future perspective. *Acta Geol. Sin.* **2016**, *90*, 1455–1470.

30. He, J.; Liang, Q.Y.; Ma, Y.; Shi, Y.H.; Xiao, Z. Geohazards types and their distribution characteristics in the natural gas hydrate area on on the northern slope of the South China Sea. *Geol. China* **2018**, *45*, 15–28.

31. Intergovernmental Panel on Climate Change. *The Fourth Assessment Report of the Intergovernmental Panel on Climate Change*; IPCC: Geneva, Switzerland, 2007.

32. *Global Anthropogenic Non-CO$_2$ Greenhouse Emissions: 1990–2020*; Office of Atmospheric Programs Climate Change Division US Environmental Protection Agency: Washington, DC, USA, 2006.

33. Sloan, E.D.; Koh, C. *Clathrate Hydrates of Natural Gases*; CRC Press: Boca Raton, FL, USA, 2007.

34. Steffensen, J.P.; Andersen, K.K.; Bigler, M.; Clausenet, H.B.; Jensen, D.D. High-resolution Greenland ice core data show abrupt climate change happens in few years. *Science* **2008**, *321*, 680–684. [CrossRef] [PubMed]

35. Wei, W.; Zhang, J.H.; Yu, R.Z.; Lin, B.B.; Chen, L.Q.; Peng, Y.; Xiao, H.P. Review on natural gas hydrate in 2017. *Sci. Technol. Rev.* **2018**, *36*, 83–90.

36. Zhang, J.X.; Ren, J.Y. Several important problems in the research of gas hydrates. *Geol. Sci. Technol. Inf.* **2001**, *20*, 44–48.

37. Mestdagh, T.; Poort, J.; De Batist, M. The sensitivity of gas hydrate reservoirs to climate change: Perspectives from a new combined model for permafrost-related and marine settings. *Earth-Sci. Rev.* **2017**, *169*, 104–131. [CrossRef]

38. Suess, E. Marine cold seeps and their manifestations: Geological control, biogeochemical criteria and environmental conditions. *Int. J. Earth Sci.* **2014**, *103*, 1889–1916. [CrossRef]

39. James, R.H.; Bousquet, P.; Bussmann, I.; Haeckel, M.; Kipfer, R.; Leifer, I.; Niemann, H.; Ostrovsky, I.; Piskozub, J.; Rehder, G.; et al. Effects of climate change on methane emissions from seafloor sediments in the Arctic Ocean: A review. *Limnol. Oceanogr.* **2016**, *61*, S283–S299. [CrossRef]

40. Wei, H.L.; Sun, Z.L.; Wang, L.B.; Zhang, X.R.; Cao, H.; Huang, W.; Bai, F.L.; He, Y.J.; Zhang, X.L.; Zhai, B. Perspective of the environmental effect of natural gas hydrate system. *Mar. Geol. Quat. Geol.* **2016**, *36*, 1–13.

41. Wang, S.H.; Song, H.B.; Yan, W. Environmental Effects of Natural Gas Hydrate. *Bull. Miner. Pet. Geochem.* **2004**, *23*, 160–165.

42. Farrimond, P. Massive dissociation of gas hydrate during a Jurassic oceanic anoxic event. *Nature* **2000**, *406*, 392–395.

43. Dickens, G.R.; O'Neil, J.R.; Rea, D.K.; Rea, D.K.; Owen, R.M. Dissociation of oceanic methane hydrate as a cause of the carbon isotope excursion at the end of the Paleocene. *Paleoceanography* **1995**, *10*, 965–971. [CrossRef]

44. He, Y.; Tang, C.P.; Liang, D.Q. The potential risks of drilling in marine gas hydrate bearing sediments and the corresponding strategies. *Adv. New Renew. Energy* **2016**, *4*, 42–47.

45. Ning, F.L.; Liu, L.; Li, S.; Zhang, K.; Jiang, G.S.; Wu, N.Y.; Sun, C.Y.; Chen, G.J. Well logging assessment of natural gas hydrate reservoirs and relevant influential factors. *Acta Pet. Sin.* **2013**, *34*, 591–606.

46. Koh, C.A.; Sum, A.K.; Sloan, E.D. State of the art: Natural gas hydrates as a natural resource. *J. Nat. Gas Sci. Eng.* **2012**, *8*, 132–138. [CrossRef]

47. Shipp, R.C. Gas Hydrates as a Geohazard: What Really Are the Issues? Shell International Exploration and Production Inc.: Houston, TX, USA, 2014. Available online: https://www.energy.gov/sites/prod/files/2014/04/f14/GH-as-Geohazards_MHAC-Galveston_27-mar-2014_final_0.pdf (accessed on 27 March 2014).

48. Yie, J.L.; Yin, K.; Jiang, G.S.; Tang, F.L.; Dou, B. Key technique and countermeasures for natural gas hydrate drilling. *Explor. Eng.* **2003**, *5*, 45–48.

49. Lu, J.S.; Li, D.L.; He, Y.; Liang, D.Q.; Xiong, Y.M. Research status of sand production during the gas hydrate exploitation process. *Adv. New Renew. Energy* **2017**, *5*, 394–402.

50. Bai, Y.H.; Li, Q.P.; Zhou, J.L.; Liu, Y.H. The Potential Risk of Gas Hydrate to Deepwater Drilling and Production and the Corresponding Strategy. *Pet. Drill. Tech.* **2009**, *37*, 17–21.

51. Wu, N.Y.; Liang, J.Q.; Wang, H.B.; Su, X.; Song, H.B.; Jiang, S.Y.; Zhu, Y.H.; Lu, Z.Q. Marine gas hydrate system: State of the Art. *Geoscience* **2008**, *22*, 356–362.

52. Bu, Q.T.; Hu, G.W.; Ye, Y.G.; Liu, C.L.; Li, C.F.; Wang, J.S. Research Progress in Natural Gas Hydrate Accumulation System. *Adv. New Renew. Energy* **2015**, *3*, 435–443.

53. Suzuki, K.; Schultheiss, P.; Nakatsuka, Y.; Ito, T.; Egawa, K.; Holland, M.; Yamamoto, K. Physical properties and sedimentological features of hydrate-bearing samples recovered from the first gas hydrate production test site on Daini-Atsumi Knoll around eastern Nankai Trough. *Mar. Pet. Geol.* **2015**, *66*, 346–357. [CrossRef]

54. Yang, S.X.; Liang, J.Q.; Liu, C.L.; Sha, Z.B. Progresses of gas hydrate resources exploration in sea area. *Geol. Surv. China* **2017**, *4*, 1–8.

55. Milkov, A.V. Worldwide distribution of submarine mud volcanoes and associated gas hydrates. *Mar. Geol.* **2000**, *167*, 29–42. [CrossRef]

56. Milkov, A.V.; Sassen, R. Preliminary assessment of resources and economic potential of individual gas hydrate accumulations in the Gulf of Mexico continental slope. *Mar. Pet. Geol.* **2003**, *20*, 111–128. [CrossRef]

57. Sun, Y.B.; Zhao, T.H.; Cai, F. Gas hydrate resource assessment abroad and its implications. *Mar. Geol. Front.* **2013**, *29*, 27–35.

58. Collett, T.S. *Gas Hydrate Petroleum Systems in Marine and Arctic Permafrost Environments*; Gcssepm Proceedings: Houston, TX, USA, 2009; pp. 6–30.

59. Veluswamy, H.P.; Kumar, R.; Linga, P. Hydrogen storage in clathrate hydrates: Current state of the art and future directions. *Appl. Energy* **2014**, *122*, 112–132. [CrossRef]

60. Zhou, S.W.; Chen, W.; Li, Q.P.; Zhou, J.L.; Shi, H.S. Research on the solid fluidization well testing and production for shallow non-diagenetic natural gas hydrate in deep water area. *China Offshore Oil Gas.* **2017**, *29*, 1–8.

61. Schoderbek, D.; Martinet, K.L.; Howard, J.; Silpngarmlert, S.; Hester, K. North slope hydrate field trial: CO_2/CH_4 exchange. In Proceedings of the OTC Arctic Technology Conference, Houston, TX, USA, 3–5 December 2012.

62. Guang, X.J.; Wang, M.S. Key production test technologies for offshore natural gas hydrate. *Pet. Drill. Tech.* **2016**, *44*, 45–51.

63. Wang, Y.; Feng, J.C.; Li, X.S.; Zhang, Y.; Han, H. Experimental Investigation on Sediment Deformation during Gas Hydrate Decomposition for Different Hydrate Reservoir Types. *Energy Procedia* **2017**, *142*, 4110–4116. [CrossRef]

64. Oyama, A.; Masutani, S.M. A review of the methane hydrate program in Japan. *Energies* **2017**, *10*, 1447. [CrossRef]

65. Song, Y.C.; Yang, L.; Zhao, J.F.; Liu, W.G.; Yang, M.J.; Li, Y.H.; Liu, Y.; Li, Q.P. The status of natural gas hydrate research in China: A review. *Renew. Sustain. Energy Rev.* **2014**, *31*, 778–791. [CrossRef]

66. Liu, C.L.; Meng, Q.G.; Li, C.F.; Sun, J.Y.; He, X.L.; Yang, S.X.; Liang, J.Q. Characterization of natural gas hydrate and its deposits recovered from the northern slope of the South China Sea. *Earth Sci. Front.* **2017**, *24*, 41–50.

67. He, T.; Lu, H.L.; Lin, J.Q.; Dong, Y.F.; He, J. Geophysical techniques of reservoir monitoring for marine gas hydrate exploitation. *Earth Sci. Front.* **2017**, *24*, 368–382.

68. Zhang, X.; Du, Z.F.; Luan, Z.D.; Wang, X.J.; Xi, S.C.; Wang, B.; Li, L.F.; Lian, C.; Yan, J. In situ Raman detection of gas hydrates exposed on the seafloor of the South China Sea. *Geochem. Geophys. Geosyst.* **2017**, *18*, 3700–3713. [CrossRef]

69. Liu, Z.X.; Feng, F.; Wang, J.B.; Hua, H.Y.; Wang, R.X. Application of Fuzzy Synthesis Assessment Method on Classification of Rock Mass in Mine. *J. Wuhan Univ. Technol.* **2014**, *36*, 129–134.

70. Chen, J.; Wang, Y.H.; Lang, X.M.; Fan, S.S. Energy-efficient methods for production methane from natural gas hydrates. *J. Energy Chem.* **2017**, *24*, 552–558. [CrossRef]

71. Kurihara, M.; Ouchi, H.; Narita, H.; Masuda, Y. Gas production from methane hydrate reservoirs. In Proceedings of the 7th International Conference on Gas Hydrates (ICGH), Edinburgh, UK, 17–21 July 2011; Volume 1721.

Energies **2018**, *11*, 2384

72. Liu, C.L.; Li, Y.L.; Zhang, Y.; Sun, J.Y.; Wu, N.Y. Gas hydrate production test: From experimental simulation to field practice. *Mar. Geol. Quat. Geol.* **2017**, *37*, 12–26. [CrossRef]

73. Ju, X.; Liu, F.; Fu, P. Vulnerability of Seafloor at Shenhu Area, South China Sea Subjected to Hydrate Dissociation. In Proceedings of the GeoShanghai International Conference, Shanghai, China, 27–30 May 2018; pp. 54–62.

MDPI

St. Alban-Anlage 66

4052 Basel

Switzerland

Tel. +41 61 683 77 34

Fax +41 61 302 89 18

www.mdpi.com

Energies Editorial Office

E-mail: energies@mdpi.com

www.mdpi.com/journal/energies

www.ingramcontent.com/pod-product-compliance
Lightning Source LLC
Chambersburg PA
CBHW051726210326
41597CB00032B/5619